Phonon Scattering
in
Condensed Matter

Phonon Scattering
in
Condensed Matter

Edited by

Humphrey J. Maris

Brown University
Providence, Rhode Island

PLENUM PRESS · NEW YORK AND LONDON

Library of Congress Cataloging in Publication Data

International Conference on Phonon Scattering in Condensed Matter, 3d, Brown
 University, 1979.
 Phonon scattering in condensed matter.

 Includes indexes.
 1. Phonons—Scattering—Congresses. 2. Solid state physics—Congresses. 3. Amor-
phous substances—Congresses. 4. Semiconductors—Congresses. I. Maris, Humphrey J.
II. Title.
QC176.8.P5I57 1979   539.7′217   80-401
ISBN-13: 978-1-4613-3065-3   e-ISBN-13: 978-1-4613-3063-9
DOI: 10.1007/978-1-4613-3063-9

Proceedings of the Third International Conference on Phonon Scattering in
Condensed Matter, held at Brown University, Providence, Rhode Island,
August 28–31, 1979.

©1980 Plenum Press, New York
Softcover reprint of the hardcover 1st edition 1980
A Division of Plenum Publishing Corporation
227 West 17th Street, New York, N.Y. 10011

INTERNATIONAL ADVISORY COMMITTEE

A. C. Anderson (University of Illinois)

W. E. Bron (University of Indiana)

L. J. Challis (University of Nottingham)

E. R. Dobbs (London University)

W. Eisenmenger (Stuttgart University)

T. Ishiguro (Electrotechnical Laboratory, Japan)

J. Joffrin (Institut Laue-Langevin, Grenoble)

I. M. Khalatnikov (Institute for Theoretical Physics, Moscow)

H. Kinder (Technical University, Munich)

P. G. Klemens (University of Connecticut)

R. L. Melcher (IBM, New York)

V. Narayanamurti (Bell Laboratories, New Jersey)

K. Weiss (Philips Laboratories, Eindhoven)

LOCAL ORGANIZING COMMITTEE

H. J. Maris (Brown University)   Chairman

C. Elbaum (Brown University)

J. D. N. Cheeke (University of Sherbrooke)

Preface

The Third International Conference on Phonon Scattering in Condensed Matter was held at Brown University, Providence, Rhode Island from August 28-31, 1979. The previous conferences in this series were held at Nottingham in 1975, and in France at Paris and Ste Maxime in 1972.

Until about 15 years ago phonon scattering was studied almost exclusively by measurements of thermal conductivity. This approach has the severe limitation that the result obtained for the phonon scattering rate is actually the average of the scattering for all of the phonons in the sample. Thus, no distinction can be made between phonons of different polarization, direction of propagation, or energy. During the 1960's several significant developments occurred. The most important of these was the application by Von Gutfeld and Nethercot of the "heat-pulse" method, previously used only in liquid helium, to the investigation of phonons in crystals. This approach makes possible the study of the propagation and scattering of phonons of known polarization and propagation direction. The early heat-pulse experiments used phonon generators which produced phonons having a broad distribution of energies and, in addition, the phonon detectors were sensitive to phonons of all energies. Since that time much research has been carried out to devise "monochromatic" phonon generators and detectors, and several successful techniques have been developed. These techniques include the use of superconducting tunnel junctions, spin-phonon spectrometers, and the direct generation of phonons by infra-red lasers. This combination of the original heat-pulse technique with the new methods of detection and generation has made the study of phonon scattering a very active area of research in the last few years, and it was the purpose of the conference to bring together specialists in this field. Areas of particular emphasis at the conference were the study of phonon scattering in amorphous materials, phonon transmission across interfaces (the Kapitza resistance problem), phonon scattering by free and bound electrons in semiconductors, and phonon propagation in elastically anisotropic materials (phonon focusing).

The conference was attended by 145 delegates from 18 countries. Ten invited papers and 95 contributed papers were read. In addition to the formal presentation of papers the scientific program included discussion sessions on the Kapitza resistance and on amorphous materials. These were chaired by K. Weiss and S. Hunklinger respectively. The social program included a traditional New England clambake held at the Haffenreffer Estate on Narragansett Bay.

These proceedings include all papers presented at the conference. Questions and comments made after each paper were tape recorded, and an edited version of these is included in these proceedings. Only those questions or comments that were judged to add significantly to the usefulness of this volume have been retained.

The conference was supported by a grant from the National Science Foundation. Additional financial support was provided by the Materials Research Laboratory at Brown University (also supported by the National Science Foundation), and by the Valpey-Fisher and Matec Corporations. The conference was sponsored by the International Union of Pure and Applied Physics. I should also like to thank the international advisory committee for their helpful suggestions regarding invited speakers, and C. Elbaum and J.D.N. Cheeke for their valiant labors as members of the local committee. Special thanks go to E. Zigas, C. Huber, J. J. McLaughry, R. Zeller, R. Mayer, F. Bucholtz, and A. Beierle for their various contributions to the success of the conference.

Humphrey J. Maris

Contents

SPIN-PHONON INTERACTIONS

Session Chairmen: W. E. Bron and A. M. de Goer

PHONON-PHONON INTERACTIONS

Session Chairman: P. J. King

HELIUM

Session Chairman: Y. Narahara

KAPITZA RESISTANCE

Session Chairmen: A. C. Anderson, J. U. Trefny, and A. Ikushima

DEFECTS

Session Chairman: V. W. Rampton

PHASE TRANSITIONS

Session Chairman: R. O. Pohl

NEW TECHNIQUES

Session Chairmen: F. de la Cruz and E. R. Dobbs

SEMICONDUCTORS

Session Chairmen: D. V. Osborne, K. Suzuki, and J. A. Rayne

METALS AND SUPERCONDUCTORS

Session Chairman: J. P. Maneval

PHONON SCATTERING IN METALLIC GLASSES

J. L. Black

Physics Department
Brookhaven National Laboratory
Upton, New York 11973

INTRODUCTION

The purpose of this article is to review some recent theoretical and experimental developments in the study of metallic glasses at temperatures near or below 1K. In this temperature regime, it appears that practically all glasses, whether metallic or insulating, behave in a similar fashion. The fact that such similarities occur, despite substantial structural differences between metallic and insulating glasses[1], constitutes a major theoretical challenge. This challenge, however, is not directly addressed in what follows. Instead, we shall concentrate upon the evidence for universal behavior and upon the theory which is necessary to understand this evidence. It turns out that most of this evidence involves a comparison of phonon scattering in metallic glasses with its counterpart in insulating glasses.

The theoretical framework for describing the low-temperature properties of amorphous solids is provided by the tunneling model of Anderson et al[2] and Phillips[3]. This model is a phenomenological prescription for calculating the effects of local rearrangements within the glassy structure. In the simplest version, these atomic rearrangements produce low-energy excitations which are quantum-mechanical two-level systems (TLS). These form a dilute (concentration of TLS active below 1K is less than 10^{-5}) collection of $S=\frac{1}{2}$ pseudospins which interact with the strain fields caused by long-wavelength phonons. It is an established fact that this model describes most low-temperature properties of <u>insulating</u> glasses in considerable detail[4]. From a fundamental point of view, this success supports the idea that the insulating-glass "ground state" actually consists of many states of roughly the

same energy and that some small fraction of these states are
accessible to each other within experimental time scales. We are
then led to wonder whether these principles apply to metallic
glasses, and (if so) whether the conduction electrons affect the
behavior of the TLS.

EVIDENCE AND PUZZLES

The natural way to look for TLS in metallic glasses involves
those experiments which give convincing proof in insulating
glasses. In this list of experiments[4] we find the linear specific
heat, the T^2 thermal conductivity, the lnT sound velocity, the
saturation of the sound attenuation, the ability of one ultrasonic
pulse to diminish the absorption of a second pulse, and generation
of phonon echoes by coherently excited TLS. In metallic glasses,
direct electronic contributions to the specific heat and thermal
conductivity overwhelm the TLS effect unless special methods of
eliminating the electrons can be devised. By using such methods,
Matey and Anderson[5] and Graebner et al[6] have reported evidence for
TLS in several bulk metallic glasses, while von Löhneysen and
Steglich[7] have done the same for amorphous metallic films. On the
other hand, ultrasonic experiments are not masked by direct electronic contributions in competition with the TLS. Bellessa and
coworkers[8] have observed the distinctive logarithmic temperature
dependence of the sound velocity in numerous metallic glasses.
Most significantly Doussineau et al[9] and Golding et al[10] have seen
convincing evidence of saturation of the ultrasonic attenuation
in the bulk metallic glass PdSiCu. Taken together, these experiments are strong support for the existence of TLS in metallic
glasses.

Upon closer examination, however, a number of puzzles appear.
As shown in Figure 1, the phonon flux required to saturate the
absorption is several orders of magnitude greater in a metallic
glass than in its insulating counterpart at the same temperature
and frequency. Secondly, consider the attenuation which remains
after the resonant attenuation has been fully saturated. A comparison between metal[10,11] and insulator[12] of the temperature dependence of this additional absorption, usually called relaxational[12], is shown in Figure 2. In the metallic glass, this absorption rises at a lower temperature and obviously has a much weaker
T dependence than the insulating glasses. A third point of difference is that various effects involving two or more ultrasonic
pulses separated by very small time intervals are completely
absent in metallic glasses[10]. In the insulating glasses, these
effects include phonon echoes[13] and the saturation of one pulse by
another[4]. Finally, a number of discrepancies in the magnitude

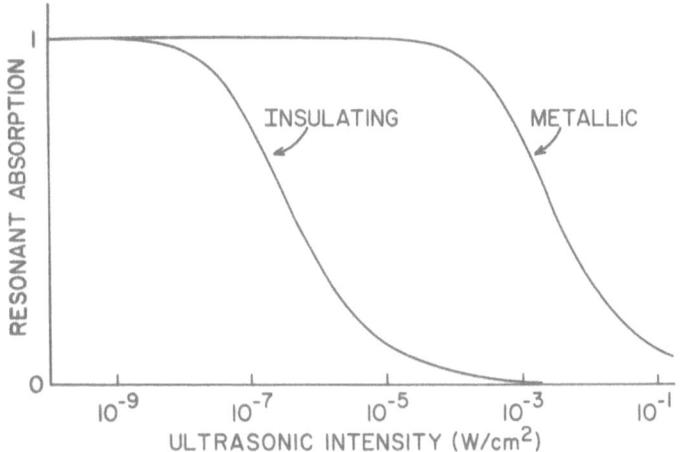

Figure 1.   Curves showing typical metallic vs insulating behavior in the saturation of ultrasonic attenuation at temperatures
near 10mK and frequencies near 1GHZ.   Absorption is normalized to
its low-power value.

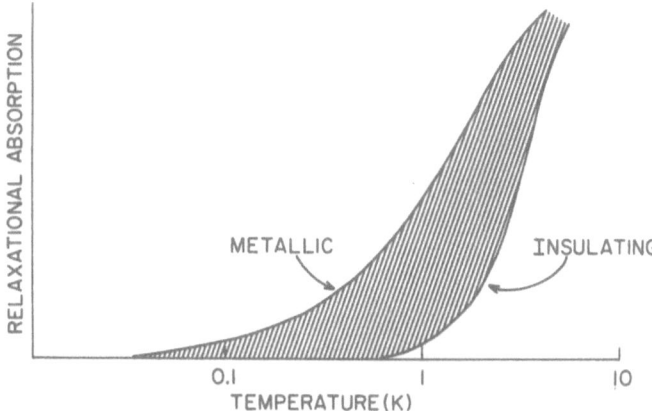

Figure 2.   Illustration of the different temperature dependence
in the relaxational attenuation in a metallic glass[10] (PdSiCu) as
compared with a typical insulating glass[12] (e.g. SiO_2).

and temperature/frequency dependence of the resonant absorption in metallic glasses have been pointed out by Matey and Anderson[14] and by Araki et al[15].

Table 1 presents a summary of the experimental situation in metallic glasses with an emphasis on the comparison with insulators. It is apparent that there are some very substantial differences, which raises the question of whether a TLS in a metal is intrinsically different than a TLS in an insulator. As we shall see below, making such an intrinsic distinction is not justified. Most of the apparent differences in Table 1 can be understood by taking into account effects of the conduction electrons upon the TLS.

THE ELECTRON/TLS INTERACTION

In insulating glasses, the TLS are described by

$$H_o = \sum_i E_i S^i_z + \sum_i (2M_i S^i_x + D_i S^i_z) \; e_i + H_{phonon}, \tag{1}$$

where E_i is the energy splitting of the TLS at site i, whose state is described by the Pauli matrices $2S^i_\alpha (\alpha=x,y,z)$. The strain field e_i modifies the TLS environment, leading to the coupling tensors M_i and D_i. Similarly, it is expected that the TLS will be coupled to conduction electrons in a metallic glass[16-18]. Figure 3 shows a schematic representation of the atomic displacements, d_i, associated with a single TLS whose atomic cores are located at positions $X_\ell = R^0_\ell \pm \frac{1}{2} d_i$. Associated with each of these cores is an atomic pseudopotential which causes the electrons to see a potential whose Fourier transform has the form

$$V_{L,R}(q) = \frac{1}{v} u(q) \rho_{L,R}(q) \tag{2}$$

where u(q) is the pseudopotential, v is the atomic volume, and the TLS density operator is given by

$$\rho_{L,R}(q) = \sum_{\ell=1}^{N'} e^{iqX_\ell} \tag{3}$$

The N' atoms comprising the TLS may be in the "left" state (L) or "right" state (R). As shown in Figure 3, the existence of these two positions for each core gives rise to a difference potential V_L-V_R, which is the origin of the important coupling to the TLS. If we now express X_ℓ in terms of the pseudospin operators of Eq. (1), we obtain interaction terms of the form[18]:

Table 1.   Classification of Metallic Glass Experiments
            (With References in Parentheses)

INSULATOR-LIKE BEHAVIOR	UNUSUAL BEHAVIOR
Linear Specific Heat (6)	Size of Saturation Threshold (9,10)
T^2 Thermal Conductivitiy (5,6,7)	T-Dependence of Relaxational Absorption (10,11)
Ln T Sound Velocity (8)	Absence of Two-Pulse Saturation (10)
Saturation of Resonant Absorption (9,10)	Absence of Phonon Echoes (10)
	Quantitative Discrepancies in Resonant Absorption (14,15)

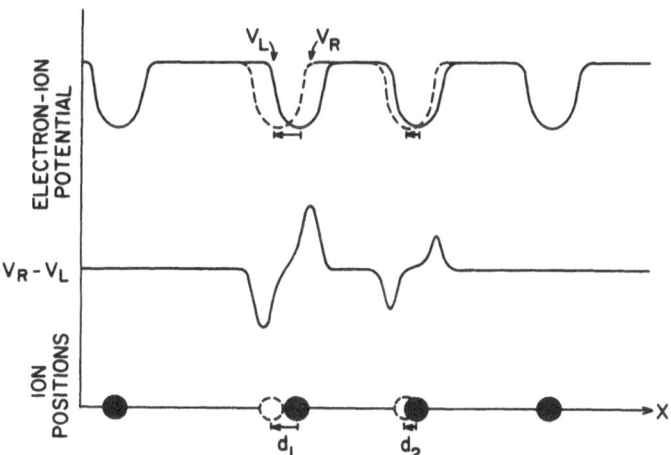

Figure 3.   Schematic drawing of the origin of the electron/TLS
coupling. The displacement of the ions in a TLS from the "L"
position to the "R" position leads to a difference potential,
$V_R - V_L$, which is felt by the electrons.

$$H_1 = \frac{1}{N} \sum_{i,k,q} e^{iqr_i} [v_{||}^i(q)s_{\mathbf{z}}^i + v_{\perp}^i(q) \ s_x^i] \ c_{k+q}^+ c_k \ , \tag{4}$$

where N is the total number of atoms, c_k is an electron operator, and r_i denotes the center of the TLS from which X_ℓ is measured. The coupling matrix elements are given by[18]

$$\left.\begin{matrix} v_{||}^i(q) \\ \\ v_{\perp}^i(q) \end{matrix}\right\} = \frac{2iu(q)}{v} \sum_\ell e^{iqx_\ell^o} \ \sin(qd_\ell/2) \ \left\{\begin{matrix} \Delta^i/E^i \\ \\ \Delta_o^i/E^i \ , \end{matrix}\right. \tag{5}$$

where Δ is the well asymmetry and Δ_o is the tunneling energy for the TLS[4].

Given the existence of this coupling, a number of interesting possibilities then present themselves. For example, what types of temperature-dependent contributions to the electrical resistivity arise from Eq. (4)? Can the TLS contribute (as phonons do) to the attractive electron-electron interactions which lead to superconductivity? Can the electrons produce RKKY-type "spin-spin" interactions among the TLS? Does a Korringa-type decay process significantly alter the dynamics of the TLS? For the present discussion of phonon scattering, it turns out that this last question is the most relevant.

PHONON SCATTERING REVISITED

The "spin lattice" relaxation rate T_1^{-1} controls the return to thermal equilibrium of a TLS which has been disturbed, for example, by an ultrasonic pulse. In insulating glasses, this rate is governed by exchange of energy with the thermal phonon bath[4]. In metals, electronic excitations provide an alternative, extremely efficient mechanism. This "Korringa process" follows from a Golden Rule calculation[10] based upon Eq. (4):

$$T_1^{-1} = \frac{\pi}{4\hbar} (\rho v_{\perp})^2 \ E \ coth \ (\beta E/2) \tag{6}$$

where ρ is the electronic density of states per atom at E_F and E is the TLS energy splitting. A rough estimate from Eqs. (5) and (6), using u/v = 2eV, $k_F d$ = 0.1, and ρ = .5 eV yields a T_1 in the nanosecond regime at T = 10mK and E/h = 1GHz. This value is to be compared with a T_1 in the millisecond regime for an insulating glass at the same temperature and energy splitting. It is this tremendous difference which explains many of the apparent puzzles listed in Table 1.

Consider, for example, the saturation curves shown in Figure 1. The point at which saturation occurs is determined by the

competition between the driving ultrasonic field e_i, which reduces
S_z, and the relaxation rate T_1^{-1} which restores it toward $<S_z>$. By
demanding that these two processes balance at threshold, it follows[4,10] that the critical ultrasonic intensity is proportional to
T_1^{-2}. Quantitative agreement with the observed threshold in
PdSiCu glass has been obtained from Eq. (6) using $\rho V \approx 0.2$, which
is reasonable for the values of u/v, $k_F d$, and ρ mentioned above[10].
In fact, this value of the coupling also gives a straightforward
explanation for the lack of multipulse saturation and coherent
effects in metallic glasses. The TLS simply returns to thermal
equilibrium long before the second pulse arrives.

Further evidence for this fast relaxation comes from the
anomalous relaxational absorption illustrated in Figure 2. This
type of absorption occurs[12] as the TLS responds to a harmonically
varying modulation of its energy splitting resulting from the term
proportional to D in Eq. (1). This response is governed by the
relationship between the frequency ω of the modulating strain
field and the response rate T_1^{-1} of the TLS, leading to an absorption proportional to $<A(A^2 + 1)^{-1}>$, where $A = \omega T_1$. Generally
this type of absorption remains small at low temperatures, where
A is large, and first becomes significant when A=1. As illustrated in Figure 2, this condition is met at a significantly
lower temperature in a metallic glass. Furthermore, the linear E
dependence of T_1^{-1} in a metal (compared with E^3 for an insulator)
leads to a more gradual temperature dependence in the metallic
glass. All of these results are in reasonable quantitative agreement with the data[10,11] for PdSiCu when the value $\rho V = 0.2$ is
used[10].

A conclusive test of these ideas would be to "remove" the
conduction electrons by cooling the metallic glass below the
superconducting transition at T_c. Recent calculations by Black
and Fulde[19] show that the electronic contribution to T_1^{-1} is
exponentially activated for $T<<T_c$ and increases with infinite
slope for $T \leq T_c$. The predicted consequence for the relaxational
absorption of Figure 2 is a dramatic crossover from the insulating
to the metallic curve as T increases toward T_c or as a magnetic
field is applied. Similar effects should be observable in the
saturation threshold and in multipulse ultrasonic experiments.

CONCLUSIONS

We have reviewed the evidence that TLS exist in metallic
glasses and have shown that most of the ultrasonic puzzles can be
understood in terms of a fast decay process via the conduction
electrons. Of the remaining puzzles, it is interesting to note
that the experiments of Araki et al[15] almost certainly fall into
a previously unexplored region ($T_1^{-1}>>\omega$) of resonant attenuation
in glasses, where new effects may be expected. Furthermore, it
has become apparent that the ultrasonic evidence for a new inter-

action in the form of Eq. (4) gives new impetus to the search for
effects of the TLS upon the electronic properties of metallic
glasses[16-18,20]. All of these new effects and possibilities are interesting in themselves, but they should not be allowed to obscure
the significance of the fact that a model originally designed for
insulating glasses[2,3] can be supplemented without essential modification to explain low-temperature experiments in metallic glasses.

REFERENCES

1.   D. Weaire, Contemp. Phys. 17, 173 (1976).
2.   P.W. Anderson, B.I. Halperin, and C.M. Varma, Phil. Mag. 25,
      1 (1972).
3.   W.A. Phillips, J. Low-Temp. Phys. 7, 351 (1972).
4.   S. Hunklinger and W. Arnold in Physical Acoustics, ed. by
      R.N. Thurston and W.P. Mason, (Academic Press, N.Y., 1976)
      12, 155.
5.   J.R. Matey and A.C. Anderson, J. Non-Cryst. Sol. 23, 129
      (1977).
6.   J.E. Graebner, B. Golding, R.J. Schutz, F.S.L. Hsu, and H.S.
      Chen, Phys. Rev. Lett. 39, 1480 (1977).
7.   H. v.Löhneysen and F. Steglich, Z. Phys. B 29, 89 (1978).
8.   G. Bellessa, J. Phys. C 10, L285 (1977).
9.   P. Doussineau, P. Legros, A. Levelut and A. Robin, J. Phys.
      (Paris) Lett. 39, L265 (1978).
10.   B. Golding, J.E. Graebner, A.B. Kane, and J.L. Black, Phys.
      Rev. Lett. 41, 1487 (1978).
11.   B. Golding, J.E. Graebner, and W.H. Haemmerle in Proc. of
      Int. Conf. on Lattice Dynamics, ed. by M. Balkanski (Flam-
      marion, Paris, 1977), p. 348; P. Doussineau, A. Levelut,
      G. Bellessa, and O. Bethoux, J. Phys. (Paris) Lett. 38,
      L483 (1977).
12.   J. Jäckle, L. Piche, W. Arnold, and S. Hunklinger, J. Non-
      Cryst. Sol. 20, 365 (1976).
13.   J.E. Graebner and B. Golding, Phys. Rev. B 19, 964 (1979).
14.   J.R. Matey and A.C. Anderson, Phys. Rev. B 17, 5029 (1978).
15.   H. Araki, G. Park, A. Hikata, and C. Elbaum, proceedings of
      this conference.
16.   R. Cochrane, R. Harris, J. Ström-Olson, and M. Zuckerman,
      Phys. Rev. Lett. 35, 676 (1975).
17.   J. Kondo, Physica 84B, 40, 207 (1976).
18.   J.L. Black, B.L. Gyorffy and J. Jäckle, Phil. Mag. (in press,
      Aug. 1979).
19.   J.L. Black and P. Fulde, Phys. Rev. Lett. 43, 453 (1979).
20.   J.L. Black and B. Gyorffy, Phys. Rev. Lett. 41, 1595 (1978).

Acknowledgements - The author wishes to thank P. Fulde, B. Golding,
J.E. Graebner, G. Gyorffy, and J. Jäckle for useful discussions.
Work supported by the Division of Basic Energy Sciences, Depart-
ment of Energy, Contract No. EY-76-C-02-0016.

DISCUSSION

R. C. Dynes: One should be able to choose a weak electron-phonon coupling material, and the saturation experiments should be substantially easier in that kind of material than in a strong electron-phonon coupling material.

J. L. Black: Yes, but we don't really know what the displacement vectors are.

R. C. Dynes: Have you estimated the contribution of the electron-tunneling level system to superconductivity? Your relaxation times are very fast, so I would expect it to make a measurable contribution.

J. L. Black: You can expect to measure and calculate λ in the BCS theory. That is proportional to the square of this coupling constant, and proportional to the two-level density of states. Because the two-level density of states is much smaller than the electron density of states λ turns out to be rather small. We estimate that it's only on the order of 10^{-3}. The spin-lattice relaxation is very fast because it's proportional to the electronic density of states, which is overwhelming compared to the two-level density of states. That's why I don't think that in materials with normal electron-phonon coupling strengths this mechanism would give a substantial contribution to the T_c of a superconductor.

R. C. Dynes: People have been tunneling into amorphous superconductors for ten years now and there's substantial weight at very low temperatures. That might be a good way to separate out the phonons from these states.

W. Eisenmenger: In your last model in which you described crucial tests for your ideas for electron coupling to the lattice systems I would like to ask about the experiment. It's very important that you exclude the direct interaction of the ultrasonic waves with the electrons which have almost the same shape. How strong are these effects?

J. L. Black: That's actually very small because the electron mean free path is so small in these materials. You can estimate what this direct electron-phonon contribution to the attenuation should be, and it's on the order of 10^{-2} dB cm^{-1}. This wouldn't be anything near what we're seeing here.

J. A. Rayne: Could you tell me how your calculations of the relaxation time are affected by the fact that these are highly absorbing systems, and that the plane wave states really aren't plane waves?

J. L. Black: What we really need to know is what the electronic
wave functions are, and if there's a lot of amplitude of those wave
functions near the two-level system then we are going to get a bigger coupling constant. If it happens to be that the two-level systems are where the electronic wavefunctions aren't, then we'll get
a decrease in the coupling constant. So, since I don't know what
the two-level system states actually look like, and I don't know
what the electronic states in amorphous materials are, I can't say
any more.

C. Elbaum: You said the relaxation time calculated gave you good
fits when $\omega\tau_1$ was greater than 1 at low temperatures. Do you care
to comment on what happens at higher temperatures?

J. L. Black: At higher temperatures it becomes very sensitive to
what one assumes for the density of states for the two-level systems. This is because when $\omega\tau$ becomes $\ll 1$, it is important to
take into account that there may be a broad distribution of relaxation times. What B. Golding and I did was make a particular
assumption which makes it look pretty lousy at high temperatures
because we wanted to fit basically four experiments with one set
of parameters.

ECHO PHENOMENA IN DISORDERED SOLIDS

Brage Golding and John E. Graebner

Bell Laboratories Murray Hill, New Jersey 07974

INTRODUCTION

It is characteristic of structurally disordered solids that many of their properties at, and below, liquid helium temperatures are dominated by localized excitations generally known as tunneling systems. A successful description[1,2] of many of the thermal[3] and heat transport[3,4] properties of glassy insulators (and even some metals)[5,6] has emerged from the realization that the tunneling systems are effective scatterers of phonons. The tunneling centers possess two energy levels and transitions between levels may be caused by the absorption and emission of resonant phonons. At sufficiently low temperatures, typically below 0.1 K, the tunneling systems may possess long phase memory times and this is reflected by the observation of coherent resonance phenomena such as phonon echoes.[7] As we shall emphasize in this review, a systematic study of echo behavior can reveal many previously obscured details of the interaction of phonons with tunneling centers. Tunneling systems also possess electric dipole moments, thereby providing an additional means, electric field excitation, for generating echoes.[8-10] A combination of phonon and electric echo techniques constitutes a very powerful approach to studying the low temperature properties of disordered solids.

TUNNELING SYSTEM DYNAMICS

In crystalline solids it is energetically favorable for atoms to occupy sites on a particular periodic lattice. In non-crystalline solids, because large static potential fluctuations are present, there may exist two nearly equivalent sites[1,2] separated by small distance d and barrier height V. If $k_B T > V$, atomic jumping rates between sites are dominated by thermally activated processes; if $k_B T < V$, no activation energy is required as a particle may tunnel through the barrier with probability $\Delta_0/\hbar = \Omega_0 \exp(-\lambda)$, where $\lambda = \hbar^{-1} d(2mV)^{1/2}$ and Ω_0 is the particle's zero-point frequency. An energy level doublet results from the double potential well structure, and the tunneling process can be assisted by absorption or emission of a quantum of energy $E = (\Delta_0^2 + \Delta^2)^{1/2}$, where Δ is the potential energy difference between the two configurations. Because of the random local environments in a disordered solid, both Δ_0 and Δ fluctuate from site to site and a continuous spectrum of E develops which gives rise to a density of states $P(E)$ which is nearly independent of E for $E/k_B \lesssim 10k$. Its magnitude is also remarkably invariant among a variety of disordered substances.[3,5,11]

Owing to the formal equivalence of quantum-mechanical two-level systems we may write the Hamiltonian of a single tunneling system as

$$H = H_0 + H_{int} = \frac{1}{2} E \sigma_z - [M \sigma_x e(t) + \mu' \sigma_x F(t)], \qquad (1)$$

where the σ's are 2×2 Pauli matrices, and e and F are classical strain and electric field amplitudes, respectively. The quantities M and μ' are, respectively, elastic deformation potential and electric dipole coefficients, coupling the two-level systems to long-wavelength perturbing fields. In the tunneling model[1,2,12] M is strictly a function of Δ_0 and Δ, a distinction we shall ignore throughout this paper. A three-dimensional unit pseudo-polarization vector $\vec{P}$ can be defined in terms of linear combinations of rotating-frame density matrix elements (u, v, w) and the state of the two-level system, or an ensemble of two-level systems, can be visualized by viewing $\vec{P}$ in this space. $\vec{P}$ may correspond to an actual electric polarization in an electric echo experiment, a stress in the phonon echo situation, or a magnetization in the NMR or EPR case. Rotations of $\vec{P}$ are generated by application of pulses of resonant radiation of duration τ with resulting rotation angle

$$\theta = \begin{cases} (M/\hbar) e \tau & \textit{acoustic excitation} \\ (\mu'/\hbar) F \tau & \textit{electric excitation} \end{cases} \qquad (2)$$

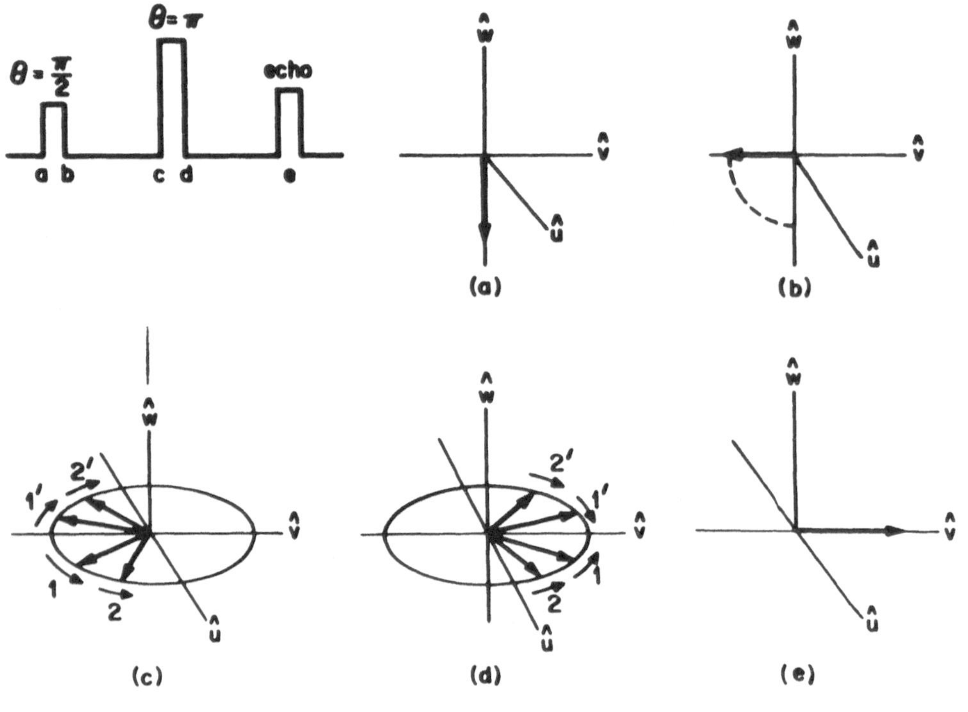

Fig. 1   Pulse sequence and pseudo-polarization vector rotations in a rotating frame describing the generation of a two-pulse echo in an inhomogeneously broadened resonant system.

The echo generation process can be understood by considering the rotations of $\vec{P}$ under a $(\theta_1, \theta_2) = (\pi/2, \pi)$ pulse sequence as shown in Fig. 1.[13] With the systems initially in thermal equilibrium $(w = w^o = -1, u = v = 0)$ a $\pi/2$ pulse places the ensemble into a non-stationary state (b) with maximum induced *transverse* polarization. This polarization rapidly decays to zero (c) as a result of the inhomogeneous nature of the excited systems. The second pulse rotates each two-level system's vector about $\hat{u}$ (d), and since the precessional sense is unchanged, all individual polarization vectors are momentarily rephased along $-\hat{v}$ (e) at a time after θ_2 equal to the θ_1, θ_2 separation, τ_{12}. At this time all systems are again oscillating in phase and a coherent pulse of radiation, the echo, is emitted.

In an echo experiment, not all excited systems will contribute to the signal since various processes may cause the phases of some systems to have become randomized.

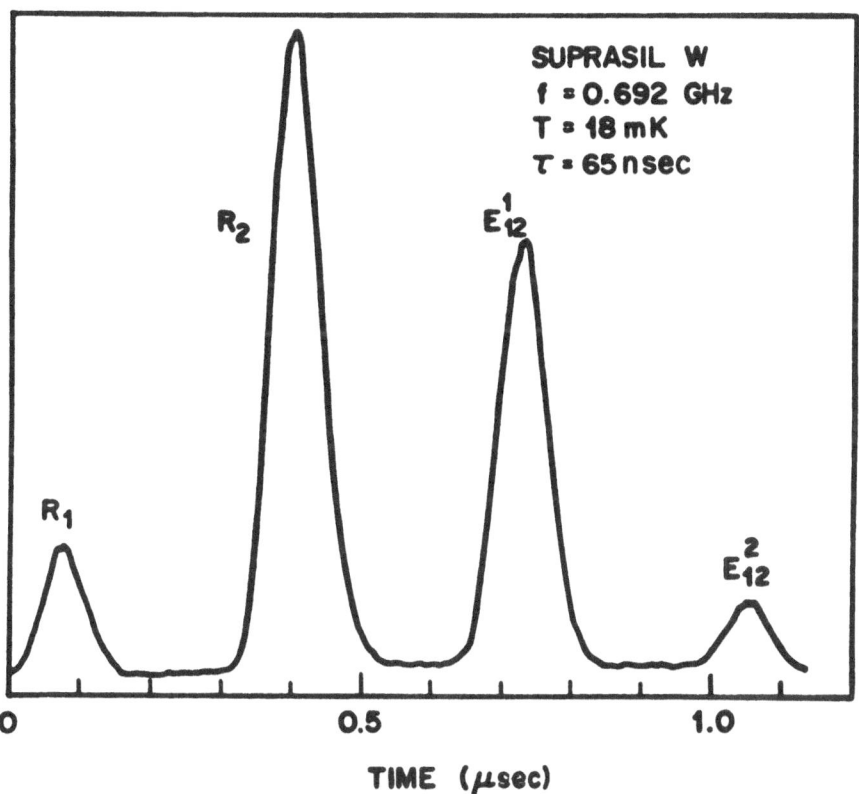

SUPRASIL W
$f = 0.692$ GHz
$T = 18$ mK
$\tau = 65$ nsec

R_2

E_{12}^1

R_1

E_{12}^2

0   0.5   1.0

TIME (μsec)

Fig. 2   Spontaneous phonon echo in a pure silica glass. R_1 and R_2 are acoustic reflections observed after one round trip through the sample, while E_{12}^1 is the echo. Acoustic pulse width is 65 nsec, and the pulse separation is 330 nsec.

This may occur from pure phase-disrupting processes, with characteristic time T_ϕ, or by longitudinal relaxation, with time T_1. The measurement of the echo decay as a function of τ_{12} yields T_2', the homogeneous relaxation time, where $T_2'^{-1} = T_\phi^{-1} + \frac{1}{2} T_1^{-1}$. It is possible to measure T_1 directly in echo experiments, either by generating a three-pulse stimulated echo,[14] or by detecting the recovery of $\vec{P}$ by generating two-pulse echoes from an initially inverted system[15] ($\hat{w} = +1$). Thus, echo methods allow a complete experimental determination of the relaxation times of tunneling systems.

Using echoes, it is possible to determine the coupling strength between the exciting field and resonant system,[7] i.e. M and μ' in Eq. (1). This results from the parametric dependence of echo amplitude E on θ_1 and θ_2. The relationship

$$E(\theta_1, \theta_2) = E_0 \sin \theta_1 \sin^2(\theta_2/2) \tag{3}$$

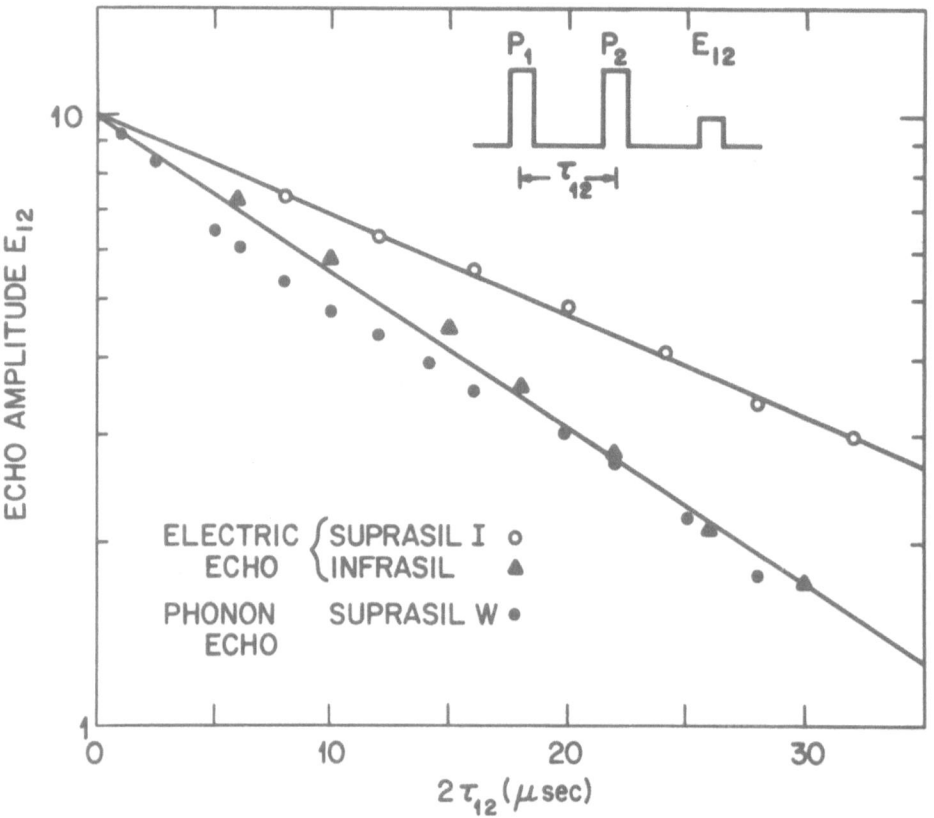

Fig. 3   Decay of phonon and electric two-pulse echo amplitudes in silica glasses vs. pulse separation τ_{12}. Echo decay is the same for intrinsic tunneling systems in OH-free glasses, Infrasil and Suprasil W, but is slower for the OH-associated systems in Suprasil I (1200 ppm OH).

is valid for a homogeneous line and can be generalized for the inhomogeneous case.[16] In both situations, however, it is true that a maximum in E occurs for a $(\pi/2, \pi)$ sequence, or if $\theta_1 = \theta_2$, for a $(2\pi/3, 2\pi/3)$ sequence. Therefore, an accurate experimental determination of the field amplitude and τ for which the echo maximum occurs allows Eq. 1 to be solved for either M or μ'.

PHONON ECHOES IN SiO$_2$

Fused Silica Glass

The first coherent resonance phenomenon observed in a glass was the phonon echo[7] in a pure (Suprasil W) fused silica. Fig. 2 shows an experimental tracing of phonon

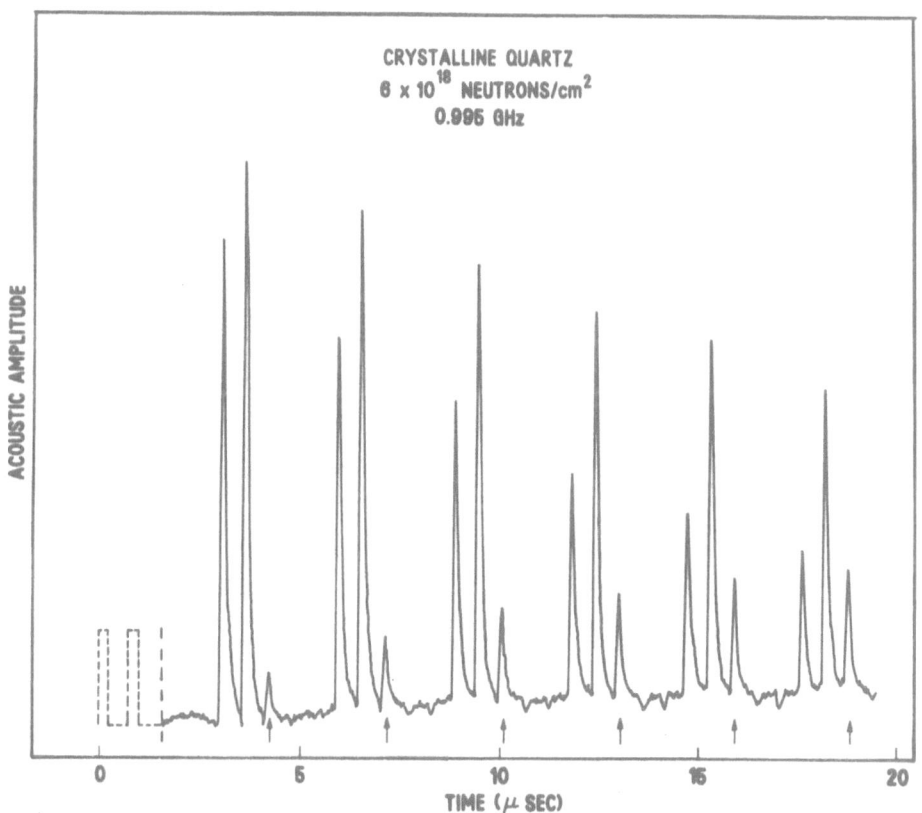

Fig. 4   Acoustic reflections and phonon echoes generated by a two-pulse sequence in neutron-irradiated quartz. Echoes are indicated by arrows. Each sequence of reflections and echoes are separated by 3 μsec, the round-trip transit time of the longitudinal waves in quartz. Note the increase in echo amplitude with propagation distance.

echoes generated by the two acoustic pulses at the left which have propagated 1.27 cm. The two generating pulses initially possessed equal amplitudes (slightly greater than R_2) but after propagating this distance R_1 is highly attenuated, illustrating the coherent absorption of energy from the pulse required to produce the situation depicted in Fig. 1b. The echo E_{12}^1 is larger than R_1 and nearly as large as R_2, illustrating that a substantial fraction of the energy utilized in exciting the tunneling systems is re-radiated into echoes. E_{12}^2 is a secondary echo generated by R_2 and E_{12}^1.

The decay of the two-pulse phonon echo is shown in Fig. 3. Assuming exponential decay, a homogeneous lifetime $T_2' \approx 16$ μsec at 18 mK is obtained. Since $T_1 \gg T_2'$, the echo decay results from pure dephasing processes. A description of dephasing due to phonon-mediated interactions between non-resonant tunneling systems has been shown to give a good description of this experiment.[17] Thus, echoes have provided strong quantitative evidence for the coupling of tunneling systems through the phonon field.[18, 19]

Neutron-Irradiated Quartz (NIRQ)

As discussed above, a characteristic feature of the glassy state is the existence of tunneling systems at very low energies. However, one may ask whether it is possible for such low energy states to exist in *crystalline* solids into which only a slight amount of disorder has been introduced.[11, 20] The initial observation[21] of non-linear acoustic propagation in a crystal quartz sample lightly irradiated with neutrons suggested that echo studies of NIRQ might be informative.[22] Fig. 4 shows the result of propagating two acoustic pulses in NIRQ at 20 mK. Since the absorption is low, it is possible to see many sets of reflections, each separated by a double-transit time of 3 μsec. Note that

TABLE I. Results of Phonon Echo Experiments on SiO_2

	α-Quartz	Fused Silica	NIRQ ($6\times10^{18}n/cm^2$)
$10^{-5}V_L$ (x-axis, cm/sec)	5.75	5.85	5.75
Density (g/cm^3)	2.65	2.20	2.65×0.999
$10^{-7}\bar{P}M_l^2$ (erg/cm^3)		20	1.3
M_l (eV)		1.5 ± 0.4	1.2 ± 0.4
T_1 (μsec)		150[a]	250[b]
T_2' (μsec)		16[a]	7[b]
$10^{-30}\bar{P}$ (erg^{-1}cm^{-3})		35	5

[a]At 18 mK and 0.692 GHz

[b]At 15 mK and 0.995 GHz

although the generating pulses *decrease* monotonically in amplitude at each pass, the echo amplitude *grows* monotonically. This increase as a function of propagation distance is related to another coherent phenomenon, self-induced-transparency.[23-26]

There is evidence that suggests that the resonant centers produced in quartz by irradiation are the same as the tunneling systems in silica glass. Table I shows a comparison of parameters for the two substances. $\bar{P}M_l^2$ has been obtained from sound velocity measurements. We see that the longitudinal coupling parameters M_l are the same, within experimental error, in both substances indicating the same center is involved. The result is also obtained that $\bar{P}$ for NIRQ is only a factor of 7 smaller than in the glass. This is surprising since the irradiation dosage (6×10^{18} neutrons/cm^2) was sufficiently small to leave the NIRQ sample essentially indistinguishable from unirradiated quartz as evidenced by its sound velocity, density,[27] piezoelectric activity,[21] and by the existence of sharp Bragg x-ray reflections. It has been argued that neutron irradiation in quartz produces thermal spikes which yield glassy regions.[28] Such regions, if they exist, must constitute an extremely small fraction of the total volume of the crystal, and cannot be the only location of the resonant systems. This is readily seen by comparison of the density change on irradiation, 0.1%, with the density difference between crystalline and glassy SiO_2, 20% as shown in Table I. If 1/7 of the irradiated crystal had the density of glassy SiO_2 one would expect to see an order of magnitude greater change in ρ on irradiation. Since ρ in NIRQ is only slightly reduced from quartz, we argue that tunneling systems can exist in the inhomogeneously strained matrix surrounding displacement regions. Therefore, it appears that complete disorder is not required for the formation of low-energy tunneling centers.[20]

ELECTRIC ECHOES

Echoes can be generated in glasses by electric field excitation. The electric echo was first observed[8] in silica glass and several detailed studies of their properties have been reported.[9, 10, 29] Recent experiments[10] in a series of pure and OH-doped silicas have shown that at least two distinct types of tunneling systems may exist, or coexist, in SiO_2. The intrinsic center possesses a weak, but finite, electric dipole moment and a strong phonon coupling. This is the center studied by phonon echoes[7] in Suprasil *W*. Fig. 3 shows the nearly identical two-pulse echo decay for phonon and electric excitation under similar experimental conditions in two forms of silica with negligible OH content, Suprasil *W* and Infrasil. On the other hand, an OH-associated center may also exist, and it is characterized by an electric dipole moment 6 times greater than the intrinsic species, but a phonon coupling only about half as large. The weaker phonon coupling leads, for example, to a distinctly longer dephasing time T_2' as shown in Fig. 3 for a Suprasil *I* glass containing 1200 ppm OH.

BACKWARD-WAVE ECHO

A different type of echo phenomenon, the backward-wave phonon echo, has been observed[30] in several glasses and related to the presence of OH impurities. This echo is generated by wave-vector reversal of a propagating phonon pulse by a uniform electric field pulse at the same frequency. Unlike the echoes discussed above, the two-pulse backward wave echo does not require for its existence long-lived coherence in the tunneling systems. The echo decay is governed by the acoustic attenuation of the medium rather than the relaxation times of the tunneling systems. Since it is a parametric process, it requires an appreciable non-linearity of the medium and is typically observed at excitation fields orders of magnitude greater than those used to generate resonant echoes.

CONCLUSIONS

We have described how phonon and electric echo studies have proven to be a valuable method for studying the interaction of phonons with tunneling systems in structurally disordered solids. One can view the tunneling systems as local centers which can be manipulated by external resonant fields to probe their environment. As a further example, we have shown that the two-level systems may be introduced into nominally crystalline substances by particle bombardment. Thus, a much broader class of materials, defective crystals, should prove to be amenable to study by phonon and electric echoes.

[1]   P. W. Anderson, B. I. Halperin, and C. M. Varma, Philos. Mag. *25* : 1 (1972).

[2]   W. A. Phillips, J. Low Temp. Phys. *7* : 351 (1972).

[3]   R. C. Zeller and R. O. Pohl, Phys. Rev. B*4*: 2029 (1971).

[4]   M. P. Zaitlin and A. C. Anderson, Phys. Rev. Lett. *33* : 1158 (1974); Phys. Rev. B*12*: 4475 (1975).

[5]   J. E. Graebner, B. Golding, R. J. Schutz, F. S. L. Hsu, and H. S. Chen, Phys. Rev. Lett. *39* : 1480 (1977).

[6]   J. R. Matey and A. C. Anderson, Phys. Rev. B*16*: 3406 (1977).

[7]   B. Golding and J. E. Graebner, Phys. Rev. Lett. *37* : 852 (1976); J. E. Graebner and B. Golding, Phys. Rev. B*19*: 964 (1979).

[8]   B. Golding, J. E. Graebner, and W. H. Haemmerle, in "Amorphous and Liquid Semiconductors", W. E. Spear, ed., CICL University of Edinburgh, Edinburgh (1977) p. 367.

[9]   L. Bernard, L. Piché, G. Schumacher, J. Joffrin, and J. E. Graebner, J. Phys. (Paris) Lett. *39* : L126 (1978).

[10] B. Golding, M. V. Schickfus, S. Hunklinger, and K. Dransfeld (to be published).

[11] R. B. Stephens, Phys. Rev. B*8*: 2896 (1973); Phys. Rev. B*13*: 852 (1976).

[12] J. Jäckle, Z. Phys. *257* : 212 (1972).

[13] See, for example, C. P. Slichter, "Principles of Magnetic Resonance", Springer-Verlag, Berlin (1978).

[14] E. L. Hahn, Phys. Rev. *80* : 580 (1950).

[15] A. Abragam, "The Principles of Nuclear Magnetism", Oxford University Press, London, (1961).

[16] W. B. Mims, in "Electron Paramagnetic Resonance", S. Geschwind, ed., Plenum Press, New York, (1972).

[17] J. L. Black and B. I. Halperin, Phys. Rev. B*16*: 2879 (1977).

[18] J. Joffrin and A. Levelut, J. Phys. (Paris) *36* : 811 (1975).

[19] S. Hunklinger and W. Arnold in "Physical Acoustics" W. P. Mason and R. N. Thurston, eds., Vol. XII, Academic Press, New York (1976), p. 155.

[20] L. F. Lou, Solid State Commun. *19* : 335 (1976).

[21] C. Laermans, Phys. Rev. Lett. *42* : 250 (1979).

[22] B. Golding, J. E. Graebner, W. H. Haemmerle, and C. Laermans, Bull. Am. Phys. Soc. *24* : 495 (1979).

[23] B. Golding, IEEE Trans. Sonics and Ultrasonics *SU*24 : 692 (1977).

[24] J. E. Graebner and B. Golding in "Proceedings of the International Conference on Lattice Dynamics", M. Balkanski, ed., Flammarion Press, Paris (1978), p. 464.

[25] S. L. McCall and E. L. Hahn, Phys. Rev. *183* : 457 (1969).

[26] E. L. Hahn, N. S. Shiren, and S. L. McCall, Phys. Lett. A*37* : 265 (1971).

[27] M. Wittels and F. A. Sherrill, Phys. Rev. *93* : 1117 (1954).

[28] See, for example, D. S. Billington and J. H. Crawford, "Radiation Damage in Solids", Princeton University Press, Princeton (1961).

[29] L. Bernard, L. Piché, G. Schumacher, and J. Joffrin, J. Low Temp. Phys. *35* : 411 (1979).

[30] N. S. Shiren, W. Arnold, and T. G. Kazyaka, Phys. Rev. Lett. *39* : 239 (1977).

****

DISCUSSION

R. O. Pohl: I was wondering if you know something about the nature of neutron-induced defect states in the silica.

B. Golding: If we really knew something about that we thought we would answer a number of important questions. Upon looking at the literature as closely as we could we found there is not really very much known about the nature of the neutron damage in oxide materials. One of the recurrent ideas is that a high-energy neutron creates some kind of molten region. I think one can dispense with that idea because one would have to have such a large fraction of the material filled up with these glassy regions to get such a large density of states. We don't think that we have melted anything at all. In fact, I would suggest that as a result of some rather careful X-ray studies which are in progress the nature of the effect is a clustering of interstitials into small localised regions perhaps on the scale of tens of A, perhaps as large as 50Å. Of course, that would depend upon the radiation dosage, but at these levels it's probably small local clusters of interstials.

R. O. Pohl: You did mention at the end of your talk that we are now looking at tunneling defects in crystalline system, but you have just described that you don't understand the nature very well. I would like to point out that thanks to the work of V. Narayanamurti and J. P. Harrison, just to mention a few names, we do have a fairly good idea of tunneling defects in crystalline solids - not in quartz, but in alkali halides. I wonder if you have thought about doing the sort of work which you have started in the quartz also in the alkali halides.

B. Golding: I would certainly agree with what you are saying. Someone should probably do that. We're not going to do that.

E. R. Dobbs: Have you got evidence whether the echoes increase linearly with radiation? Have you any evidence whether they disappear with annealing the radiation damage?

B. Golding: That is something we are intending to do.

FURTHER STUDY OF THE "GLASSY" LOW TEMPERATURE

PROPERTIES OF IRRADIATED CRYSTALLINE QUARTZ

C. Laermans[x] and B. Daudin[xx]

x Katholieke Universiteit Leuven, Afd. Vaste Stof en
  Hoge Druk-Fysica, 3030 LEUVEN, Belgium.
xx Service des Basses Températures, Centre d'Etudes
  Nucléaires de Grenoble, 85X, 38041 GRENOBLE-CEDEX,
  France.

The low temperature anomalous thermal, acoustic and dielectric
properties of amorphous materials are explained by assuming the existence of low energy excitations which can be successfully described
as systems in which a tunneling process between two nearly degenerate
energy levels can occur[1], or more simply as two-level systems (2LS)[2].
Recently it has been seen that a quartz crystal, after light fast-
neutron-irradiation, shows similar hypersonic properties as glasses[3].
The further observations of phonon echoes in the same sample showed
that the acoustic deformation potential coupling to the defect centers, responsible for the echo and their lifetemes T_1 and T'_2, are
remarkably similar to those found in vitreous silica[4]. A density of
states one order of magnitude smaller than in the glass can account
for all the observations[4]. It is also remarkable that fast neutron-
irradiation of the crystal introduces the 2LS, while on the other
hand the irradiation of the glass reduces the number of such systems[5,6]. The purpose of the work reported here was to extend the
study of defective quartz crystals. 9 GHz hypersonic attenuation
measurements in and without a magnetic field and thermal conductivity
measurements were carried out in a neutron-irradiated and an electron-irradiated sample.

9 GHz hypersonic attenuation measurements were carried out in a
neutron-irradiated sample in a magnetic field of 0 to 1.5 T, as a
function of acoustic intensity at 1.5 K and as a function of temperature (between 1.5-4.2 K). The temperature and acoustic intensity
dependence were found to be the same in a magnetic field as the results previously reported without field[3]. The experiments were
carried out on the same sample as used before[3,4], which was slightly

defective and still long range ordered after a fast neutron dose of
6.10^{18}n/cm^2. The magnetic field was directed along the x-axis of the
sample in one experiment and in a direction which makes an angle of
45° with the axis in another experiment.

Similar experiments were carried out in an electron-irradiated
sample. This sample was irradiated with 2 MeV electrons up to a dose
of 3.10^{19}e/cm^2. It was rotating during the irradiation to ensure
uniform treatment and kept at about 50°C. The measurements did not
reveal any temperature or acoustic intensity dependence of the hypersonic attenuation, and therefore no evidence for the presence of 2LS.

The thermal conductivity measurements of the same samples are
plotted in fig. 1 together with the data for a sodasilicate glass
sample and the results of Zeller and Pohl[7] for a virgin quartz
sample. The data for neutron-irradiated quartz are an extension down

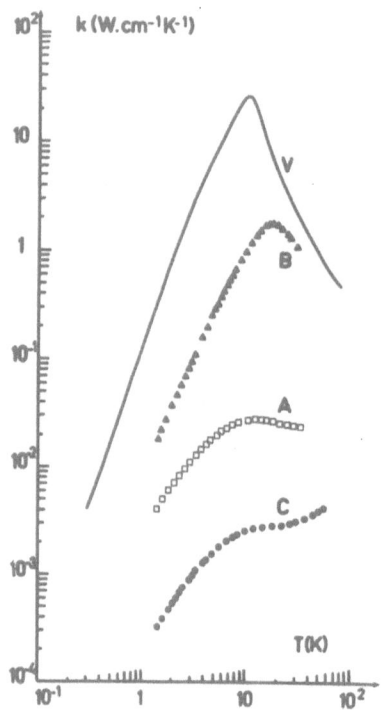

Fig. 1 : Thermal conductivity data.
      V : quartz crystal ; 5x5x40 mm (after Zeller and Pohl)[7]
      A : neutron-irradiated quartz ; 3 mm diameter, 8.4 mm length
      B : electron-irradiated quartz ; 3 mm diameter, 6 mm length
      C : sodasilicate glass ; 3 mm diameter, 6 mm length

to lower temperatures of Berman's[8] measurements. They are consistent
with Berman's data, taking into account the possible difference in
irradiation circumstances. It is seen that the thermal conductivity
of this slightly defective crystal has a behavior which qualitatively resembles very much that of the glass. At the lowest temperatures
the data are consistent with the T^2 behavior so well known for glasses. Furthermore it is found that the magnitude of the thermal conductivity in this low temperature region is consistent with a number
of 2LS one order of magnitude less than in vitreous silica, as already
found before[4]. Here it is taken into account that the coupling
strength between the 2LS and the acoustic phonons is similar to that
in vitreous silica[4]. In the electron-irradiated sample, the thermal
conductivity data give no evidence for the presence of 2LS, in agreement with the hypersonic observations.

    While there is evidence that neutrons cause single defects and
clusters of defects[9], electrons give rise only to single defects[10].
For the used neutron and electron doses, estimates show that the
number of single defects in both samples is of the same order of
magnitude : $10^{19}/cm^3$. If this is the case, the absence of any measurable effect in the electron-irradiated sample would indicate that
2LS, which are believed to cause the acoustic anomalies in the neutron-irradiated sample, are not related to the single defects.

    We are grateful to M. LOCATELLI who made the thermal conducti-
vity experiments possible. We also thank Mrs. A.M. DE GOËR, MM.
BONJOUR and CALEMCZUK for fruitful discussions. One of us (C.L.)
whishes to thank the C.E.N. Grenoble for its kind hospitality and
the Belgian N.F.W.O. and I.I.K.W. for financial support.

1. P.W. Anderson, B.I. Halperin, C.M. Varma, Philos. Mag. 25, 1,
   (1972) ; W.A. Phillips, J. Low Temp. Phys. 7, 351 (1972).
2. Hunklinger and W. Arnold in "Physical Acoustics", Ed. W.P. Mason
   and R.N. Turston (Academic Press, New York, 1976) 12, 155.
3. C. Laermans, Phys. Rev. Lett. 42, 250 (1979).
4. B. Golding, J. Graebner, B. Haemmerle, C. Laermans, Bull. Am.
   Phys. Soc. 24, 495 (1979).
5. C. Laermans, L. Piché, W. Arnold and S. Hunklinger in "Phys. of
   Non Crystalline Solids", Ed. G.H. Frischat (Trans. Tech. Aeder-
   mannsdorf, Switzerland, 1977), 562.
6. T.L. Smith, P.J. Anthony and A.C. Anderson, Phys. Rev. B., 17,
   4997 (1978).
7. R.C. Zeller and R.O. Pohl, Phys. Rev. B., 4, 2029 (1971).
8. R. Berman, Proc. Roy. Soc. A 208, 90 (1951).
9. P.G. Klemens, Proc. Roy. Soc. A 208, 109 (1951).
10. D.S. Billington and J.H. Crawford, "Radiation Damage in Solids",
   (Princetown, New Jersey, 1961).

DISCUSSION

J. W. Vandersande: I'd like to point out that in your thermal conductivity data on neutron-irradiated quartz if you had clusters that
would show up as a change in scattering from geometrical to Rayleigh
scattering - your results do not show this. But you did stop at 1°K,
and if you had gone lower, you might actually have seen the transition. If you don't get this changeover, it's unlikely that you have
clusters. You just have point defects or very small clusters.

C. Laermans: I don't agree with your implication.

A. C. Anderson: John Gardner measured the thermal conductivity of
irradiated quartz crystals to much lower temperatures and the temperature dependence is T^2 all the way.

EFFECT OF THERMAL TREATMENTS ON THE LOW TEMPERATURE SPECIFIC HEAT

OF VITREOUS SILICA

J.C. Lasjaunias, G. Penn, M. Vandorpe

Centre de Recherches sur les Très Basses Températures,
C.N.R.S., BP 166 X, 38042 Grenoble-Cedex, France

During the last few years there have been several experimental attempts to test if the low temperature anomaly of the specific heat of glasses could be sensitive to special physical treatments in order to clear up the structural origin of the excitations responsible for this anomaly. In the case of vitreous silica, the most extensively studied material, irradiation by fast neutrons[1] produced a big variation of the mass density ($\sim$ 2 %) and a decrease of 30 % for C_p. However other treatments such as fast electrons irradiation[2], strain induced by quenching[3] and thermal treatments[4], revealed no detectable or small effects.

We have investigated the effect of thermal treatments on Suprasil-W vitreous silica. In comparison to ref. 4, all our measurements were done on the *same* silica sample, and the temperature range extended to lower temperatures : 30 or 80 mK, depending on the size of the specimens. The results for the initial sample (refered to as sample 1 in ref. 5) and the techniques used have been previously published[5,6]. Several modifications have been made for these measurements such as the use of doped Si thermometers instead of carbon ones.

The procedure was the following : a piece of the initial sample[7] in form of a rod ($\phi \sim$ 2 cm, $\ell \sim$ 5 cm) was successively stabilized at 1100°C for 7 days (then air-quenched) and at 1300°C for 6 hours (then water-quenched). Such durations were required to enable the structural equilibrium to be established.[8] The density was measured by the Archimedes method in water : no variation was detected between the initial value and after the first treatment, ρ = 2.203 $\pm$ 0.0005 g/cm^3. The second treatment resulted in an increase to 2.205 g/cm^3. Thereafter this bulk sample was cut into

slabs ($\phi \sim 2$ cm, e $\lesssim 1$ mm) and submitted to the next stabilization
at 1300°C for 24 hr (then water-quenched). Traces of surface crys-
tallinity of the cristobalite phase were removed by chemical or mechanical treatment and each slab tested by X-rays. The mean density
was found to be $2.207_5 \pm 0.0005$ g/cm^3 : an increase of about 10^{-3} in
comparison with the precedent value (we have verified indeed that
the variations of density both at 1100 and 1300°C were greater for
the samples of small size, as if quenching was more effective than
for the bulk rods).

Results are shown in figure 1. We distinguish two temperature
ranges :
a) Below about 0.5 K – The effect of the thermal treatments is
rather slight above 0.1 K. For the bulk sample annealed at 1100°C
the data are shifted upwards from the initial values by about 30 %
at 40 mK and by 10 % from 0.1 to 0.5 K. The next treatment at 1300°C
modifies very slightly the specific heat up to 0.1 K which joins
again the initial values at 0.2 K. No further change is observed
for the slabs annealed at 1300°C and measured down to 80 mK (we have
not reported these data on the figure for clarity).
b) Between 0.5 and 2 K – In this range a direct correlation
appears between the specific heat and the thermal treatments : C_p
decreases systematically when the temperature of stabilization increases and the effect is strongly enhanced for the sample in the
form of slabs. For a sample of IR Vitreosil annealed at 1400°C C_p
was found to be reduced by about 10 % from a sample annealed at
1000°C for temperature between 1.3 and 4 K, as already pointed out
by White and Birch.[9] However for Suprasil-W, a bump in the curve
$C_p(T)$ exists at these temperatures, which is not observed in the
other types of vitreous silica, including Suprasil with hydroxyl
content[5]. We don't know the origin of this bump, but a similar reduction of this excess, whereas the values below 0.5 K remained un-
modified, has been obtained by irradiation with fast electrons[2].
The irradiation resulted in the creation of a density of free spins
of order of 10^{17}/cm^3.

In conclusion the effects observed in the two temperature ranges are very different, probably involving separate origins.
One may suppose that the increase of density of the samples treated
at 1300°C, when in the form of slabs, revealed the presence of inhomogeneities in the bulk form. These internal strains indeed gave rise to cracks during the quenching. Also there is a large difference
in the strain density between the two bulk 1100°C and 1300°C samples.
In consequence the distribution of the two-level systems responsible
for the anomaly below 0.5 K appears to have little sensitivity to
the density and the presence of strains.
On the other hand, the effect above 0.5 K which is well related to
the thermal treatments seems not entirely caused by the variations
of the density, especially when compared to the results of White
and Birch. Another mechanism, for example, the annealing of a defect,
is perhaps involved.

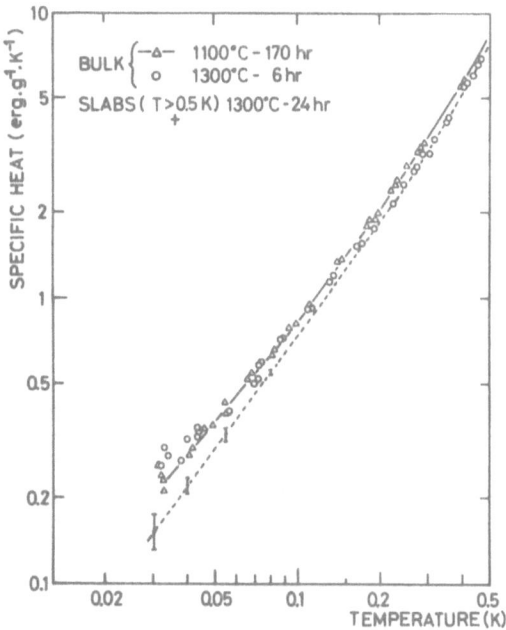

Fig. 1. Specific heat of Suprasil-W
silica before heat treatments
from data of ref. 5 and recent
results above 1 K (---), and
after the successive stabili-
zations as indicated in the
insert (symbols Δ, o, +).
For the slabs, only
results above 0.5 K
(symbol +) are reported.

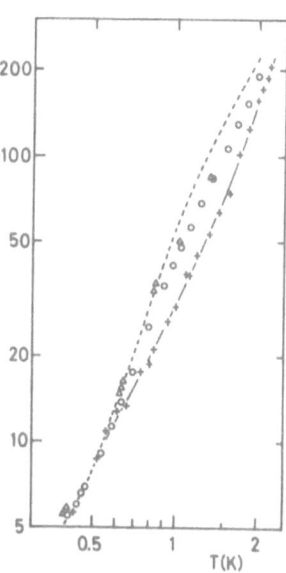

REFERENCES

1. T.L Smith, P.J. Antony, and A.C. Anderson, Phys. Rev. B17, 4997 (1978).
2. H.v. Löhneysen and B. Picot, J. Physique 39, C6-976 (1978).
3. L.E. Wenger, K. Amaya, C.A. Kukkonen, Phys. Rev. B14, 1327 (1976).
4. R.L. Fagaly and R.H. Bohn, J. of Non-Cryst. Solids 28, 67 (1978).
5. J.C. Lasjaunias, A. Ravex, M. Vandorpe, S. Hunklinger, Solid State Comm. 17, 1045 (1975).
6. J.C. Lasjaunias, B. Picot, A. Ravex, D. Thoulouze, M. Vandorpe, Cryogenics 17, 111 (1977).
7. An annealing of 24 hr at 1100°C was previously made by the producer Heraeus-Schott.
8. G. Hetherington, K.H. Jack, J.C. Kennedy, Phys. and Chem. of Glasses 5, 130 (1964).
9. G.K. White and J.A. Birch, Phys. and Chem. of Glasses 6, 85 (1965).

ELECTRICAL POLARIZABILITY OF PHONON SCATTERING STATES IN GLASSES[*]

Arup K. Raychaudhuri

Physics Department, Cornell University

Ithaca, N. Y. 14853

Tunneling states[1] in glasses should be electrically polariz-able. Hence one might expect that the thermal conductivity would change in an electric field. In two previous studies, no effect was seen[2,3]. In the first experiment, however, the dielectric polarizability of the glass was not known, and the sensitivity of the experiment was reduced because of the relatively high tempera-tures used, $T > 2°$ K. In the second experiment, the experiment was performed at lower temperatures ($T > 0.15°$K), and the polarizability was measured. The latter, however, was found to be quite small, which again should decrease the sensitivity of the experiment. In the present experiment, we have chosen a glass of very high polar-izability, and have also measured as low as $0.2°$ K. The experimen-tal technique was the same as that used by Stephens[3]. The soda-silica glass sample had the composition 75 wt% SiO_2, 12 wt% Na_2O, with the remainder B_2O_3, K_2O, and Al_2O_3. The relative change in thermal conductivity, defined as $\Delta \Lambda / \Lambda = (\Lambda(E,T) - \Lambda(0,T))/\Lambda(0,T)$ in an applied field as high as 53 kV/cm, is shown in Fig. 1. As in the previous studies, no systematic change was detectable. In zero field, the thermal conductivity was found to scatter within the range indicated by the two curves in Fig. 1, arising mostly due to thermal drift. Average changes in the electric field are marked as data points, with the error bars indicating the r.m.s. scatter. Above 0.3 K, the maximum relative change is $|\Delta \Lambda / \Lambda| \leqslant 0.2\%$, with r.m.s. standard deviation of $\pm 0.2\%$. At 0.2 K, a maximum $(\Delta \Lambda / \Lambda)$ $\leqslant 0.8\%$, was observed with r.m.s. standard deviation $\pm 0.4\%$. The dielectric constant ϵ (T) at low temperatures can be explained with the tunneling model, assuming a uniform density of states[4], n_o:
$\Delta \epsilon = \epsilon(T) - \epsilon(T_0) = -\frac{8\pi}{3} n_o \mu^2 \ell n(T/T_0)$ (cgs units), where T_0 is some arbitrary reference temperature, and μ is the dipole moment. From the data, we determine $n_o \mu^2 = 8 \times 10^{-4}$. Since n_o is unknown, we

29

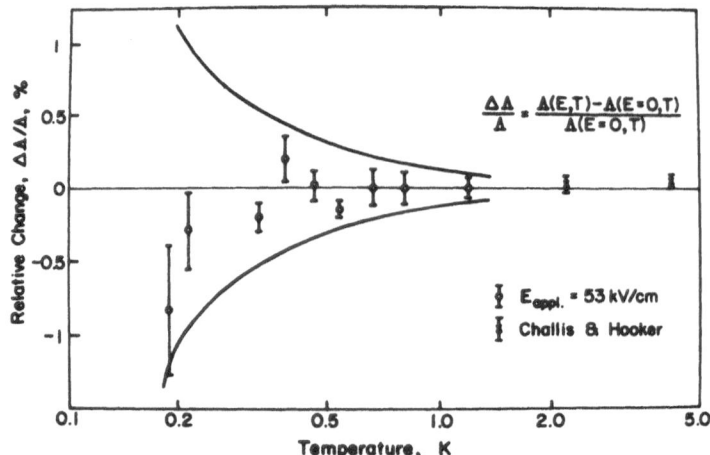

Fig. 1   The relative change of thermal conductivity in an applied
         field.   The two solid curves give the range of scatter of
         the zero field Λ (T).   The data points are average change
         and the bars indicate r.m.s. scatter.

can put a <u>lower limit</u> on μ by using the total density of states
obtained from specific heat measurements[3], $n_0 = 7.7 \times 10^{32}$ erg^{-1}
cm^{-3}.   From this we obtain $\mu \gtrsim 1$ Debye.   If the same dipole moment
were associated with the phonon scattering states, the applied field
would cause a change in the energy of the resonant scatterers by
as much as $uE = 5K\ k_B$ (local field corrected).   This would be about
ten times larger than the energy of the dominant heat carrying
phonons at $T = 0.2$ K.

      In the tunneling model, the states which scatter the phonons
most strongly are symmetric (tunnel splitting $\approx$ total energy).
If we assume that the phonon scattering states are polarizable and
carry an average dipole moment ~ 1 D, then they will cease to do
so when a field is applied.   An absence of change in Λ (E,T) therefore can only be possible if an equal number of states with correct
tunnel splitting becomes symmetric by the application of the field.
This however cannot be the case in the frame work of tunneling
model.   The only explanation of the null effect can therefore be
that the majority of the phonon scattering states are not polariz-
able.   From $|\Delta \Lambda / \Lambda| \lesssim 1\%$ and using a lower limit of $\mu \sim 1D$, we
estimate that more than 95% of phonon scattering states are not
dipole active.

      Hence, we conclude that the states seen in thermal conductiv-
ity are different from those seen in dielectric measurements and

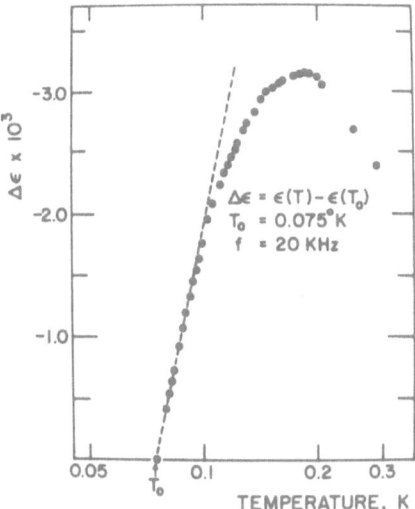

Fig. 2 Change in dielectric constant of the soda silica based
glass with temperature. The slope of the dashed line
gives $n_o \mu^2$ (see text).

their interactions are not straightforward. It is extremely
unlikely that the phonon active states are connected with the
motions of individual atoms or molecules, since such are likely to
carry a dipole moment.

* Work supported by the National Science Foundation, Grant
# DMR 78-01560

References
1. J. L. Black, B. I. Halperin, Phys. Rev. B16, 2879 (1977).
2. L. J. Challis, C. N. Hooker, J. Phys. C5, 1153 (1972).
3. R. B. Stephens, thesis - Cornell Univ., report # 2304 of the
        Materials Science Center (1974).
4. M. von Schickfus, S. Hunklinger, J. Phys. C9, L439 (1976).
****
DISCUSSION

L. J. Challis: Dr. Golding was telling us earlier that he has
measured the dipole moments of the phonon scattering states. They
weren't in fact so small.

B. Golding: I think that what I was talking about earlier this
morning is somewhat different from what you're saying. I was talking
about induced dipoles whereas you are talking about permanent di-

poles. Nevertheless one can show from other dielectric measurements
that associated with these states is also a permanent dipole and that
even in the absence of impurities a pure silica glass is essentially
a paraelectric substance. I don't think that invoking arguments
about those states which are effective scatterers of phonons is
really going to change the argument very much. There are certain
of these states which show up in resonance experiments which are
strongly coupled to the fields, but I think you can show by calculations of the thermal conductivity and specific heat that these
distributions are probably very similar, in both the intrinsic levels
and in the impurity levels.

HEAT PULSE EXPERIMENTS AT VERY LOW TEMPERATURES IN VITREOUS SILICA

J.E. Lewis* and J.C. Lasjaunias

Centre de Recherches sur les Très Basses Températures,
C.N.R.S., Cedex 166, 38042 Grenoble, France
*Permanent address: Physics Department, State Univer-
sity of New York, Plattsburgh, NY 12901, USA

According to the most widely accepted model[1] of the amorphous
state, which postulates the existence of localized two level tunneling defects, the specific heat at very low temperatures should be
time dependent, increasing as the logarithm of the experimental time
scale. This important prediction has not yet been unequivocably
confirmed, with two negative results (no time dependence) reported[2]
for time scales of the order of 1-100 microseconds and only our
initial work[3] on SiO_2-Suprasil W at $T < 200$ mK on a time scale ~30mS
to 1 S agreeing with the predictions of the generalized tunneling
model. However, our more recent work on SiO_2 using a different
addenda mounting does not agree with the tunneling model nor with
the solution of the heat equation assuming constant specific heat
(as did the previous negative results). Rather, a time dependent
specific heat is implied, but one that decreases with time, in direct contradiction to the tunneling model.

Two types of well characterized[4] SiO_2 have been measured,
Suprasil II (1200 ppm OH) and Suprasil W (< 1.5 ppm OH and 0.5 ppm
metal ions), both having similar dimensions ~10 cm long and 2 cm
diameter, using two different addenda mountings. The A mounting
consists of a 7 mm wide annular electrical heater placed at one
end of the sample, providing a 0.1 mS pulse of thermal energy, and
the resulting temperature transient $\theta(t)$ is measured by a small ion
doped silicon resistance thermometer greased to a further annular
band situated in the first half of the sample's length. Both bands
were lightly greased to the sample and secured by tight nylon
threads, giving excellent thermal contact. The B mounting is an
attempt to more closely respect a one dimensional heat equation. A
planar heater greased to the front face of the sample provides a
more uniform initial heating. The thermometer band is now 2 mm wide
and sunk into a shallow groove ground in the sample, giving a more

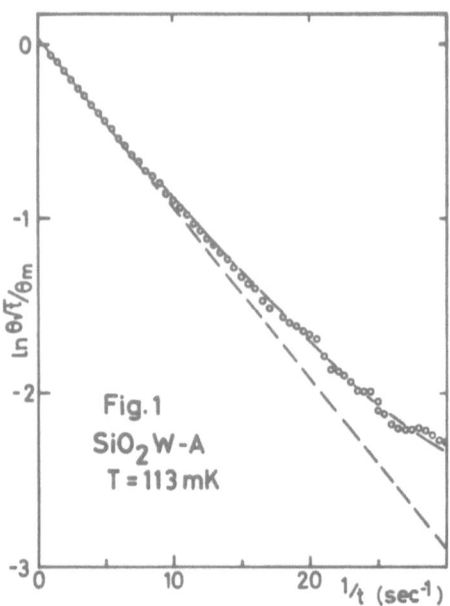

Fig. 1.   Typical temperature transient with A mounting.

precise location for the temperature. Various criteria[5] indicate
that for both the A and B mounts, with the pulse and addenda used,
the flux of heat in the sample is accurately described by a one dimensional heat equation during the experimental time scale and that
for $t < 1$ S the effect of the heat leak to the refrigerator at the
far end of the sample is negligible. The sample is pulsed repetitive-
ly and signal averaging techniques recover the signal from the noise,
as the relative temperature rise is only several percent.

Fig. 1 shows a typical temperature transient obtained in SiO_2
W-A at T = 113 mK plotted as ln $\theta\sqrt{t}/\theta_m$ vs. 1/t, where θ_m is the
maximum temperature of the profile. It has a clearly upward curva-
ture indicating that the sample specific heat is increasing with
time. The dashed line is the behavior expected for a constant
specific heat and is only achieved at long time scales. At short
time scales the profile rises faster than predicted on classical
diffusion theory, the smooth curve being the behavior predicted by
the generalized two level tunneling model[6]. Further analysis gives
the predicted temperature dependence of two critical model parameters, the defect's density of states and their fastest relaxation
time.

However, when measured in the same temperature range with the
B mounting a radically different profile is obtained for the same
sample (Fig. 2). Now the arrival of the thermal pulse is retarded

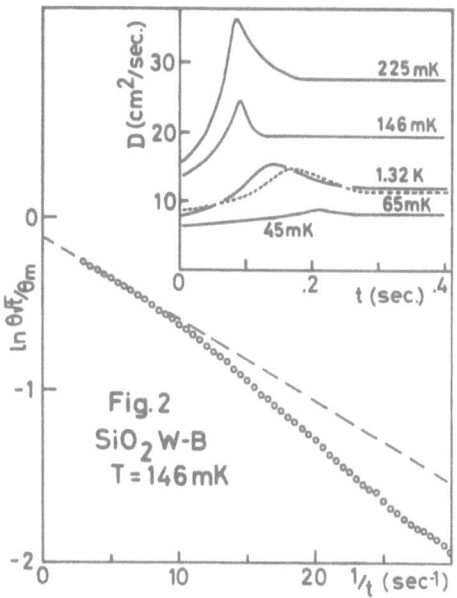

Fig. 2.   Typical temperature transient with B mounting.

and only reaches its long time scale behavior at t~.1 S.   This re-
tardation effect is found for <u>both</u> the A and B mounts at pumped He
temperatures,~1 K.   For the Suprasil II sample the general results
are similar: T < 200 mK, A mounting- slight upward curvature, though
less than for Suprasil W; T< 200 mK, B mounting- retarded heat
pulse; T~1 K, both A and B mounts- pronounced retarded effect.
This retarded effect at time scales~1-100 mS has been reported in
vitreous silica (Suprasil I and W) between 2.5 and 15 K, using a
seemingly 1 dimensional geometry and virtually no addenda (thermo-
couple sensors).
    Although the plotted transient lies on two linear sections, only
that at long time scales obeys the heat equation with constant specific heat as the slope m and intercept b = ln c must obey the condition $-c^2/m$ = 2e = 5.44.   The retarded profiles have been analyzed
using the standard theory[8] for time dependent diffusivities (Fig. 2,
inset).   An initially low value of diffusivity rises to a maximum at
intermediate times, then falls quickly to its constant long time
scale value, which can imply that the specific heat is initially decreasing with time.   The time of the maximum and its relative value
are quite temperature sensitive.
    We believe that, for the time dependent effect in agreement with
the tunneling model, observed with the A mounting, a geometric origin
can be ruled out, as not seen at 1 K, it arises progressively only at

the lowest temperatures. But it is also hard to ascribe geometric/
addenda effects as the absolute cause of the retarded profiles, especially those at 1 K, where so little dependence on addenda is seen,
and we are forced to examine their physical origin. However, it is
difficult to formulate any acceptable model that would generate the
required time dependence of the diffusivity either from a time dependent specific heat and/or thermal conductivity, or even from non-
linear effects. We will repeat these heat pulse experiments at
T > 1 K using a tightly controlled geometry-thin wall tubing. We expect an unequivocal demonstration of an intrinsic physical origin of
this retarded effect. If so, it will be an increasing problem to
explain its cause.

REFERENCES

1. P.W. Anderson, B.I. Halperin, and C.M. Varma, Phil. Mag. 25,
    1 (1972). W.A. Phillips, J. Low Temp. Phys. 7, 351 (1972).
2. W.M. Goubau and R.A. Tait, Phys. Rev. Lett. 34, 1220 (1975).
    R.B. Kummer, V. Narayanamurti, and R.E. Dynes, Phys. Rev.
    Lett. 40, 1187 (1978).
3. J.W. Lewis, J.C. Lasjaunias and G. Schumacher, J. de Physique
    39, C6-967 (1978).
4. M. Vandorpe, J.C. Lasjaunias, and R. Maynard, J. de Physique,
    39, C6-973 (1978). J.C. Lasjaunias, A. Ravex, M. Vandorpe,
    and S. Hunklinger, Sol. State Comm., 17, 1045 (1975).
5. S. Alterovitz, G. Deutscher, and M. Gershenson, J. App. Phys.,
    46, 3637 (1975). J.A. Cape and G.W. Lehman, J. App. Phys.,
    34, 1909 (1963).
6. R. Maynard and R. Rammal, J. de Physique, 39, C6-970 (1978).
7. W. Block and M. Meissner, Conf. on Glasses and Spin Glasses,
    Aussois, (1977).
8. J. Crank, Ch. IX, "The Mathematics of Diffusion," Oxford U.
    Press (1956).

INELASTIC PHONON SCATTERING IN GLASS

W. Dietsche and H. Kinder

Physik-Department E 10, TU München

8046 Garching, West-Germany

A plot of the thermal conductivity vs. temperature for glasses shows a distinct "plateau" at temperatures between 1 K and 10 K. This indicates that strong phonon scattering processes set in at phonon frequencies above about 100 GHz. Several theories have been proposed.[1,2,3] All, however, have been subject to some debate about the parameters necessary to fit the data.[3,4]

We used superconducting tunnel junctions to make the first spectroscopic measurements on the phonon scattering in glass. We determined the phonon mean free path and ascertained the qualitative nature of the scattering process. A SiO_2 glass film was evaporated by electron beam so as to cover one half of the face of a cylindrical Si substrate crystal. Two Sn generator junctions[5] were prepared simultaneously, one of them on the glass film, the other one closely aside on the bare Si surface, for reference. The common detector was an Al-PbBi junction on the far side of the Si crystal.

In the first experiment we used this junction as a detector with a voltage tunable detection threshold.[6,7] In Fig.1, the measured phonon signal is shown as a function of the generator frequency. The thresholds are indicated by the dashed lines. The upper trace of each pair corresponds to the reference generator while the lower one corresponds to the junction on the glass film. A step like increase is observed whenever the frequency crosses the threshold. The height of the step is proportional to the number of phonons which reach the detector with their

original frequency. Thus we can determine the relative
intensity of phonons transmitted elasticly through the
glass by dividing the step height of the "glass" trace
(I) by that of the "reference" trace (I_O). The phonon
mean free path (1) can be calculated from the relation
I/I_O=exp(-d/1) where d is the glass thickness. In Fig.2
1/1 is plotted vs. the phonon frequency. The data were
taken with two film thicknesses 1.6 μm and 0.74 μm. In
the case of the latter the temperature was varied from
1.05 K to 1.25 K with no discernible effect. The straight
line in Fig.2 corresponds to a least squares fit which
yielded 1=(10±1) μm (f/100 GHz)$^{-2.9\pm.3}$

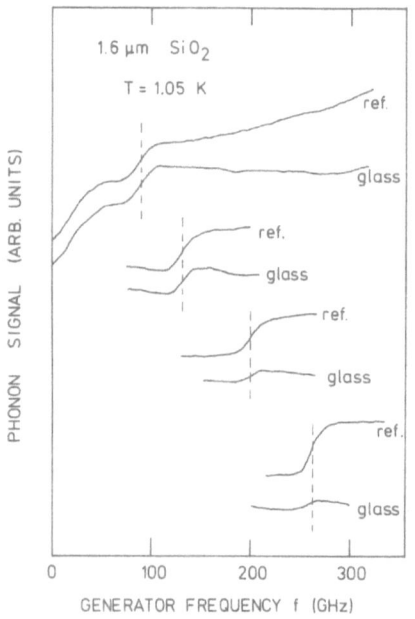

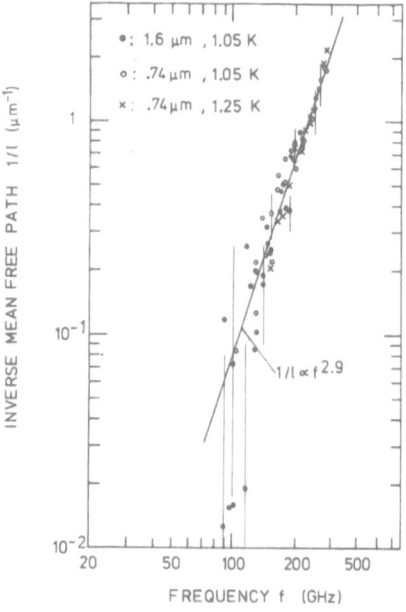

Fig. 1: Phonon signal vs.
frequency with various detector thresholds (dashed
lines). Phonons are either
transmitted through the
glass film (lower traces)
or reach the detector di-
rectly (upper traces).

Fig. 2: Inverse mean free
path vs. phonon frequency
calculated from the rela-
tive step heights in Fig.1.
Straight line is a least
squares fit.

The phonon loss by the glass may be either due to
elastic backscattering into the helium bath or due to
inelastic scattering to frequencies below the set detector thresholds. To distinguish between the two possibilities we determined the spectrum of the phonons trans-

mitted through the glass film. This was made possible by using the Al-PbBi detector in the phonon spectrometer mode.[7] In this case, we used recombination phonons (∿290 GHz) produced by biasing the generator junctions at $2\Delta/e$. Fig.3 shows the measured spectrum as emitted by the reference generator as well as that of the "glass" generator.

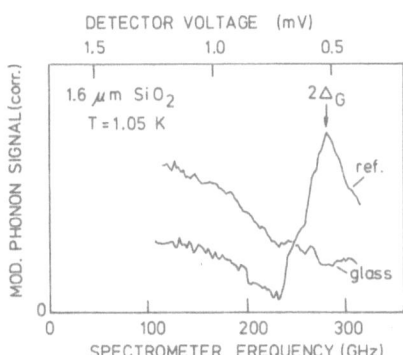

Fig.3: Spectra of phonon as generated (ref) and after being transmitted through a glass film (glass)

Comparing the two spectra one finds that the frequency distribution is considerably changed after transmission through the glass film. At high frequencies the intensity is much reduced while at low frequencies it is increased. This indicates that high frequency phonons are converted into low frequency ones. Thus we conclude that the scattering of high frequency phonons in glass is predominantly inelastic.

REFERENCES
1. R. C. Zeller and R. O. Pohl, Phys.Rev.B4:2029(1971).
2. M. P. Zaitlin and A. C. Anderson, Phys.Stat.Sol.(b) 71:323(1975).
3. A. J. Leadbetter, A. P. Jeapes, C. G. Waterfield and R. Maynard, J.de Physique, Paris, 38:95(1977).
4. D. P. Jones, N. Thomas, and W. A. Phillips, Phil. Mag.B38:271 (1978).
5. H. Kinder, Phys.Rev.Lett.28:1564 (1972); Z.Phys. 262:295 (1973).
6. W. Dietsche, Phys.Rev.Lett.40:786 (1978).
7. W. Dietsche, Bremsstrahlung and Recombination Phonons studied by Al-PbBi Heterojunction Spectrometer, this volume.
                              ****

DISCUSSION

H. J. Maris: What is the temperature dependence of this scattering process?

W. Dietsche: The data are taken at ambient temperature 1.2 and about 1.0°K.   It's not a very large variation, but with this variation there was no difference in the mean free path.

H. J. Maris: Those are both temperatures significantly less than the energy of the phonons.

A. Long: I agree that we have seen strong evidence for inelastic processes, but I wouldn't rule out the possibility of there being strong elastic scattering in the same energy range.   I'd like to discuss that tonight at the discussion session.

R. O. Pohl: Can you tell me what you would expect if you were to evaporate a poly-crystalline film of the same thickness and repeat the experiment?   Are there any experiments along those lines?

W. Dietsche: We haven't done any experiments yet, but we would like to do experiments, for example, with amorphous germanium.

PHONON DENSITY OF STATES AND SPECIFIC HEAT OF AMORPHOUS AND CRYSTALLINE SELENIUM

R. CALEMCZUK and R. LAGNIER

Centre d'Etudes Nucléaires de Grenoble, S.B.T./L.C.P.

85 X - 38041 GRENOBLE-CEDEX (France)

In the harmonic approximation, the thermodynamic functions are related to the phonon density of states $g(\omega)$(PDS) by a linear integral equation. In particular, for the specific heat $C(T)$ the equation is :

$$C(T) = \int_{0}^{\infty} g(\omega)(\omega^2/T) \exp(\omega/T) \left|1-\exp(\omega/T)\right|^{-2} d\omega \qquad (1)$$

Usually, calculations of $C(T)$ are done by assuming some physically reasonable PDS [1][2]. This procedure enables to get some information about the main features of the PDS.

Montroll [3] showed that equation (1) can be formally inversed to give the PDS from the experimental values of $C(T)$. Korshunov et al.[4] pointed out that Montroll's solution is not stable and, by using a theory of ill posed problems, they performed calculations of PDS which are satisfactory in a large range of frequencies, but fail in the low frequency region.

In this communication we present a new method for computing the PDS from specific heat data. The PDS of glassy and crystalline Se are presented and discussed.

METHOD

It is well known that at low frequencies $g(\omega)$ behaves like ω^2 and that it vanishes at some high frequency. Then, the phonon specific heat is proportional to T^3 at very low temperature and it approches to a saturation value at high temperature. The family of functions

$$c_n(T) = \left[T/(T+T_0)\right]^n \;;\; n \geq 3 \qquad (2)$$

which satisfy the low and high temperature behaviour, is well sui-
ted to develop C(T). We write then :

$$C(T) = \sum_{n=3}^{N} a_n c_n (T) \qquad (3)$$

It is possible to fit any function C(T) with a very few terms
(N $\sim$ 10) of (3). The parameter T_0 and the coefficients a_n are determined by minimizing, through an orthogonal polynomials method,
the mean square deviation betwen (3) and the experimental value
of C(T).

To obtain the PDS we calculate analytically the functions
$g_n(\omega)$ which are defined by the equation (1) for the "specific
heats" $c_n(T)$. This is made by developing $c_n(T)$ in powers of T for
which the inversion of equation (1) is well known. Owing to the
linearity of (1), we obtain :

$$g(\omega) = \sum_{n=3}^{N} a_n g_n (\omega) \qquad (4)$$

To check the efficiency of our algorithm we considered the
same model case than Korshunov et al[4], namely a PDS very similar
in shape to that of an FCC metal. We first calculated the corresponding specific heat on a grid of 1 K steps between 0.1 and 300 K.
With this input data we calculated the approximate PDS. Both the
exact and the approximative PDS are shown in fig.1a. We obtain a
better agreement than Korshunov et al.in the low frequency range
whereas the performances are similar at higher frequencies.

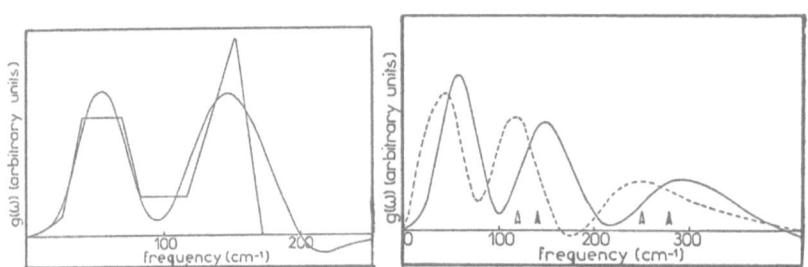

Fig.1:Phonon density of states :(a) model case; (b) Selenium ;
full curve: crystal, dashed curve : glass.
Full and open arrows indicats respectively the positions of infrared peaks [9] of glassy and trigonal Se.

SPECIFIC HEAT AND PDS OF Se

We measured the specific heat of a sample of 99.999 Se (Materials Research) in two different-stabilized and quenched-glassy states and also in the trigonal crystalline state. Measurements were performed between 4 and 300 K with the apparatus and procedure which are described elsewhere[5]. Our results are in good agreement with previous ones[6]. In contrast with other glassy systems like glycerol[7], the specific heat of both glasses do not differ very much. So, we only calculated the PDS for the crystal and one of the glasses.

Below 4 K we extrapolated C(T) for the crystal through a T^3-law. For the glass the corresponding extrapolation did not take into account the contribution which is usually attributed to the two level systems. The PDS thus obtained are shown in Figure 1b.
For both crystal and glass the PDS shows 3 structures each of which contain roughly 1/3 of the total degrees of freedom. For the crystal this result is in good agreement with dispersion curves adjusted to spectroscopic data [8]: the structures correspond to the acoustic and the optical modes of Se.
The PDS corresponding to the glass is shifted to the lower frequencies in agreement with the infrared data [9] (see fig.1b). The enhancement in the density of low frequency states accounts for the anomalous excess of specific heat in amorphous Se.

REFERENCES –

[1] M. RAY-LAFON and E. BONJOUR, Mol.Cryst.and Liquid Cryst.29, 191 (1973)

[2] J.C. LASJAUNIAS and R. MAYNARD, J.Non-Cryst.Sol.6, 101 (1971)

[3] E.W. MONTROLL, J.Chem.Phys.10, 218 (1942)

[4] V.A. KORSHUNOV and V.P. TANANA, Fiz.Metal.Metalloved, 42, 455 (1976)

[5] R. LAGNIER, J. PIERRE and M.J. MORTIMER, Cryogenics (June 1977) 349

[6] S.S. CHANG and A.B. BESTUL, J.Chem.Thermodynamics, 6, 325 (1974)

[7] R. CALEMCZUK, R. LAGNIER and E. BONJOUR, J. Non Cryst. Sol. (to be published)

[8] R.M. MARTIN, G. LUCOVSKY and K. HELLIWELL, Phys.Rev.B, 13

[9] R.S. CALDWELL and M.Y. FAM, Phys.Rev. 119, 664 (1959)
****

DISCUSSION

H. J. Maris: One way you might approach that problem would be to
calculate the eigenfunctions of the integral operator in Eq. (1), and
then if the operator has reasonable properties (I think it's a
symmetric operator) you would be able to go from the specific heat
to a density of states by a projection procedure. The eigenfunctions
would be universal quantities, i.e. nothing to do with any particular crystal so they would only have to be determined once.

R. Calemczuk: You can find an inverse operator by the Fourier
transform method, but there are problems of the stability of the
solutions. You must add some information about the solution you
want.

PHONON SCATTERING IN GLASSES AND HIGHLY DISORDERED CRYSTALS*

A. K. Raychaudhuri, J. M. Peech[+], R. O. Pohl

Physics Department, Cornell University

Ithaca, N.Y.   14853

A thermal conductivity varying as T^δ below 1 K, with the exponent δ close to, but less than 2, appears to be common to all noncrystalline solids[1]. Here, we studied the following questions: How closely does this power law actually describe the data, and can the same power law also occur in suitably disordered crystals? In answer to the first question, we show in the upper part of Fig. 1 for two different glasses ($_{exp}$ -Λ)/Λ, where Λ_{exp} are the experimental data points, and Λ is a least squares fit to the data below $\sim$0.6 K. The scatter of the points is less than 5% in this temperature range; at higher temperatures, the data deviate systematically as the conductivity approaches the plateau region. For the nitrate glass sample shown, Λ= 1.62 x 10^{-4} T $^{1.87}$(W cm^{-1} K^{-1}); the uncertainty of the exponent is $\pm$0.01; for the prefactor, it is $\sim$10% because of uncertainty in the geometry of the sample. For three other samples of the same glass the exponents were found to be (1.84 $\pm$ 0.01), (1.86 $\pm$ 0.01), and (1.87 $\pm$ 0.01). The pre-factors also agreed to within the experimental uncertainty. For the WO_3-doped borate glass shown in Fig. 1, Λ = 1.047 x 10^{-4} $T^{1.907}$ (W cm^{-1} K^{-1}); for an undoped sample, Λ = 1.24 x 10^{-4} $T^{1.910}$, again with the uncertainty in the exponent of $\pm$ 0.01. Based on a careful analysis of all data available, we conclude that for all glasses for which accurate data are available the exponent ranges between 1.8 and 1.9; it depends somewhat on the material, while the sample dependence is smaller than the experimental accuracy. The pre-factors differ by as much as a factor of 10 for different materials. With increasing disorder, a crystalline solid becomes amorphous, and should display the thermal conductivity characteristic for the amorphous state. As a first example, we show the thermal conductivity of a metamict mineral, a titanate containing $\sim$10 wt% thorium and uranium. Their radioactive decay (α - particles) made the

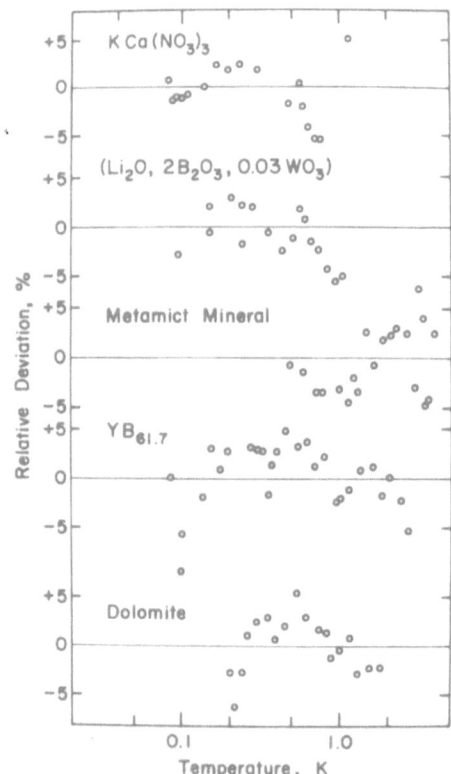

Fig. 1 Quality of the power law fit to the thermal conductivity of
    amorphous and of high disordered crystalline solids.

crystal amorphous, as determined by x-ray diffraction.[2] Its
thermal conductivity, shown in Fig. 2, is described by the power
law $\Lambda=6.5 \times 10^{-5}$ $T^{1.78}$ (W cm^{-1} K^{-1}). The quality of the fit is
shown in Fig. 1. The second example is crystalline, $YB_{61.7}$ and
YB_{66}.[5] Its crystal structure is known. It has been concluded that
the number of yttrium ions is approximately one half the number of
possible sites; the ions seem to be randomly distributed among these
sites.[6] The low temperature thermal conductivity is again of the
form characteristic for the non-crystalline state, $\Lambda=1.44 \times 10^{-4}T^{1.84}$
(W cm^{-1} K^{-1}). The quality of the fit is shown in Fig. 1. Finally,
we have measured a sample of a polycrystalline natural rock, dolomite ($CaMg(CO_3)_3$). It undoubtedly contains a large number of impurities. Its conductivity is clearly not that of an amorphous
solid, see Fig. 2. Note, however, that its conductivity is also
well described by a power law, $\Lambda=5.37 \times 10^{-5}$ $T^{2.64}$ (W cm-1 K-1),
see Fig. 1. In conclusion, the power law actually describes the
thermal conductivity of amorphous solids to a remarkably high

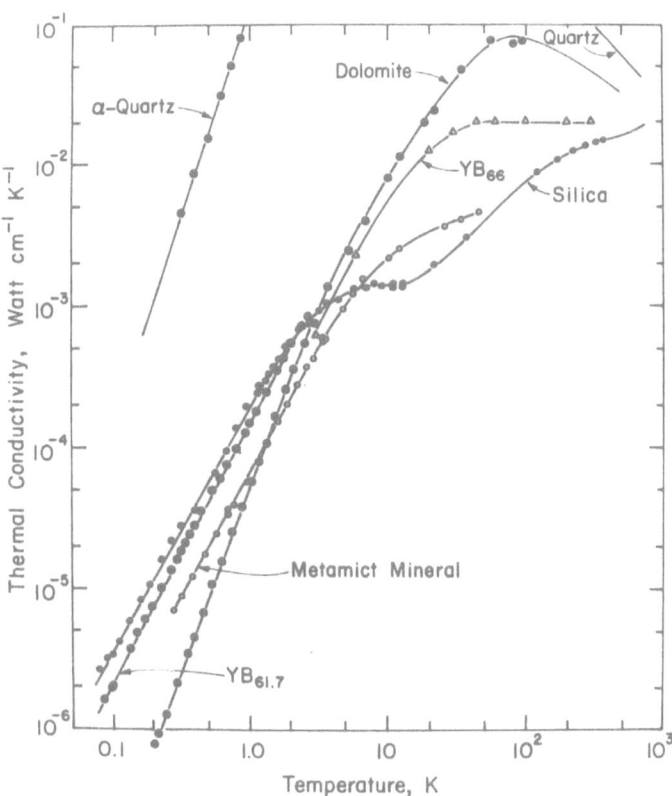

Fig. 2 Thermal conductivity of disordered crystals. YB_{66} after
Ref. 3. The sample of $YB_{61.7}$ was provided by G. A. Slack.
The metamict mineral, labelled R-13 in Ref. 2, was provided
by Dr. R. C. Ewing. Dolomite is a natural rock.

degree of accuracy. It is possible to produce the same power law
in a single crystal (the Y-B compound). It is not clear what
causes the amorphous thermal conductivity in this case. Finally,
a power law has also been found in another disordered solid,
dolomite, although the exponent is quite different. Conceivably,
a study of the power law in disordered crystalline solids may help
to unravel the mystery of the origin of the power law in amorphous
solids.

* The work on glasses was supported by the National Science
  Foundation, Grant No. DMR 78-01560. The work on disordered
  crystals and rock was supported by the U.S. Nuclear Regulatory
  Commission, Contract No. NRC-04-78-261.
+ Deceased

References
1. R. O. Pohl, G. L. Salinger, Ann. NY Acad. Sci. 279, 150 (1976).
2. R. C. Ewing, A. J. Ehlmann, Can. Mineral. 13, 1 (1975).
3. G. A. Slack, D. W. Oliver, and F. H. Horn, Phys. Rev. B 4, 1714 (1971).
4. G. A. Slack, D. W. Oliver, G. D. Brower, and J. D. Young, J. Phys. Chem. Solids 38 45 (1977).

DISPERSION AND THE THERMAL CONDUCTIVITY OF SiO_2

D.P.Jones, J.Jäckle and W.A.Phillips

Cavendish Laboratory
Madingley Road
Cambridge CB3 OHE, England

The thermal conductivity κ of amorphous materials has a characteristic temperature dependence quite unlike that of crystalline
solids. κ varies roughly as T^2 below 1 K, shows little temperature
dependence in a 'plateau' region between 1 and 10 K and increases
again at higher temperatures.[1] This behaviour is illustrated for
vitreous silica in the figure. The low temperature behaviour is
well understood, and is the result of scattering from two-level
systems.

Scattering in the plateau region is not understood, though any
mechanism which gives a sufficiently rapid decrease of phonon free
path ℓ with phonon frequency will fit the data. The usual approach
is to use a Debye spectrum for all the phonons, and to derive
numerical values for the strength of the scattering using theory as
a guide to the frequency and temperature dependences of the scattering mechanism of interest. None of the existing explanations
is satisfactory, and the criticisms can be summarised by saying
that in all cases the calculated scattering is too weak to explain
the magnitude of κ at the plateau. For example, the rate of Rayleigh scattering from density fluctuations can be calculated from
light scattering data, but is too small by almost two orders of
magnitude.[2] As a second example, if it is assumed that the plateau
is a result of scattering from two-level systems[3], their total
number is too large.[4] Further, the heat capacity of amorphous
arsenic in the temperature range 0.3 to 1 K agrees with that calculated from the measured velocities of sound[5], indicating that
the density of two level systems is much less than in silica.
Since the magnitude of κ at the plateau is similar in a-As and
silica[4], it would appear that scattering from two level systems
is not the cause of the plateau.

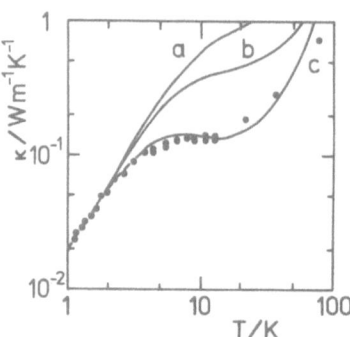

Fig.1.   The thermal conductivity of vitreous silica.   Solid
         circles: experimental results.[1]   Curves a, b, c:
         calculations as described in the text.

As all the unsuccessful explanations of the plateau have been
based on the Debye model, it is worth examining the validity of this
approximation.   In fact, experiments show clearly that the Debye
model is not accurate above about 1 K in amorphous solids, as the
measured heat capacity C is greater than that calculated from the
sound velocities.   This indicates an increased density of vibrational states which in the crystal would be the result of dispersion,
and which changes the analysis of κ.   In this paper the effect is
calculated for the example of Rayleigh scattering from density
fluctuations.

The calculation[2] treats the vibrational states as plane waves
with well defined wavevectors k_α, where α is the polarization index.
Treating the density fluctuations as a perturbation, the elastic
scattering rate $\tau_{\alpha\alpha'}^{-1}(\omega)$ from branch α to branch α' is given by

$$\tau_{\alpha\alpha'}^{-1}(\omega) = \frac{\pi}{2} S_o V_o g_{\alpha'}(\omega)\ I_{\alpha\alpha'}\ \omega^2 \tag{1}$$

Here S_O is the long wavelength limit of the structure factor, V_O is the atomic volume, $g_{\alpha'}(\omega)$ is the density of states, and the $I_{\alpha\alpha'}$ are constants which depend on the Gruneisen parameters for the elastic constants, but are of order unity. κ is calculated by

$$\kappa = \sum_{\alpha} \int C_\alpha(\omega) v_\alpha^2 \tau_\alpha(\omega) d\omega \quad . \tag{2}$$

where $C_\alpha(\omega)$ is the contribution to the heat capacity of the modes of frequency ω in branch α and v_α is their group velocity. For the Debye model, where the scattering rate τ_α^{-1} is proportional to ω^4, the calculated curve is shown as a) in the figure. Scattering by two-level systems is included using as the density of these states the slowly increasing function of energy needed to fit the $T^{1.8}$ dependence below 1 K.

This calculation is extended to a non quadratic density of states by using the dispersive crystal to define the phonon modes: only the transverse branches are dispersive so that only their densities of states are enhanced. It can be seen from eq. 1 that scattering into transverse branches will be increased by a factor of $g_t(\omega)/g_D(\omega)$, where $g_D(\omega)$ is the Debye density of states. In the non-dispersive case $\tau_{\alpha\ell}^{-1}$ is smaller than $\tau_{\alpha t}^{-1}$, so that the contribution of longitudinal modes to κ is reduced by approximately this factor. For transverse modes the increased $C_t(\omega)$ almost exactly compensates for the decrease in τ_t in eq.2. and the only effect of dispersion is to reduce the group velocity factor v_t^2, which also turns out to be approximately proportional to $1/g_t(\omega)$. The contributions to κ of both longitudinal and transverse modes of frequency ω are therefore reduced by the factor $g_D(\omega)/g_t(\omega)$. Since $g_t(\omega)$ dominates the total density of states, this factor can be deduced from the heat capacity data after subtracting a contribution from two level systems. A two parameter fit to the heat capacity of silica accurate to within 5% between 3 and 15 K was used in eqs. 1 and 2 to calculate κ. The magnitudes of S_O and the $I_{\alpha\alpha'}$ were unchanged from the non-dispersive calculations. The result is plotted as b) in the figure. This curve gives the maximum possible value of κ in SiO$_2$: both S_O and $g_t(\omega)$ are taken from experiment and the calculation does not introduce any unknown fitting parameter. The calculation success-fully predicts a plateau, and also shows why the plateau is closely related to the peak in C/T^3, a correlation which seems to hold in all disordered solids. However, κ is still a maximum factor of three too large: instead of a discrepancy of almost two orders of magnitude between calculated and fitted values of the Rayleigh scattering coefficient the difference is now about 15 (curve c). This remaining discrepancy may well be the result of factors, such as anisotropic fluctuations or the increase of $S(k)$ with k, that have been neglected in this calculation of the Rayleigh scattering.

REFERENCES

1. R.C.Zeller and R.O.Pohl, Phys.Rev. B4:2029 (1971)
2. J.Jäckle, 4th Int.Conf. on "Physics of Non-Cryst. Solids" (1976)
3. M.P.Zaitlin and A.C.Anderson, Phys.Rev.B12:4475 (1975)
4. D.P.Jones, N.Thomas and W.A.Phillips, Phil.Mag.38:271 (1978)
5. W.H.Haemmerle and B.Golding, Bull.Am.Phys.Soc.24:282 (1979)

ULTRASONIC MEASUREMENTS ON ALUMINOSILICATE AND BORATE GLASSES

M. J. Lin* and R. L. Thomas

Wayne State University
Detroit, MI   48202

The low temperature properties of amorphous materials exhibit a number of features which are nearly independent of composition and strikingly different from their crystalline counterparts. The unusual behavior of the specific heat, thermal conductivity, and ultrasonic properties in the amorphous state have been successfully described[1,2] by theories based upon the tunneling model of Anderson, Halperin, and Varma[3], and by Philips[4].

In this paper we present the results of measurements of the temperature dependences of the velocity and attenuation of longitudinal sound waves for a number of insulating glasses over the temperature range between 1.3K and 150K, and for frequencies between 30MHz and 270MHz. Evaporated CdS thin film transducers were used, and the phase-comparison method was used to measure the temperature dependence of the sound velocity. The temperature was regulated point by point, and calibrated Pt and Ge thermometers were used for the temperature measurements. Two non-magnetic aluminosilicate glasses ($MgO-Al_2O_3-SiO_2$: 30-11-59 mole %; ZnO-: 17-15-68 mole %), two magnetic aluminosilicate glasses (MnO-: 35-10-55 mole %; CoO-: 13-25-62 mole %), and one rare-earth borate glass ($Nd_2O_3-B_2O_3$: 60-40 mole %) were studied.

The ZnO- glass has a maximum in sound velocity at about 2.5K and a broad minimum at about 110K. The minimum is very weakly frequency dependent, ranging from 104K at 29MHz to 118K at 200MHz. Similar anomalous high temperature dependence of the sound velocity has been observed for other glasses, and has been attributed[5] to frozen-in fluctuations in the elastic moduli, The low temperature dependence of the velocity is qualitatively consistent with that expected from resonant scattering and relaxation of two-level systems.

The MgO- glass also shows a peak in sound velocity at about 2.5K,
as shown in Fig. 1. No minimum was observed for this glass up to
180K. A detailed comparison has been made of the sound velocity in
the range 1.3K-4.2K with the form expected from resonant interaction
with two-level systems[1],

$$\frac{\Delta v}{v} = \frac{n_0 M^2}{\rho v^2} \ln (T/T_0), \tag{1}$$

where n_0 is the constant part of the two-level density of states, M
is the coupling energy, ρ is the mass density, and T_0 is a reference
temperature. The low temperature limiting behavior agrees with Eq.
(1) and is frequency independent from 30MHz-145MHz. Taking
$\rho = 2.61$ g/cm^3, and v = 6.27 x 10^5 cm/sec, we find $n_0 M^2 = 1.01$ x 10^8
erg/cm^3, the same order of magnitude as the coupling constants deduced by other workers[1] for vitreous silica and borosilicate glass.

The sound velocities of the MnO- glass (Fig. 1) and the CoO-
glass differ qualitatively from the non-magnetic aluminosilicates
in two respects. First, a shallow minimum at $\sim$3K is observed which
correlates with spin-glass behavior and has been discussed elsewhere[6]. Secondly, no maximum is observed down to 1.3K.

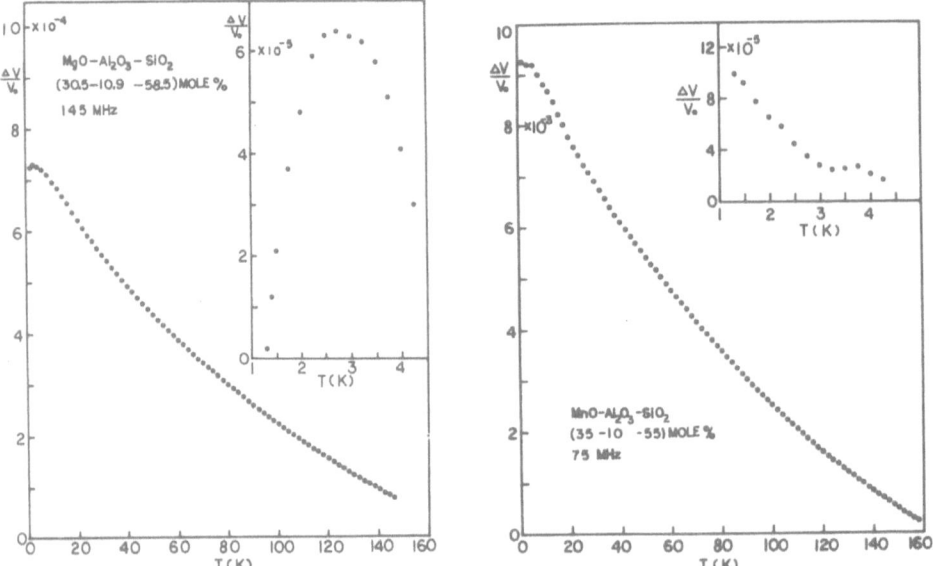

Fig. 1. Relative sound velocity as a function of temperature for
two aluminosilicate glasses.

The Nd_2O_3-B_2O_3 glass yields a peak in sound velocity at ∼4K, and frequency independent behavior describable by Eq. (1) below ∼3K. Taking ρ = 4.22 g/cm^3, and v = 3.15 x 10^5 cm/sec, we find n_0M^2 = 0.92 x 10^8 erg/cm^3.

The high temperature ultrasonic attenuation of these glasses can be compared with that expected from thermally activated relaxation of two-level systems with relaxation times given by

$$\tau(V) = \tau_0 \exp(V/kT), \tag{2}$$

where V is the potential barrier of the double well. Typical results are shown in Fig. 2, where the values of τ_0 and V have been deduced from the frequency dependence of the various relaxation peaks in the ultrasonic attenuation over the range from 30MHz – 270MHz.

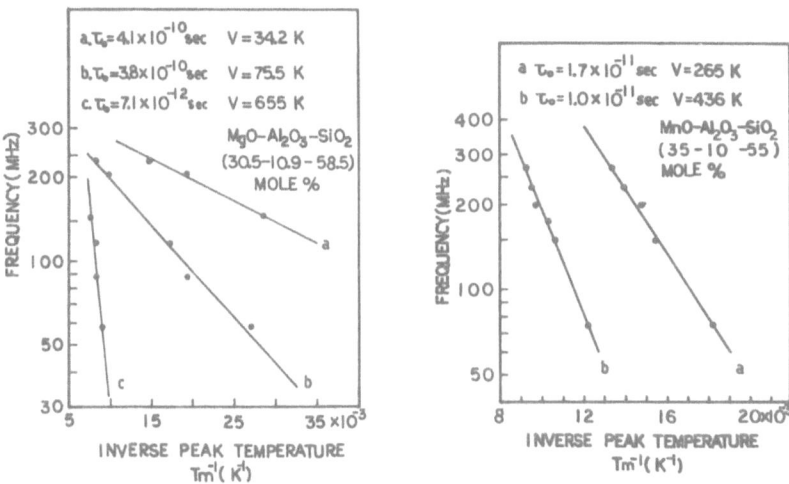

Fig. 2.   Logarithm of the measuring frequency as a function of the inverse attenuation peak temperature for the glasses of Fig. 1.

REFERENCES

*Present address, National Taiwan Normal U.
1.   S. Hunklinger and W. Arnold in "Physical Acoustics," ed. by

R. N. Thurston and W.P. Mason, (Academic Press, N.Y., 1976) 12, 155.

2.   J. L. Black, Phys. Rev. 17, 2740 (1978).
3.   P. W. Anderson, B.I. Halperin, and C.M. Varma, Phil. Mag. 25, 1 (1972).
4.   W.A. Phillips, J. Low-Temp. Phys. 7 351 (1972).
5.   M. N. Kul'bitskaya, V.P. Romanov, and V.A. Shirtilov, Sov. Phys.-Acoust. 19, 399 (1974).
6.   T.J. Moran, N.K. Batra, R.A. Verhelst, and A.M. de Graaf, Phys. Rev. B11, 4436 (1975).

ULTRASONIC MEASUREMENTS IN Na β-ALUMINA AT LOW TEMPERATURES

TEMPERATURE

Pierre Doussineau, Robert G. Leisure,* Alain Levelut
and Jean-Yves Prieur
Laboratoire d'Ultrasons,†
Université Pierre et Marie Curie (Paris VI),
Tour 13, 4 place Jussieu, 75230 Paris Cedex 05, France

Sodium β-alumina has been well-studied because of its high
ionic conductivity at elevated temperatures, but its low temperature properties are also interesting.[1] The specific heat at very
low temperatures exceeds the Debye specific heat by an amount which
is approximately linear in temperature while the thermal conductivity varies approximately as the square of the temperature. The
dielectric properties are also unusual. The high frequency
dielectric absorption exhibits saturation behavior and the
dielectric constant has a minimum as a function of temperature, the
position of which depends on the measuring frequency. These
properties are quite similar to those observed in amorphous
materials and thus have been interpreted in terms of the theory of
two level systems (T.L.S.) which was developed to explain the low
temperature behavior of amorphous materials. Sodium β-alumina
consists of spinel-type blocks of alumina which are bonded together
with oxygen atoms. The layers between these blocks contain the Na
ions. These layers usually contain excess sodium over that
corresponding to stoichiometry and excess oxygen for charge
compensation. The result is that these planes are quite disordered
and apparently account for the amorphous properties of the material.

Ultrasonic experiments are known to be useful for understanding
amorphous materials ; thus we have measured the low temperature
ultrasonic attenuation and velocity in Na β-alumina. The experiments
were performed on a 1 mm thick melt-grown single crystal. One GHz

*Permanent address : Department of Physics, Colorado State
                     University, Fort Collins,Colorado 80523,
 Associated with the Centre National de la Recherche Scientifique.

longitudinal waves were propagated perpendicular to the c-axis. The
sample was placed between two quartz delay rods and attenuation and
velocity were investigated by measuring the amplitude and phase of
the transmitted signal.

The change in phase velocity was measured from 0.1 K to 15 K .
The results are shown in Fig. 1 . The velocity increases in a
logarithmic manner with increasing temperature at the lowest temperatures, passes through a maximum at approximately 8 K and decreases
at higher temperatures. No variation of attenuation was detected as
the temperature was varied between 0.1 K and 4 K . No intensity
dependence of the acoustic attenuation was detected in the range of
acoustic flux from 10^{-5} W cm^{-2} to 10^{-2} W cm^{-2} in the temperature
range 0.1 K to 2 K .

We interpret our results in terms of a coupling of the acoustic
wave to T.L.S. with an energy density of states $n(E) = n_0$ which is

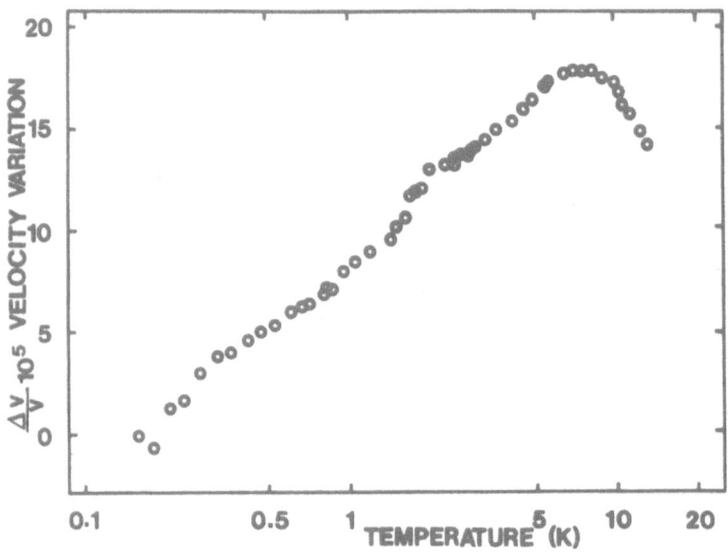

Fig. 1.   Temperature dependence of the phase velocity of 1 GHz
           longitudinal acoustic waves propagating perpendicular to
           the c-axis in single-crystal Na β-alumina. The variation
           is relative to the value at 0.16 K .

independent of energy for low energies. The resonant interaction between the acoustic wave and the T.L.S. gives rise to a temperature dependent phase velocity and attenuation.[2] The phase velocity variation is given by

$$\frac{V(T) - V(T_0)}{V(T_0)} = \frac{n_0 G_x^2}{4 \rho V^2} \ln \frac{T}{T_0} \tag{1}$$

where V is the phase velocity, T is the temperature, T_0 is some reference temperature, G_x is the coupling constant, and ρ is the density of the material. The data of Fig. 1 show the expected logarithmic temperature dependence in the temperature range 0.1 K - 5 K . The slope of the curve in this range gives $n_0 G_x^2 / 4 \rho V^2$. With $\rho = 3.23$ g cm^{-3} and $V = 1.04 \cdot 10^6$ cm sec^{-1}, we obtain $n_0 G_x^2 / 4 = 1.7 \cdot 10^8$ erg cm^{-3} . If we take $n_0 = 2 \cdot 10^{33}$ erg^{-1} cm^{-1} from specific heat data we have $G_x \simeq 0.4$ eV .

The power attenuation coefficient due to the resonant inter-action is given by

$$\alpha = \frac{\pi \omega n_0 G_x^2}{4 \rho V^3 (1 + \phi/\phi_c)^{1/2}} \tanh \left(\frac{\hbar \omega}{2 kT} \right) = \frac{\alpha_{UNS}}{(1 + \phi/\phi_c)^{1/2}} \tag{2}$$

where $\omega/2\pi$ is the frequency of the acoustic wave, ϕ is the acoustic flux in the sample and ϕ_c is a critical flux given by

$$\phi_c \simeq 2 \hbar^2 \rho V^3 / G_x^2 \Delta t\, T_2 \tag{3}$$

where T_2 is the relaxation time of the T.L.S. due to coupling with other T.L.S. Equ. (3) gives an approximate value of ϕ_c when the acoustic pulse width Δt (= 1 µsec) is much less than the T.L.S. - lattice relaxation time, a condition we expect to exist in β-alumina at low temperatures. We can estimate α_{UNS}, the unsaturated attenuation, with the value of $n_0 G_x^2$ obtained from the velocity measurement. At T = 0.1 K and $\omega/2\pi \doteq$ 1 GHz we find $\alpha_{UNS} \simeq 2$ dB cm^{-1}. This attenuation, the largest expected, is just at the limit of sensitivity of the experiment. We can also estimate ϕ_c. Using G_x obtained above and estimating $T_2 = 0.26$ µsec from the dielectric measurements we find $\phi_c = 10^{-5}$ W cm^{-2} at T = 1 K . This value, which lies at the lower limit of the acoustic flux range explored in the experiment, together with the value of α_{UNS} estimated above, show that the expected magnitude of the resonant attenuation is too small to be detected in the present experiment.

Although $\Delta V/V$ and α_{UNS} are smaller in Na β-alumina than in amorphous materials such as silica, this difference appears to be due to the high value of the acoustic velocity in former material. The quantity $n_0 G_x^2$ has approximately the same value in the two materials. Thus, the T.L.S. in Na β-alumina appear to have charac-teristics quite similar to those found in amorphous materials.

REFERENCES

1.   P. J. Anthony and A. C. Anderson, Low Temperature Dielectric
     Susceptibility of Li, Na, and Ag β-Alumina, *Phys. Rev.* B 19 :
     5310 (1979) and references therein.
2.   S. Hunklinger and W. Arnold, Ultrasonic Properties of Glasses at
     Low Temperatures, in *"Physical Acoustics"*, Vol. XII, ed. by
     W. P. Mason and R. N. Thurston, Academic Press, N.Y. (1976).

HIGH RESOLUTION STUDIES OF BRILLOUIN SCATTERING

IN AMORPHOUS MATERIALS

R. Vacher[*], H. Sussner, M. Schmidt
and S. Hunklinger

Max-Planck-Institut für Festkörperforschung,
Heisenbergstr. 1, D-7000 Stuttgart 80, FRG

The interest to study the elastic properties of amorphous materials with Brillouin scattering (BS) is manifold. First of all, interactions of phonons with defects can be studied in an energy range which is accessible neither by ultrasonic nor by thermal measurements. Furthermore, the theoretically interesting regime where the phonon energy $h\omega$ is greater than kT is already reached at liquid He temperatures, whereas conventional acoustic measurements have to be performed in the mK range.[1] In addition, BS allows the observation of absorptions where the mean free path extends over several wavelengths only. Due to the small wavelength of the probed phonons (between 0.1 and 0.5 μ), BS can be also applied to small samples of different shapes and, in particular, thin films. Even for non-transparent media, the elastic properties can still be obtained by BS from surface phonons.

In this paper, these two aspects will be illustrated by reporting BS measurements on two amorphous materials of quite different character: We first report on Brillouin measurements of sound velocity and attenuation in vitreous silica, which have been performed below 1K for the first time. In the second part, we describe the application of BS to determine the attenuation of 15 GHz surface phonons in films of amorphous silicon.

The measurements of velocity and mean free path of 35 GHz phonons in vitreous silica between 0.3 and 20K are shown in Fig. 1. The sound velocity passes through a maximum around 7K and decreases approximately logarithmically with temperature below the maximum.[2,3] Within the accuracy of our measurement, it becomes temperature independent at the lowest temperatures where $h\omega \simeq 5kT$. As reported previously,[2] the attenuation decreases upon cooling to 4K where the

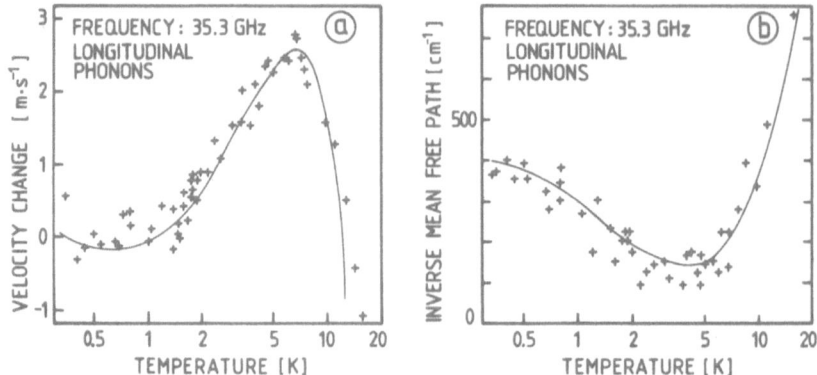

Fig. 1.   Velocity (a) and attenuation (b) of hypersounds vs. tem-
          perature in vitreous silica.

observed linewidth becomes comparable with the instrumental width
of 7 MHz.   At still lower temperatures, when hω > kT, the linewidth
increases again due to the resonant absorption of phonons by two
level systems characteristic of amorphous materials.[4,5]   At the
lowest temperature, the absorption - being observed under strictly
unsaturated conditions - tends to a constant value, as predicted by
theory.   Our measurement of the mean free path of 35 GHz phonons,
which are dominant at 0.4K, is in good agreement with that deduced
from thermal properties at this temperature,[6] assuming that the mean
free path is similar for longitudinal and transverse phonons in
this temperature range.[5]

     Whereas only qualitative agreement between theory (solid lines
in Fig. 1) and experiment is found above 10K, the low temperature
data are fully described by current theories based on the existence
of two-level systems in glasses.

     In contrast to vitreous silica, the elastic properties of
amorphous silicon have not been investigated extensively until now.
As the samples are only available in the form of thin films and in
addition are non-transparent, BS from surface phonons has been used
to probe their acoustic behaviour.   We studied two samples: The
first was prepared by bombarding a Si single crystal by Si ions,
resulting in one amorphous layer of about 1 µ thickness.   The
second was an amorphous film deposited by rf-sputtering on a
crystalline Si substrate.   This sample contained about 10 % hydrogen
impurities.   We have also studied surface waves on a Si single
crystal for comparison.   The two main results are (i) the mean free
path of the ion-bombarded sample shows the same small value as in
the single crystal, even though the corresponding velocities differ
considerably (Fig. 2a) and (ii) the sputtered sample, in contrast,

is characterized by a small sound velocity[7] and a very high sound
attenuation. The frequency dependence of this attenuation
(Fig. 2b) as well as its continuous decrease upon cooling observed
from 300 to 80K suggest relaxational processes rather than scattering of phonons by bulk or surface defects to be responsible for
this large damping.

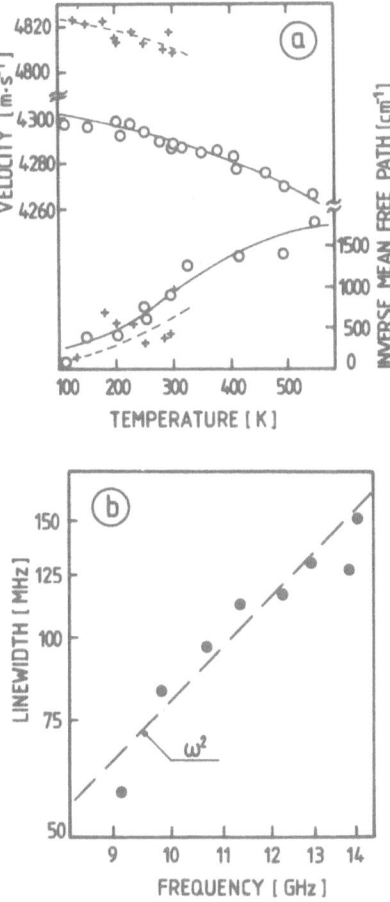

Fig. 2.   (a) Velocity and attenuation vs. temperature in
           crystalline (+) and amorphised (O) Si.   (b) Frequency
           dependence of linewidth in sputtered amorphous Si.

*Permanent address: University of Montpellier, F-34060 Montpellier (France)

REFERENCES

1. B. Golding, J.E. Graebner, and A.B. Kane, Phys. Rev. Lett. 37:1248 (1976).
2. J. Pelous and R. Vacher, J. Physique 38:1153 (1977).
3. L. Piché, R. Maynard, S. Hunklinger and J. Jäckle, Phys. Rev. Lett. 32:1426 (1974).
4. W. Arnold, S. Hunklinger, S. Stein, and K. Dransfeld, J. Non-Cryst. Sol. 14:192 (1974).
5. B. Golding, J.E. Graebner, and R.J. Schutz, Phys. Rev. B14:1660 (1976).
6. R.C. Zeller and R.O. Pohl, Phys. Rev. B4:2029 (1971).
7. W. Senn, G. Winterling, M. Grimsditch, and M. Brodsky, in "Physics of Semiconductors 1978", B.L.H. Wilson, ed., Institute of Physics Conf. Ser. No. 43:709, London (1979)

SATURATION OF THE ACOUSTIC ATTENUATION AND RELAXATION OF

THE TWO-LEVEL SYSTEMS BY THE ELECTRONS IN AMORPHOUS NiP

Pierre Doussineau and Alain Robin

Laboratoire d'Ultrasons*
Université Pierre et Marie Curie (Paris VI)
Tour 13, 4 place Jussieu, 75230 Paris Cedex 05, France

At low temperature the amorphous metals present some properties similar to those observed in amorphous insulators. The thermal conductivity varies as the square of the temperature,[1] the acoustical velocity increases logarithmically with the temperature.[2] These properties are explained by the model of the two-level systems.[3] Recently one of the most spectacular effect, the saturation of the acoustical attenuation, has been observed in the amorphous metal PdSiCu.[4,5] We present here the results of acoustic attenuation measurements in another amorphous metal $Ni_{0.78}P_{0.22}$.

The sample was electrodeposited from the bath. We used a standard pulse echo method. In the first experiment we measured the attenuation of a transverse wave at fixed temperature (T = 60 mK) as a function of the acoustical power in the sample. The results clearly show an increase of the attenuation when the power decreases (see Fig. 1). We interpret this effect as due to the resonant interaction between the acoustic wave and some localized excitations intrinsic to the amorphous state, often called two-level systems (T.L.S.). This attenuation is given by

$$\alpha_{Res} = (1 + \Phi/\Phi_c)^{-\frac{1}{2}} \, (n\gamma^2/\rho v^3) \, \pi\omega \, \tanh(\hbar\omega/2 \, kT)$$

where Φ is the acoustical flux in the sample, Φ_c a critical value given by $\Phi_c = \rho v^3 \hbar^2/2 \gamma^2 T_1 T_2$, n the constant density of states of the T.L.S., γ the coupling constant between T.L.S. and acoustical wave, ρ the density of the material, v the velocity of the acoustical wave of frequency $\omega/2\pi$, T the temperature, T_1 and T_2 respectively the

---

*Associated with the Centre National de la Recherche Scientifique.

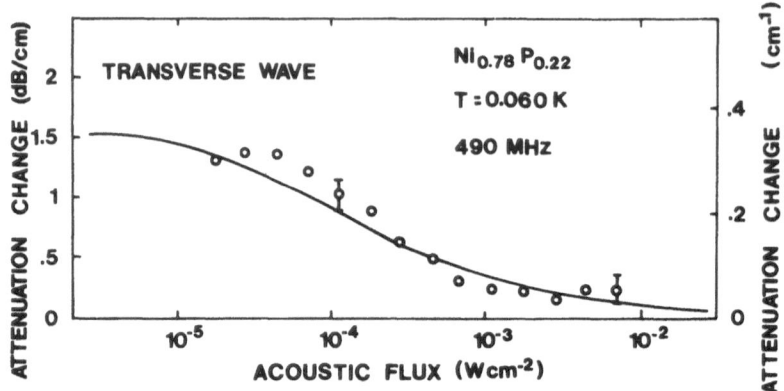

Fig. 1 — Relative acoustical attenuation as a function of the
acoustical flux in the sample. The solid line is a
calculated curve with $\alpha_R = (1 + \Phi/\Phi_c)^{-1/2}\alpha_{unsat}$.

longitudinal and transverse relaxation times of the T.L.S. From
our measurements we deduce $n\gamma^2 \simeq (2 \pm 0.4)\ 10^7$ erg cm^{-3} and
$\Phi_c \simeq 5 \cdot 10^{-4}$ W cm^{-2}. This value of the critical flux is nearly the
same as in PdSiCu but is three orders of magnitude higher than
that measured in amorphous insulators. As a consequence we infer
short relaxation times of the T.L.S. in NiP.

In the second experiment we measured at various frequencies
between 275 and 820 MHz the attenuation change in the temperature
range 0.06 up to 1.5 K and at high acoustical powers. At all the
frequencies the variation is linear at the lowest temperatures.
Moreover the slope is frequency independent. At higher temperature
the attenuation becomes frequency dependent (Fig. 2). We explain
this behavior as the relaxational attenuation when the relaxation
of the T.L.S. population perturbed by the ultrasonic wave is
produced by scattering of a conduction electron. In that case, in
the low temperature regime when $\omega T_1 \ll 1$, the attenuation is
given by

$$\alpha_{Rel} = \pi^3 n\gamma^2 (n_{E_F} K)^2 kT / 24 \hbar \rho v^3$$

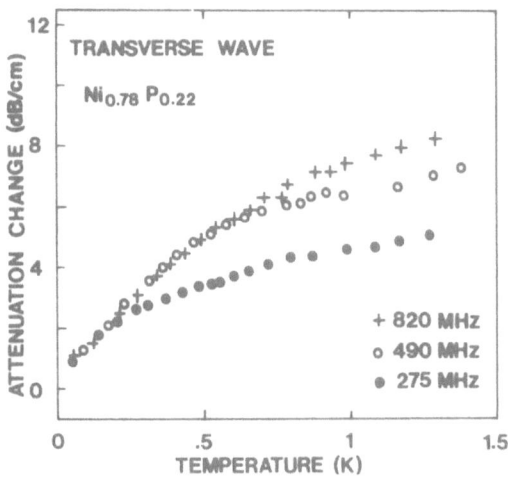

Fig. 2 — Attenuation change as a function of the temperature for
three frequencies. The three curves are shifted on a
manner such as they coincide at 0.1 K .

where K is the coupling constant between T.L.S. and electrons, n_{E_F}
is the electronic density of states at the Fermi level. Our
measurements leads to a value $n_{E_F} K \simeq 0.25$ for NiP. This $\omega^0 T$ law
is the analog of the well-known $\omega^0 T^3$ law observed for the high
power ultrasonic attenuation in amorphous insulators where the
relaxation of the T.L.S. is by the phonons.[6]

Thus our experiment confirms clearly that T.L.S. are relaxed
by conduction electrons at very low temperatures in amorphous
metals. As previously noted,[5] the behavior at higher temperatures
is more complicated because : i) the condition $\omega T_1 \gg 1$ is no
more fulfilled, ii) the relaxation by phonons must be added to
the relaxation by electrons.

REFERENCES

1. J. R. Matey and A. C. Anderson, *Phys. Rev.* **B 16** - 3406 (1977) ;
    J. E. Graebner, B. Golding, R. J. Schutz, F. S. L. Hsu and
    H. S. Chen, *Phys. Rev. Lett.* **39** : 1480 (1977)
2. G. Bellessa, P. Doussineau and A. Levelut, *J. Physique Lett.*
    38 : L-65 (1977) ;
    G. Bellessa and O. Béthoux, *Phys. Lett.* **62 A** : 125 (1977).

3. P. W. Anderson, B. I. Halperin and C. M. Varma, *Philos. Magn.*
    25 : 1 (1972) ;
    W. A. Philipps, *J. Low Temp. Phys.* **7** : 351 (1972).
4. P. Doussineau, P. Legros, A. Levelut and A. Robin, *J. Physique
    Lett.* **39** : L-265 (1978).
5. B. Golding, J. E. Graebner, A. B. Kane and J. L. Black, *Phys.
    Rev. Lett.* **41** : 1487 (1978).
6. See for example, S. Hunklinger and W. Arnold, in "*Physical
    Acoustics*", ed. by R. N. Thurston and W. P. Mason, Academic
    Press, New-York (1976), Vol. 12, p. 155.

ULTRASONIC STUDIES OF METALLIC GLASSES*

H. Araki, G. Park, A. Hikata and C. Elbaum

Brown University, Providence, R.I. 02912

New results on ultrasonic properties of amorphous PdSiCu are reported. The temperature and amplitude dependence of the ultra‑sonic attenuation and velocity changes were measured by a pulse echo method[1], between 0.3 and 10 K, using transverse waves at 10 to 90 MHz and longitudinal waves at 30 MHz. A typical temperature dependence of the ultrasonic attenuation coefficient α for transverse waves at 25.1 MHz and four different amplitudes is shown in Fig. 1. As is seen from the figure, below 4 K there is a definite negative amplitude dependence. Detailed results of the amplitude dependent part α_A at various temperatures are shown in Fig. 2. Measurements carried out at several different frequencies showed that α_A increases as the ultrasonic frequency increases[2]. For longitudinal waves, there is a much smaller negative amplitude dependence of attenuation than for transverse waves.

Fig. 1
Temperature dependence of the attenuation coeffi‑cient for 25.1 MHz trans‑verse waves at four dif‑ferent amplitudes.

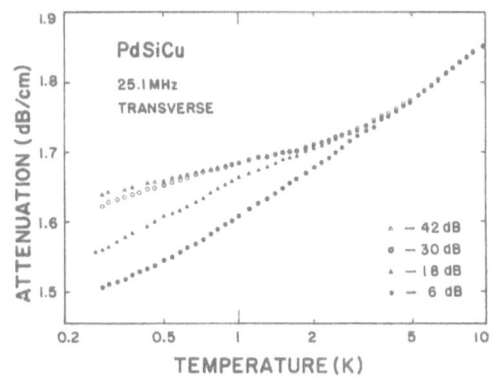

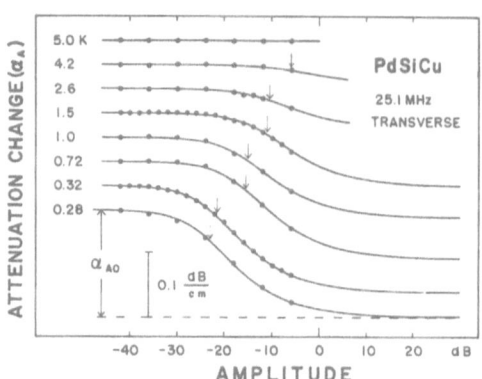

Fig. 2
Amplitude dependence of
the attenuation coeffi-
cient for 25.1 MHz trans-
verse waves at various
temperatures. Fitted
curves (eq. (1)) are also
shown.

The experimental results are compared with the predictions
of the two level system (TLS) tunneling model theory[3]. The
theory predicts for α_A the formula

$$\alpha_A = \alpha_A 0 /(1 + J/J_c)^{1/2} \qquad\qquad (1)$$

where J and J_c are the input and the critical acoustic intensity,
respectively. The data were found to agree very well with this
functional form (Fig. 2). The predicted form of α_{AO} is[4]

$$\alpha_{AO} = (\pi n M^2/2\rho v^3)\omega \ tanh(\hbar\omega/2kT) \qquad\qquad (2)$$

where n, M are the density of states of TLS and the deformation
potential, respectively, ρ is the mass density, v is the acoustic
wave velocity and the other symbols have their usual meanings.
In the present study $\hbar\omega \ll 2kT$, therefore $\alpha_{AO} \propto \omega^2/T$. In order to
compare the results with this prediction, α_{AO} obtained from the
data by fitting to eq. (1) were plotted against temperature (Fig.
3), which shows that the temperature dependence is clearly not
1/T. Similar dependences were found for the other frequencies;
for T $\gtrsim$ 1K, α_{AO} may be proportional to 1/T, but at lower temperatures it tends to a constant value. Fig. 4 shows α_{AO} for various
frequencies at 0.3 K. The frequency dependence is linear, rather
than quadratic. In the same figure we plot the result by
Doussineau et al.[5] for the same material at 720 MHz, but taken
at 0.062 K. Their result falls on the extrapolation of our data;
this strongly suggests that the temperature independent region can
be extended down to 60 mK with linear frequency dependence. (The
broken line in the figure shows α_{AO} expected at 0.3 K from eq. (2)
based on the data of ref. 5)). In order to compare these results
with dielectric glasses, experiments on vitreous silica Suprasil-
I were carried out in the same temperature and frequency range.
No amplitude dependent attenuation, within the accuracy of our
experiment, was found for a 50 dB dynamic range (highest amplitude
used is roughly the same as that for PdSiCu); this is consistent

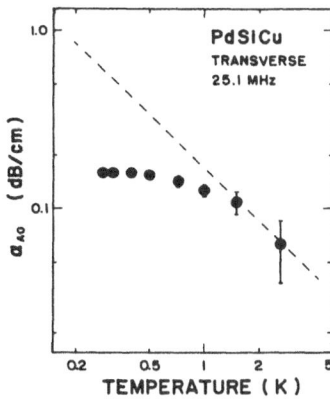

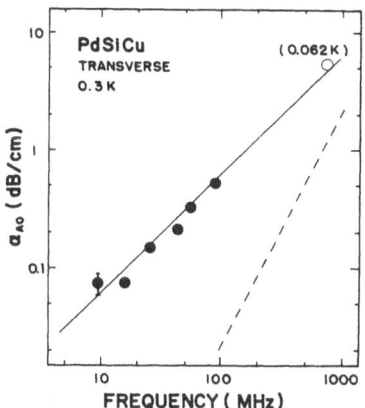

Fig. 3
Temperature dependence of α_{AO} for 25.1 MHz transverse waves. The dashed line represents 1/T dependence.

Fig. 4
Frequency dependence of α_{AO} at 0.3 K. The open circle represents the data of ref. 5). The dashed line is expected dependence from eq. 2 (see text).

with the TLS theory predictions, i.e. $\alpha_{AO} \propto \omega^2$.

In summary, some aspects of the observed ultrasonic properties of amorphous PdSiCu are consistent with TLS theory (amplitude dependent attenuation and logT dependent velocity change[2]) but there are definite discrepancies in frequency and temperature dependences.

References

1. The resolution of the equipment is better than 0.004 dB and 10^{-6} for amplitude and velocity change measurements, respectively.
2. The velocity changes $\Delta v/v$ were found to be proportional to logT at low temperatures ($0.3 \lesssim T \lesssim 1.0$ K) for both modes in agreement with previous observations. In addition, however, amplitude dependence of the slope of $\Delta v/v$ versus logT for transverse waves was found (the higher the amplitude, the smaller the slope).
3. S. Hunklinger and W. Arnold, Physical Acoustics, Vol. XII, ed. M. P. Mason and R. N. Thurston, Academic Press, N. Y. (1976), p. 155, and references within.
4. Difference of factor 2 from other papers comes from the difference in amplitude and intensity attenuation coefficient.
5. P. Doussineau, P. Legros, A. Levelut and A. Robin, J. Phys. (Paris) Lett. 39, L-265 (1978).

*This research was supported by the NSF through the Materials Research Lab. of Brown University and Grant DMR77-12249.

DISCUSSION

J. Black: In the low frequency regime in which you are operating
the widths due to the conduction electron processes would be very
much bigger than the splittings of your two-level systems, and that
might explain why you see a very weak temperature dependence and
very weak frequency dependence, although I would also expect that
you wouldn't see saturation.

S. Hunklinger: Can you give a number for nM^2 from your absorption
measurements because there is a very high coupling obviously?

H. Araki: It depends on the frequencies used. It ranges from 100
to 1000 larger than the values obtained by high frequency measurements on metallic glasses.

S. Hunklinger: How does it differ from the velocity data?

H. Araki: If we use the TLS theory we get a reasonable value for
nM^2 from the velocity data.

H. J. Maris: If you consider a tunneling system that has a very
small splitting, when that splitting becomes small enough you would
imagine that the way that that will be damped out through interaction with the conduction electrons would be more the sort of
process where you talk about the viscosity of the elctron gas
rather than individual electrons coming in and de-exciting the
tunneling state. The tunneling motion is a mechanical motion and
therefore there must be some similarity to the damping of an
ordinary sound wave by conduction electrons. So I think that for
a small enough splitting there may be some other regime.

J. Black: Are you saying that somehow you want the width to narrow
down again? It's certainly true that these other processes give
you a hugh width at these small splittings.

H. J. Maris: I guess that this would narrow it down.

J. Black: It doesn't seem how by adding another process you can
narrow it.

H. J. Maris: The attenuation of a sound wave in a metal is less
if the mean free path of the electrons is short. You then go into
the viscosity regime where the damping is proportional to ω^2.

J. Black: I don't see how the other process goes away in that
limit.

H. J. Maris: It goes away because you have to take into account the
width of the electron levels, and it's not legitimate to talk about
individual electrons.

ULTRASONIC ABSORPTION IN THE SUPERCONDUCTING AMORPHOUS METAL PdZr

G. Weiss, W. Arnold, and H.J. Güntherodt*
Max-Planck-Institut für Festkörperforschung,
D-7000 Stuttgart 80, FRG
*Institut für Physik, Universität Basel, Switzerland

ABSTRACT

Propagation of longitudinal waves at 740 MHz in superconducting amorphous $Pd_{30}Zr_{70}$ has been studied between 0.4 K and 70 K. No change of the absorption occurs when passing through the superconducting transition temperature at 2.62 K whilst a strong decrease is found for temperatures from 1.5 K down to 0.4 K. Below 1.5 K a strong magnetic field dependent enhancement of the acoustic attenuation is observed. We compare these results with a recently developed theory.

At low temperatures the thermal /1/ and acoustic properties of insulating /2/ and metallic glasses differ considerably from those of their crystalline counterparts. These differences have been attributed to the existence of tunnelling states /1/ present in amorphous materials. In particular, amorphous metals show a logarithmic temperature dependence of the sound velocity /3/, and a linear contribution to the specific heat originating from the tunnelling states as recently observed in the superconducting amorphous PdZr /4/.

In addition saturation of the ultrasonic absorption in PdSiCu at very low temperatures has been reported /5,6/. In contrast to insulating glasses, a much higher power level was required because of lifetimes of the tunnelling states, which are four orders of magnitude shorter than in insulating glasses. It has been proposed that the conduction electrons are responsible for this enhanced relaxation /6/. In a superconductor the electronic contribution should drop drastically when cooling below T_c as studied in detail in a recently developed theory /7/.

We have performed ultrasonic measurements in the superconducting amorphous metal $Pd_{30}Zr_{70}$ prepared by a modified piston-and-anvil
technique providing splats of 50 μm thickness whose structure was
carefully examined by X-rays. In order to work with a sufficiently
large sized sample, we made a stack of eleven equal platelets, each
0.05 x 1 x 10 mm^3, pressed together by a clamp of stainless steel.
Experimental details can be found elsewhere /8/.

Figure 1 shows the
measured ultrasonic
attenuation between 0.4 K
and liquid notrogen tem-
perature. From 20 K,where
a shallow maximum occurs,
the attenuation de-
creases steadily with de-
creasing temperature to
about 3 K. From thereon
the absorption remains
almost constant down to
about 1.5 K, and in par-
ticular there is no
change at the supercon-
ducting transition tem-
perature (T_c = 2.62 K).

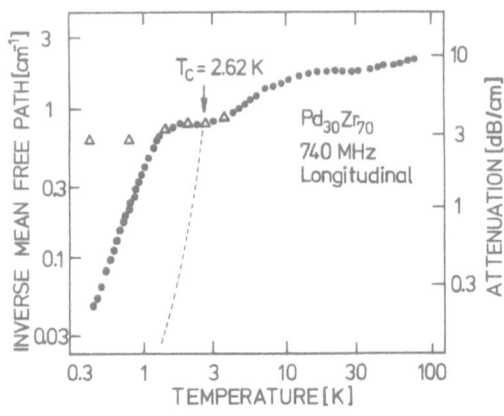

Fig. 1.

Between 1.5 K and 0.4 K, however, the attenuation decreases dramatically. At these temperatures a strong magnetic field enhancement
is observed as shown also in Fig. 1 (triangles). With increasing
field the absorption first rises linearly, and then levels off when
H_{c2} is exceeded /8/.

The observed attenuation cannot be explained by the classical
absorption mechanism arising from the viscosity of the electron gas,
because the mean free path of the electrons is only of the order of
the atomic distance in amorphous metals. In amorphous materials the
tunnelling states give relaxational and resonant contributions to
the ultrasonic absorption at low temperatures. The resonant part
can be saturated for intensities exceeding a certain critical value
determined by the relaxation rates of the tunnelling systems and by
their coupling strength M to ultrasonic phonons. This resonant interaction is responsible for a logarithmic temperature variation /2/
of $\Delta v/v$, which we observe also in PdZr between 0.4 K and 0.8 K.
From our sound velocity measurement we deduce nM^2 = $1.4 \times 10^7 erg\ cm^{-3}$,
where n is the usual density of states. From this value a small
resonant absorption contribution /2/ α_{res} = $1.3 \times 10^{-2}\ cm^{-1}$ can be
calculated, which is comparable to the accuracy of our experiment
and perhaps explains why we did not find, in a first attempt, a
saturatable contribution to the absorption.

The relaxational contribution to the ultrasonic attenuation arises from strain modulation and subsequent relaxation of the energy levels of the tunnelling states by phonons. In this process the temperature dependence of the absorption is determined mainly by the temperature dependence of the relaxation time T_1. In amorphous metals T_1 is apparently reduced drastically by the conduction electrons /6/. Because below T_c the thermally excited quasi-particles only promote the relaxation rate T_1^{-1}, it should drop rapidly as provided by /7/:

$$T_1^{-1} \sim (\rho V_\perp)^2 kT (\exp\Delta/kT + 1)^{-1} \qquad (1)$$

Here 2Δ is the superconducting energy gap, $\rho V_\perp$ is an effective coupling constant for the electron tunnelling-state interaction. Eq. (1) holds for tunnelling states with splitting $E \ll kT$. In Figure 1 the dashed line shows the theoretical prediction /7/ using the density of states of the tunnelling model and $\rho V_\perp = 0.2$, both deduced from other experiments in nonsuperconducting metallic glasses /6/. Obviously the steep fall of the absorption at T_c does not agree with our experimental results.

It might be possible to get better agreement by assuming a distribution of relaxation times. However this leads to much larger effective coupling constants $\rho V_\perp$ than 0.2. In our view further experiments at various frequencies and at lower temperatures are necessary to clarify this point. Finally, there may be another mechanism responsible for the electron tunnelling states interaction presently still unknown.

We thank our colleagues for helpful discussions and K. Dransfeld for stimulating and encouraging support.

REFERENCES

/1/ W.A. Phillips, J. Non-Cryst. Solids 31:267 (1978), and ref. contained therein.

/2/ S. Hunklinger and W. Arnold, Physical Acoustics 12:155, eds. W.P. Mason and R.N. Thurston (Academic Press, New York, 1976), and ref. contained therein.

/3/ G. Bellessa and O. Bethoux, Phys. Lett. 62A:125 (1977).

/4/ J.E. Graebner, B. Golding, R.J. Schutz, F.S.L. Hsu, and H.S. Chen, Phys. Rev. Lett. 39:1480 (1977).

/5/ P. Doussineau, P. Legros, A. Levelut, A. Robin, J. de Phys. 39:L-265 (1978).

/6/ B. Golding, J.E. Graebner, A.B. Kane, and J.L. Black, Phys. Rev. Lett. 41:1487 (1978).

/7/ J.L. Black and P. Fulde, Phys. Rev. Lett. 43:453 (1979)

/8/ G. Weiss, W. Arnold, K. Dransfeld and H.J. Güntherodt, to be published.

****

DISCUSSION

R. C. Dynes: I would like to convince myself that the superconductivity that transpired at T = 2.62 K is a bulk effect. I could very
easily explain, using several parameters of course, those results by
suggesting that your sample is inhomogeneous and that you don't
really have any direct evidence to say that the material is going
superconducting at 2.62 K.

W. Arnold: It might be true for this particular sample, but there
are measurements of the specific heat by Graebner and Golding and
they show clearly that there is only one T_c. They report a broadness
which is even stronger than ours - ours is 0.105, they report 0.01.

R. C. Dynes: I don't disagree in principle with the experiment.
I'm suggesting that in your sample you could be seeing bulk superconductivity at 1 K, but not at 2.62 K.

B. Golding: Could you describe how you measured the superconducting
transition width?

W. Arnold: We measured it by the change in susceptibility through
the Meissner effect in a separate system.

B. Golding: Was the transition width measured on the sample shown
on the viewgraph?

W. Arnold: It was on the same sample.

B. Golding: On the clamped array of plates?

W. Arnold: The plates were measured before they were compressed
in the clamps. The clamps were made out of stainless steel.

B. Golding: One possible source of inhomogeneity is that there
might be an inhomogeneous stress on these samples. It's also well
known that in these almost unstable metallic glasses it takes a very
small inhomogeneous stress to cause crystallization of the material.
I think it would be interesting if you could try to determine the
transition width in the experimental arrangement that was used.

R. I. Boughton: Do you have any estimate of what the stress on the
platelets is?

W. Arnold: These are 1 mm screws, and I wouldn't even guess.
You tighten them by hand.

SPECTRAL DIFFUSION AND ULTRASONIC ATTENUATION IN AMORPHOUS MATERIALS

W.Arnold, J.L.Black[*], and G.Weiss

Max-Planck-Institut fuer Festkoerperforschung

7000 Stuttgart 80 , FRG

Of the many low-temperature effects attributable to tunneling states (TLS) in glasses[1] , the saturation of the resonant ultrasonic attenuation has proved to be one of the richest [2,3] . This phenomenon occurs when the flux of phonons is sufficiently high to drive the TLS population out of thermal equilibrium. Up until now , there has been no complete theory to describe what is meant by a "sufficiently high" phonon flux. Consequently there has been no framework in which one could understand the temperature , frequency, and pulse-length dependences of the saturation threshold in glasses below 1 K . In this paper we present a rate-equation theory which contains all the essential processes and agrees qualitatively with experimental results , some of which are presented here.

The ultrasonic saturation threshold in glasses is determined by the competition between three processes which affect the population difference $n(\omega)$ for TLS with energy splitting $E = \hbar\omega$. These processes are (a) the lowering of $n(\omega)$ as energy is resonantly absorbed from the ultrasonic pulse , (b) the "spin-lattice" relaxation , which attempts to restore $n(\omega)$ to its thermal equilibrium value n_0 , and (c) spectral diffusion , which tends to smooth out the ω-dependent features in $n(\omega)$ caused by (a). The spectral diffusion effect arises because the TLS move in ω-space due to "spin-spin" interaction [3] . This motion is limited to a region of ω-space defined by the diffusion kernel , $D(\omega-\omega')$, which is a Lorentzian whose width is called $\Delta\omega$.

We now suggest that these processes can be included in a rate equation which is based upon a modification of Wolf's theory of saturation [4] .

$$\dot{n}(\omega) = -R_2 g(\omega-\omega_0)n(\omega) - R_1(n(\omega)-n_0) - \omega_D(n(\omega)-\bar{n}(\omega)) \qquad (1)$$

where R_2 is proportional to the acoustic power , g is a resonant
shape function of width ω_2 , ω_0 is the ultrasonic frequency ,
R_1 is the spin-lattice relaxation rate , ω_D is the rate of spectral
diffusion , and $\bar{n}(\omega)$ is an average population difference given
by $\int d\omega' n(\omega')D(\omega-\omega')$. In the simplest approximation , eq.(1) reduces to two coupled equations for $n(\omega_0)$ and $\bar{n}(\omega_0)$, and the instantaneous attenuation is proportional to $n(\omega_0,t)$. In the cw-
limit $(R_1\tau_p \gg 1$ where τ_p is the pulse length) this leads to a
saturation threshold given by

$$J_{cw} \sim (\Delta\omega) \cdot R_1 \cdot (1+\mu)^{-1} \qquad (2)$$

where $\mu = (\Delta\omega)R_1/(\omega_D \cdot \omega_2)$. For T= 0.56 K and $\omega_0/2\pi$ = 795 MHz , we
estimate that $\mu \lesssim 0.2$, so that $J_{cw} \sim (\Delta\omega)R_1$, as though we had a
homogenous line whose width were given by the spectral diffusion
width $\Delta\omega$. The magnitude /3/ of $\Delta\omega$ qualitatively explains why
the saturation threshold is as high as it is in glasses and why
it varies roughly as T^2 ($\Delta\omega \sim T$ and $R_1 \sim T$) /2/. At lower temperatures ω_D may become sufficiently slow that $J_{cw} \sim \omega_2\omega_D$, indicating that the homogenous linewidth has been reduced to ω_2 and that
the relaxation rate R_1 has been replaced by ω_D .

        This cw analysis cannot , however , be complete because
estimates of R_1 indicate that $R_1\tau_p \ll 1$ in the temperature regime
near 0.5 K . This is substantiated by the saturation curves of
Fig. 1 , which show that the saturation threshold is indeed dependent on the pulse length τ_p . Averaging our general (i.e. time
-dependent) solution of eq.(1) over one pulse length τ_p , we obtain a normalized attenuation given by

$$\langle a/a_0 \rangle_{\tau_p} = (1+A)^{-1} (1+Am(\Theta)) \qquad (3)$$

where A = J/J_{cw} , $m=\Theta^{-1}(1-e^{-\Theta})$ and $\Theta = R_1\tau_p(1+A)$. This equation
is valid when $\mu \ll 1$ and $\omega_D\tau_p \gg 1$ in accord with the conditions that
pertain to Fig. 1. It is apparent from eq.(3) that the critical
intensity is given by

$$J_c \sim J_{cw} (R_1\tau_p)^{-1} \qquad (4)$$

whenever $R_1\tau_p < 1$. This expression is in good qualitative agreement
with the results shown in Fig. 1 , which show a consistent upward
shift in the saturation threshold as τ_p becomes shorter. These
results are consistent with a spin-lattice relaxation time of
10 μsec at this temperature and frequency.

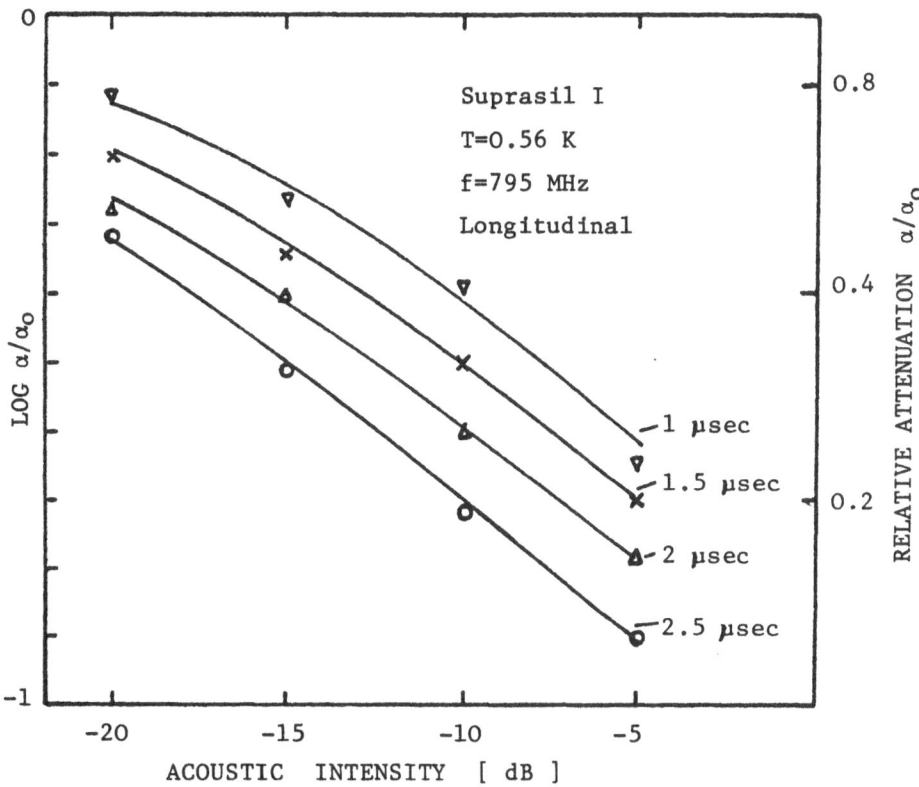

Fig.1: Ultrasonic absorption of longitudinal waves in Suprasil I
as a function of pulswidth and incident acoustic power.
Smaller pulswidths require higher power levels to saturate
the ultrasonic absorption. 0 dB corresponds approximately
to an acoustic intensity of 30 $\mu W/cm^2$. a_0 is the unsaturated
absorption value which is independent of the pulswidth.

        To summarize , we have proposed and solved a rate-equation
theory for the saturation threshold in glasses in the presence
of spectral diffusion and spin-lattice relaxation.Our results
apply in the temperature regime near 0.5 K , where spectral diffusion is sufficiently rapid that coherent effects are excluded.Our
findings show that the spectral diffusion process strongly en-
hances the saturation threshold while not significantly affecting
the pulse-length dependence. The pulse-length dependence , on the
other hand , is due to a long spin-lattice relaxation time.

REFERENCES

/1/ W.A.Philips, J.Non-Cryst.Solids 31:267 (1978) , and references
    contained therein.
/2/ S.Hunklinger and W.Arnold , Physical Acoustics 12:155 , eds.
    W.P.Mason and R.N.Thurston (Academic Press, New York , 1976)
/3/ J.L.Black and B.I.Halperin , Phys.Rev.B 16:2879 (1977)
/4/ E.L.Wolf , Phys.Rev. 142:555 (1966)

*On leave from Department of Physics , Brookhaven National Labora-
tory , Upton , New York 11973 , USA

STRESS EFFECTS IN THE SCATTERING OF THERMAL PHONONS BY JAHN-TELLER

IMPURITIES IN MgO

J.K. Wigmore, J.L. Patel and X.H. Mkhwanazi
University of Lancaster, Lancaster, U.K.,
and
J.R. Fletcher
University of Nottingham, Nottingham, U.K.

INTRODUCTION

We have observed the effect of uniaxial stress on the scattering of heat pulse phonons by $MgO:Cr^{2+}$ and $MgO:V^{3+}$. Both these impurities show a significant Jahn-Teller effect and are strongly coupled to the MgO lattice. Heat pulses are useful in their investigation since the energies of the lowest impurity states are roughly equal to those of the dominant heat pulse phonons, that is, a few cm^{-1}. The application of stress varies the energies of the levels, introduces extra splittings, and modifies the transition probabilities for phonon scattering between the states.

THE EXPERIMENTS

The heat pulse technique has already been fully described[1]. The present specimens were small rectangular parallelopipeds a few mm along each edge, immersed in liquid helium pumped to 1.5 K. Stress was applied to specimens directly through a rod loaded by weights at room temperature. Compression was applied in two separate directions, (100) corresponding to a strain of symmetry e_θ with negative sign, and (011), also e_θ but with positive sign. The phonon flux at the semiconducting avalanche bolometer was measured a few microseconds after heater excitation by means of a boxcar integrator, and plotted as a function of the applied stress determined using a wire strain gauge.

RESULTS AND INTERPRETATION

(a) $MgO:Cr^{2+}$

The data are summarised in figure 1, which includes (100) and
(011) compression results. For both positive and negative e_θ the
number of phonons reaching the bolometer decreased with increasing
stress, went through a minimum, and then increased again. In
addition, there was the suggestion of slight but reproducible
structure close to the minima.

The increased scattering away from $e_\theta = 0$ was at first sight
unexpected. The currently accepted theory of $MgO:Cr^{2+}$ gives the
lowest group of states for an unstrained site as A_1, T_1, E, T_2, and
A_2 in ascending order of energy[1]. At 1.5 K, only the A_1 ground
state should be significantly populated and from this state the only
allowed transition for e symmetry phonons is $A_1 \rightarrow$ E. The matrix
element for this transition depends only slightly on strain, and so
presumably should the phonon scattering. However, it has become
apparent during computations that if e_θ is non-zero, a further transition $A_1 \rightarrow A_2$ is also allowed and scatters phonons strongly even for
small values of e_θ. With hindsight, we note that this is because
$A_1 \rightarrow A_2$ becomes spin-allowed as e_θ is increased. The total phonon
scattering eventually decreases again as e_θ increases further, since
the $A_1 \rightarrow A_2$ splitting widens beyond the energy range of the dominant
heat pulse phonons. The calculated curve for comparison with figure
1 is rather insensitive to the actual values of the Jahn-Teller
parameters 3Γ and D. We are hopeful that the additional structure
may be assigned to level crossings of the A_1 and B_1 states (tetragonal representations).

Most impurity sites in MgO are strained naturally by amounts
comparable to those we applied. It seems likely therefore that the
$A_1 \rightarrow A_2$ transition causes significant scattering even in nominally
unstressed crystals. The data of Hasan et al.[2], for example, using
the superconducting tunnel junction technique, appear to show this
transition.

(b) $MgO:V^{3+}$

Figure 2 shows the dependence of bolometer signal on (100)
stress. The phonon scattering went through a series of maxima and
minima, which we interpret as being due to the occurrence of
frequency crossings among the V^{3+} hyperfine levels as stress was
varied. Similar features have previously been observed by thermal
conductivity with magnetic field as the variable[3].

The currently accepted theory of V^{3+} in MgO due to Ray[4] gives
the lowest state as an E doublet with a T_2 triplet a few cm^{-1}
higher in energy. Interaction with the V^{51} nucleus, spin $7/2$, splits
T_2 into three hyperfine states having F = $9/2$, $7/2$, and $5/2$. It is the

splitting of these states by strain into pairs of Kramers doublets, combined with the splitting of the ground state E doublet (which is not coupled to the nucleus in first order) that gives rise to more than thirty frequency crossings. In the absence so far of a detailed fit to the experimental curve, we have been unable to determine a value for the $E \rightarrow T_2$ splitting. However, a rough esti-mate of an average magnetoelastic coupling coefficient has been made by comparing theory with experiment for the stress at which the first frequency crossing occurred. The hyperfine splitting is known to be $A = .0042$ cm^{-1}, leading to a value for the magnetoelastic coupling of 10^3 cm^{-1} unit strain^{-1}.

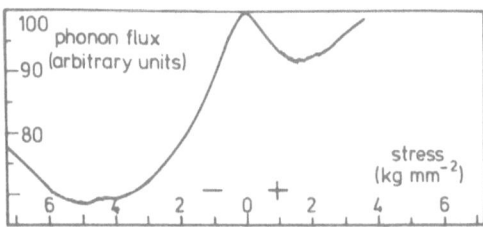

Fig. 1   Heat pulse phonon flux in MgO:Cr^{2+} as a function of applied stress, (100) and (011).

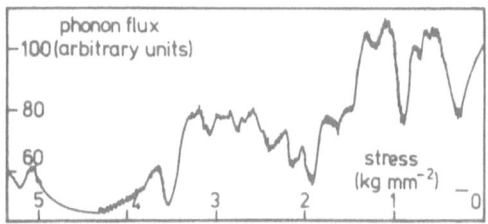

Fig. 2   Heat pulse phonon flux in MgO:V^{3+} as a function of (100) stress.

****

REFERENCES

1.   J.L. Patel and J.K. Wigmore, J. Phys. C10 1829 (1977)

2.   F. Hasan, P.J. King, D. Murphy and V.W. Rampton, Journal de Physique, 39 suppl. C6-993 (1978)

3.   L.J. Challis and D.L. Williams, J. Phys. C11 3787 (1978)

4.   T. Ray, Phys. Rev. B5 1758 (1972)

DISCUSSION

L. J. Challis: Are you saying that the hyperfine splitting constant
is strain dependent?

J. K. Wigmore: No, but I'm saying that with strain you split the
hyperfine levels, and you split the T_2 excited state. It's the
transition from the lower of the E state to one of the upper hyperfine states frequency crossing with a similar transition. So you're
tuning the hyperfine levels through each other with strain.

PHONON SPECTROSCOPY OF Mn^{3+} in MgO

F. Hasan, P.J. King, D.J. Monk, D.T. Murphy,
V.W. Rampton and P.C. Wiscombe

Department of Physics, University of Nottingham,
University Park, Nottingham NG7 2RD, England

INTRODUCTION

Superconducting tunnel junction phonon spectroscopy measurements have been made on manganese-doped MgO and four resonant absorptions, which we attribute to Mn^{3+} fitted to a current Jahn-Teller model.

Manganese readily substitutes for magnesium in MgO and occurs predominantly in the Mn^{2+} state. Small fractions of the manganese may be converted to Mn^{3+} by oxidizing heat treatment. Reducing heat treatment lowers the Mn^{3+} concentration. The $3d^4$ Mn^{3+} ion is strongly coupled to the lattice and, having a doublet electronic ground state, is expected to exhibit the Jahn-Teller effect. The iso-electronic ion Cr^{2+} has been much studied in an MgO host by methods which include acoustic paramagnetic resonance[1], thermal conductivity[2], acoustic relaxation[3] and tunnel junction phonon spectroscopy[4]. The experimental data has in each case been fitted to a model which invokes the dynamic Jahn-Teller effect, and values for the tunnel splitting, δ, the spin-orbit parameter D and the sign of anharmonicity parameter B have been deduced. Despite the great volume of work the agreement is far from satisfactory in detail.

Much less work has been done on the iso-electronic Mn^{3+}:MgO system, although recently Challis et al.[5] have inferred energy level separations of ~ 5 cm^{-1} and 21 cm^{-1} from thermal conduction measurements. These values suggest that this system is similar to Cr^{2+}.

EXPERIMENTAL

The results described here were obtained at 1 K from single crystals of MgO cut for phonon propagation along the <110> axis.

Lead-bismuth generator junctions and dirty aluminium detector junctions were used, making available a spectroscopic range of ~ 4 cm^{-1}
to 24 cm^{-1}.   The transmission spectrum taken from a sample obtained from Spicers Ltd., containing 840 ppm in the 'as received' state
is shown in Figure 1.   In addition to non-spectroscopic structure
resulting from the tunnel junctions, the figure indicates a number
of broad resonant absorptions.   Four of these, those centred at
4.8, 8.4, 14.3 and 21.4 cm^{-1}, were attributed to Mn^{3+}.   This spectrum was obtained using low frequency sine modulation.   Pulse techniques reveal strong diffusive scattering, the phonon mean free path
being much smaller than the sample dimensions.   Information on mode
or directional properties is therefore not forthcoming from these
measurements.

       Reducing heat treatment produces a spectrum with lines at 5.2,
9.6, 12.5 and 22 cm^{-1} typical of Cr^{2+}.   This type of treatment is
expected to enhance Cr^{2+} at the expense of Mn^{3+}.   A sample of MgO
containing titanium and manganese produced the four lines we attribute to Mn^{3+} when oxidized to enhance the Mn^{3+}.   The lines at 4.8
and 21.0 cm^{-1} are in excellent agreement with the lines attributed
to Mn^{3+} by Challis et al.[5].

DISCUSSION

       In fitting the four absorption lines to the model of Fletcher
and Stevens[6], the selection rules for phonon transitions and the low
temperature at which these experiments are conducted must be borne
in mind.   The latter implies that transitions most probably originate from the ground state or the first excited state.   A better fit,
with the more probable transitions probabilities, is found when
$B < 0$.   The best fit we have obtained to-date, given by $D = 3.7$ cm^{-1},
$\delta = 18$ cm^{-1}, $B < 0$, is shown in Figure 2.   These parameters provide
a transition of 4.75 cm^{-1} originating from the ground state and a
transition of 8.24 cm^{-1} from the $^1T_1$ to T_2 levels.   This transition
is expected to be a little weaker than the 4.75 cm^{-1} transition since
the depopulated first excited state is involved.   The $A_1 \to A_2$ transition provides an absorption at 14.34 cm^{-1}.   This transition is
forbidden in cubic symmetry, but is allowed under the influence of
the static tetragonal distortions known to be present.   The higher
frequency transition of 23.7 cm^{-1} predicted by the model is only in
fair agreement with the experimental line at 21 cm^{-1}.   However,
higher levels ignored in the Fletcher and Stevens model may shift
this transition by the small amount needed to provide a better fit
to the data as could the influence of static strains.   We feel that
the Fletcher and Stevens model provides a satisfactory fit to the
lines we attribute to Mn^{3+} and it is interesting to note that this is
not the case for the lines attributed to Cr^{2+} from tunnel junction
spectroscopy.   We find the acoustic resonance behaviour of Mn^{3+} to
be dominated by strain-dependent features, and to be similar to Cr^{2+}.
It is probable that the acoustic resonance properties of both sys-
ems are not strongly dependent on D or δ.

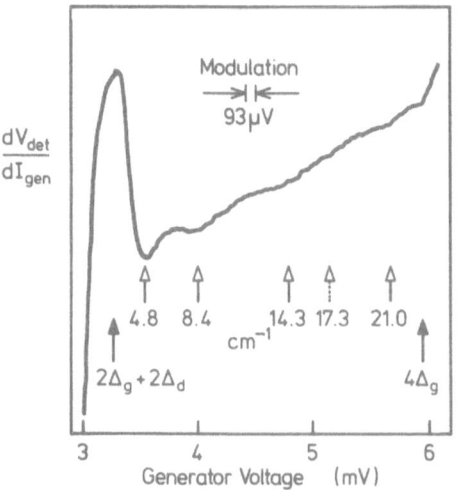

Fig. 1.   Phonon Transmission through Mn:MgO in the As-Received State.

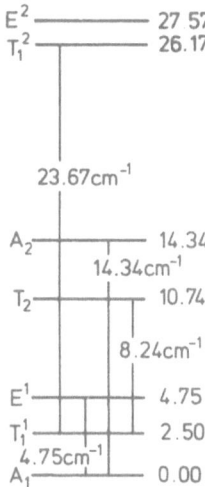

Fig. 2.   The Energy Level Diagram of
          Mn:MgO using the Jahn-Teller
          Parameters:

          $\delta = 18$ cm^{-1}, D $= 3.7$ cm^{-1}

          for B < 0.

REFERENCES

1. Marshall, F.G. and Rampton, V.W. (1968), J. Phys. C: Solid
    State Phys., 1, 594.
2. Challis, L.J., de Göer, A.-M., Guckelsberger, K. and Slack,
    G.A. (1972), Proc. Roy. Soc. London, A330, 29.
3. Lange, J.N. (1973), Phys. Rev. B, 8, 5999.
4. Hasan, F., King, P.J., Murphy, D. and Rampton, V.W. (1968), J.
    de Physique, Colloque C6, 39, 993.
5. Challis, L.J., Ghazi, A.A. and Maxwell, K.J. (1979), J. Phys.
    C: Solid State Phys., 12, 303.
6. Fletcher, J.R. and Stevens, K.W.H. (1969), J. Phys. C: Solid
    State Phys., 2, 444.

ACOUSTIC PARAMAGNETIC RESONANCE OF Cr^{2+} IN CaO

V. W. Rampton, P. C. Wiscombe and J. R. Fletcher

Department of Physics
University of Nottingham, University Park,
Nottingham NG7 2RD, England

Cr^{2+} ions in dielectric hosts have been widely studied. The ion is strongly coupled to the crystal lattice and is subject to the Jahn-Teller effect. Strong acoustic paramagnetic resonance (APR) absorptions are observed as well as large effects seen in thermal conductivity and acoustic relaxation attenuation. A detailed theory fitting the experimental results has been given for Cr^{2+} in Al_2O_3 (Bates 1978), but a detailed fit has not yet been achieved for Cr^{2+} in MgO (e.g. Lange 1976). We have made APR experiments on chromium-doped CaO, which has a structure identical with MgO, and in which the dopant Cr^{2+} ions are in an octahedral environment. The experimental results show significant differences from the results obtained with Cr^{2+} in MgO. A preliminary theoretical interpretation, including only APR results with the magnetic field along [111], has been given by Bates et al. (1979).

Specimens were cut from a sample of chromium-doped CaO given to us by Professor B. Henderson. Experiments were made using longitudinal ultrasonic waves at frequencies between 9.3 GHz and 9.7 GHz propagating along the [010] direction of the specimen. The experiments were made at a temperature of 1.8 K. A number of resonance absorptions were observed, all having axial symmetry about a fourfold axis of the crystal. An isofrequency plot of the peaks of these resonances is shown in Figure 1, though it should be noted that the resonance at lowest magnetic field sometimes has a double peak. Over most of the range of angles, the curves are consistent with the expressions:

$$h\nu = 2g_{\shortparallel}\beta H_z + \varepsilon$$
$$h\nu = 2g_{\shortparallel}\beta H_z$$

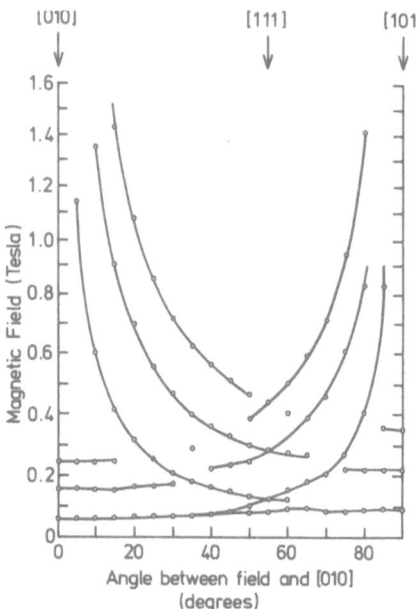

Figure 1.   Isofrequency plot of APR of Cr^{2+} in CaO.

$$h\nu = 2g_{,,}\beta H_z - \varepsilon$$

where H_z is the component of magnetic field along one of the <100>
directions of the crystal. We find $g_{,,}$ = 2.05 ± 0.05 and ε/h =
5.66 GHz. The limits of error on $g_{,,}$ are derived from the central
line of the three. Because of the small range of frequencies used,
it is possible to fit the other two lines with a range of values of
$g_{,,}$ and ε. It can be seen from Figure 1 that the isofrequency curves
do not fit the expressions given in the vicinity of cross-overs
where two resonances might be expected at the same magnetic field.
In a number of instances one, or both, resonances is shifted slightly
from the expected position.

The largest attenuation is produced by the resonance appearing
at the lowest magnetic field, whereas the one at highest magnetic
field is very weak. It should also be noted that the resonances
having axis of symmetry parallel to the direction of propagation of
the longitudinal ultrasonic waves are slightly stronger than those
with other orientations.

The Jahn-Teller theory of Cr^{2+} ions on octahedral sites has been given by Fletcher and Stevens (1969) (see also Bates et al. 1979) and is there applied to MgO.   The results presented here for CaO show some specially significant features.   The resonances show tetragonal symmetry which might imply that the Cr^{2+} ions are on tetragonally distorted sites, but two facts indicate that this view must be modified.   First, the shift of resonant field at crossovers, which implies a coupling between states having diffent symmetry axes.   Secondly, the appearance of strong resonance absorptions with symmetry axis parallel to the ultrasound direction, which implies some mixing of states from other distortion directions. Thus, for Cr^{2+} in CaO we find the results can be fitted using a Jahn-Teller theory and a slow rotation model in contrast to the fast rotation model used for MgO.   We believe the anharmonicity parameter, B, is large and the tunnelling splitting is smaller than the Zeeman splitting, or the second-order spin-orbit splitting.   APR transitions are possible when states corresponding to different tetragonal Jahn-Teller distortions are mixed by tunnelling.   Thus, at any point on the isofrequency plot the sites giving rise to a particular resonance are those for which the combined effects of magnetic field, spin-orbit coupling and strain produce approximately equal energies for different distortion axes.

REFERENCES

Bates, C. A., 1978, J. Phys. C: Solid State Phys., 11, 3447-60.

Bates, C. A., Maynard, C., Rampton, V. W. and Shellard, I. J., 1979, J. Phys. C: Solid State Phys., in the press.

Fletcher, J. R. and Stevens, K. W. H., 1969, J. Phys. C: Solid State Phys., 2, 444-56.

Lange, J., 1976, Phys. Rev. B, 14, 4791-802.

****

DISCUSSION

J. K. Wigmore: Are you saying this is a static Jahn-Teller effect?

V. W. Rampton: No. We're merely saying that the tunnel splitting is extremely small while the anharmonicity parameter β is very large so that the three holes are then separated from each other and one has to think of the states as being essentially the states one would get in the three wells, but mixed very slightly by the tunneling. This mixing allows transitions which otherwise wouldn't occur at all.

P. J. King: Are you saying this is almost in the limit of the static Jahn-Teller effect?

V. W. Rampton: Yes, I think so, if the static Jahn-Teller effect means anything. Some theorists don't like to consider a static Jahn-Teller effect.

LOW TEMPERATURE THERMAL CONDUCTIVITY OF $TmVO_4$ and $TmAsO_4$ CRYSTALS

B. Daudin, A.M. de Goër and S.H. Smith[*]

S.B.T., Centre d'Etudes Nucléaires de Grenoble
B.P. 85 X, 38041 Grenoble Cedex, France
[*]Clarendon Lab., Parks Road, Oxford OX1 3PU, U.K.

Rare-earth vanadates and arsenates undergo cooperative Jahn-Teller transitions at low temperatures and have been extensively studied both experimentally and theoretically[1] but heat transfer has been investigated only in few cases[2]. We have measured the thermal conductivity K(T) of $TmVO_4$ and $TmAsO_4$ single crystals from 80 mK to 100 K. The crystals were grown by the flux method[3] and cut as parallelepipeds with the largest dimension ($\leqslant$ 10mm) along c-axis. Special mounting techniques have been developed to allow K(T) measurements using the stationary heat flow method on such small samples[4]. The absolute accuracy of K is thought to be $\simeq$ 10%.

The results of the measurements on the $TmVO_4$ crystals are shown in fig. 1(a). The thermal conductivities of the three samples are nearly the same, with a sudden change of slope at the structural transition temperature T_D = 2.15 K. Just below T_D, K varies as $T^{5.5\pm0.5}$ and below 0.8 K as $T^{1.5\pm0.5}$. The effect around T_D was previously observed[2] with a slightly larger slope $T^{7.1\pm0.5}$. The results for the two $TmAsO_4$ samples are similar, and those for sample (A) are shown in fig. 1(b). Near T_D = 6 K there is again a sudden change of slope ; we find that K varies as $T^{12\pm1}$ below T_D. Also a minimum is observed at 4.5 K and the temperature variation below 0.8 K is about $T^{2.2}$. Preliminary experiments with applied magnetic field H up to 70 kG parallel to c-axis have been carried out on $TmVO_4$ sample (B) above 1.3 K. Typical results are shown in fig. 2 at selected fixed temperatures.

It is probable that the strong decrease of K(T) observed below T_D in both systems in zero field is related to the appearance of resonant phonon scattering at Tm^{3+} ions : the splitting 2W of the

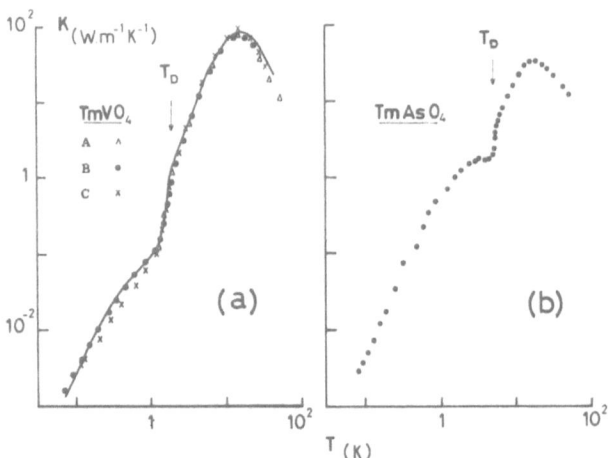

Fig. 1 - Thermal conductivity as a function of temperature.

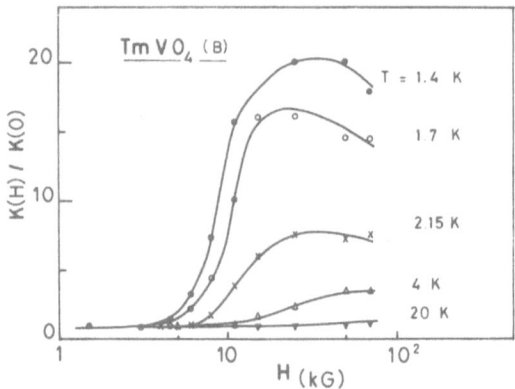

Fig. 2 - The ratio K(H)/K(O) as a function of magnetic field.

ground doublet is strongly temperature dependent and is given by
the implicit equation[1] :

$$W/kT_D = \tanh (W/kT)$$

A quantitative fit of the TmVO$_4$ results has been achieved using
Callaway's model[5], with a total inverse relaxation time :

$$\tau^{-1}(\omega) = \frac{v}{L} + A\omega^4 + G\omega + \tau_N^{-1} + \tau_U^{-1} + \tau_1^{-1} + \tau^{-1}$$

where the first five terms describe the usual scattering processes[6],
τ_1^{-1} the scattering (supposed elastic) between the ground state doublet
and the first excited state at 53.8 cm^{-1}, and τ_R^{-1} the phonon scattering at the split doublet below T_D, with the semi-empirical expression:

$$\tau_R^{-1}(\omega) = D_0 \, \omega^2 \, \exp - [(\omega - 2W/\hbar)^2/2\sigma^2]$$

The temperature dependence of W is included in the calculation but it
was found necessary to suppose a minimum value of 2W = 1 K above T_D
to get a good fit. D_0 and σ are adjustable parameters. One of the best
fits obtained for TmVO$_4$ (B) is illustrated in fig 1(a)(full line) and
the corresponding parameters (in cgs units) are given in table below.

	(v/L) cal.	(v/L) exp	A exp	G	D_0	σ_0 (K)	D_1
TmVO$_4$ (B)	3.6 10^6	2 10^6	7.10^{-45}	4.5 10^6	1.10^7	3.8	3.5 10^7
TmVO$_4$ (C)	1.9 10^6	2 10^6	7 10^{-45}	3.5 10^6	1.6 10^7	3.5	3.5 10^7
TmAsO$_4$	3 10^6	3 10^6	1 10^{-44}	4 10^6	3.10^8	1.4 3	10^8

The non zero value of 2W above T_D could be due to the effect of random strains induced in the crystals by mounting and cooling[7]. The values of the parameters are physically reasonable with the exception of
G, which should correspond to a dislocation density $\simeq 10^{10}$ cm^{-2}. The
origin of this term is so far not understood. A similar analysis was
done for TmAsO$_4$ and a good fit was obtained with 2W = 0 above T_D.
Finally the results with applied magnetic field below T_D are qualitatively in agreement with the known variation of 2W with H.

REFERENCES
1 G.A. Gehring and K.A. Gehring, Rep. Prog. Phys., 38 : 1 (1975).
2 M.W.S. Parsons and H.M. Rosenberg, 1st Int. Conf. on Phonon Scattering in Solids, ed. by M.J. Albany (CEN-Saclay, 1972), p. 326.
3 S.H. Smith, G. Garton, B.K. Tanner and D. Midgley, J. of Material
   Science, 13 : 620 (1978) and references therein.
4 B. Daudin and M. Locatelli, Note SBT 510/79 and to be published.
5 J. Callaway, Phys. Rev., 113 : 1046 (1959)
6 A.M. de Goër, J. de Phys., 30 : 389 (1969)
7 P.J. Becker, M.J.M. Leask and R.N. Tyte, J. of Phys.C,5 : 2027
   (1972).

DISCUSSION

H. Kinder: This is a famous example, where you cannot explain the
results just in terms of scattering because when you approach the
cooperative Jah-Teller phase transition you also have a softening
of the sound velocity just in the c-direction. So I think you have
to take into account also the softening of the sound velocity which
also will give you a depression of the heat conductivity without any
additional scattering.

A. M. de Goer: But I think this could be operative in the details
of the neighborhood of the transition. If I had put the fit as
K/T^3, you would have seen that it is not a good fit of the details
of the K/T^3 curve. So probably what you suggest would be important
in this region, but not over the whole region. Also, the softening
of the sound velocity occurs in just one direction.

H. Kinder: But that's just the direction ———

A. M. de Goer: But we have all phonons of all directions so we
have some mean value to take into account, so I don't know if it's
very important.

J. E. Rives: One possible way to eliminate some of the arbitrari-
ness or ambiguity is to calculate the field dependence.

A. M. de Goer: Yes, we have tried. But very probably it will not
be consistent with this first analysis because we have a very big
increase of thermal conductivity, and as I have mentioned, the
dislocation term is very important, so we must certainly change
some things in the analysis.

W. Eisenmenger: Have you any idea on the numbers for the mean free
path besides this problem of sound velocity? I have the feeling
that there is a very high concentration of ions - the material is
full of these Jahn-Teller systems. Don't you think that these
phonons are completely filtered out in your heat conduction experiment?

A. M. de Goer: No. They are an important part of the scattering,
but it is very difficult to say if they are completely stopped.

J. P. Harrison: If there is a splitting of that doublet due to the
stress field, this could show up in heat capacity.

A. M. de Goer: Yes, I know that the heat capacity results for some
mixed crystals with thulium and luterium have a huge peak at the
transition temperature. You have some tail and in the case of the
mixed crystal this tail has been explained by splitting of levels by
strain distribution. In the pure crystal I am not sure.

ACOUSTIC FARADAY AND COTTON-MOUTON EFFECTS IN PARAMAGNETIC CRYSTALS

J. W. Tucker

Physics Department,
Sheffield University,
Sheffield S3 7RH, U.K.

In an insulating crystal containing paramagnetic ions the presence of the spin-phonon interaction leads to the existence of coupled spin-phonon modes whose polarization vectors may depend not only on the space group of the insulating crystal concerned and the local symmetry of the paramagnetic ion, but also upon the presence or otherwise of an external magnetic field. Of particular interest are those host crystals where degenerate transverse phonons can propagate in high symmetry directions. In the presence of the paramagnetic ions the corresponding degeneracy of the spin-phonon modes may be lifted by the application of a magnetic field. In particular, for suitable orientations of the magnetic field two effects, the acoustic Faraday effect and the acoustic Cotton-Mouton effect can occur. Hitherto, only the acoustic Faraday effect has been observed experimentally, for Ni ion pairs in MgO. However, during a recent theoretical study of the Faraday and Cotton-Mouton effects in crystals containing iron-group ions which we have made, (to be published) numerical estimates show that both effects should be observable. As an example, we present in this paper results for one such system, Ni^{2+} ions in potassium magnesium fluoride, which suggests that both effects should be readily observable in the microwave frequency range. This is a particularly simple system in that the paramagnetic ions are in sites of octahedral symmetry and may be described by a very simple spin Hamiltonian containing just a Zeeman term with an effective spin of unity. The complete Hamiltonian is

$$H = \sum_{qj} \hbar\omega_{qj} a^+_{qj} a_{qj} - \sum_n \hbar\omega_o S^z_n + \sum_{qjn\gamma} \hbar\varepsilon^\gamma_{qj} T^\gamma_n \phi_{qj} \exp(i q \cdot R_n)$$

where the notation of Care and Tucker (1976), hereafter referred to as I, has been adopted. In the diagrammatic treatment of spin–phonon modes (paper I) the phonon Green function A, regarded as a matrix with respect to the polarization indices, satisfies the matrix equation $A=A^o+CA$ where C is related to the polarization matrix M of the generalised cumulant spin Green function. The problem here is to evaluate A for the geometry appropriate to the Faraday and Cotton–Mouton experiments.

The acoustic Faraday effect can occur for transverse phonons propagating parallel to the magnetic field which is assumed to be directed along a four–fold crystal axis. The appropriate spin–phonon coupling coefficient for this geometry may be obtained from eq.(25) of I. The assumption there of elastic isotropy is satisfied in potassium magnesium fluoride to within 5%. In lowest order one includes in M just M_o and this leads to excitations that are undamped. For the chosen direction of q the matrix A is diagonalised by choosing polarization vectors $e(1)=(\hat{x}+i\hat{y})/\sqrt{2}$ and $e(2)=(\hat{x}-i\hat{y})/\sqrt{2}$ representing two transverse circularly polarized waves. For a fixed frequency the poles of the appropriate phonon Green function give the wave vectors at which the system responds. These are found to be

$$q_{1,2}=(\omega/v_t)(1\pm B/(\omega\pm\omega_o\mp B))^{\frac{1}{2}}$$

with

$$B=(2N_sG_{44}^{2}/\hbar\rho Vv_t^{2})(\sinh\beta\omega_o/(1+2\cosh\beta\omega_o))$$

Since the two circularly polarized waves travel with different velocities, a transverse linearly polarized wave propagating along the z–axis will have its plane of polarization rotated as it propagates. The Faraday rotation/unit length is just $\frac{1}{2}(q_1-q_2)$.

The second geometry of interest is when q is directed along the x–axis (i.e. perpendicular to the magnetic field). In this case

$$A_{yy}=A_{yy}^o/(1-C_{yy}-C_{xy}C_{yx}/(1-C_{xx}))$$

and A_{xy} and A_{yx} are non-zero. However in most situations the residues
of A^{xy} and A^{yx} at the dominant pole of A_{yy} are very small (as we
have verified for the value of parameters chosen below). It is
therefore sufficient to ignore the off-diagonal components of A and
thus the last term in the denominator of A_{yy}. Under these conditions
two transverse waves can propagate, one with its polarization vector
parallel to the y-axis and the other along the z-direction. The
corresponding wave vectors at which the system responds at a
frequency ω are

$$q_y = (\omega/v_t)(1-4B\omega_o/(4\omega_o^2-\omega^2))^{-\frac{1}{2}}; \quad q_z = (\omega/v_t)(1-B\omega_o/(\omega_o^2-\omega^2))^{-\frac{1}{2}}$$

Thus if one launches a transverse wave into the crystal along the
x-axis with the polarization vector initially in the y-z plane at
45° to the y-axis, an elliptically polarized wave with its main axes
at 45° and 135° to the y-axis will result. The ellipticity defined
as the ratio of the length of the minor to major axis is given by
$\tan(\delta/2)$ where δ is the Cotton-Mouton phase shift. This shift per
unit length, δ', is just $q_y - q_z$.

For Ni^{2+} in $KMgF_3$ the measured spin-phonon coupling constants,
G_{11} and G_{44}, are 58 and 39 cm^{-1} respectively, and the longitudinal
and transverse velocities along the cubic axes are 6.47 and
3.94×10^5 cm/sec. Also the density ρ is 3.188 gm/cc and the lattice
parameter is 3.973 Å. As a demonstration of the size of the effects
we take the frequency, ω, to be in the microwave frequency range
(say, 9.4 GHz) and assume ω_o is a factor 1.5 larger. (This in fact
requires an applied magnetic field of 4.4 kG.) For these values,
we find that at a temperature of 4.2 K and a nickel ion concentration of 1%, the Faraday rotation and the Cotton-Mouton phase shift
are 1.99 and 1.12 rad/cm respectively. Even though these angles are
sufficiently large to be observed, their observation can be destroyed
if the attenuation of the excitations is large. Damping of the excitations due to the finite width of the spin levels can be included
phenomenologically into the theory by replacing ω_o by $\omega_o + i\Gamma_s$ in
the denominator of M_o. Here Γ_s is the single ion linewidth due to
processes other than the spin-phonon interaction. For real frequencies, the poles of the phonon Green function then lead to complex
values of q with an imaginary part α_s. In addition there is the
intrinsic linewidth due to the spin-phonon interaction itself. In
the phonon regime this is obtained by including M_{12} into M (see I
for details). Again an imaginary part, α_i, of q results. For
realistic values of Γ_s we find in fact that $\alpha_s \gg \alpha_i$ but that both
have a negligible effect on the observation of the Faraday and
Cotton-Mouton effects for the frequencies we have chosen. Ni doped
$KMgF_3$ should thus be a suitable material for an experimental

investigation of these effects.

Reference

Care, C. M., and Tucker, J. W., 1976, Phonon scattering by paramagnetic ions with $S>\frac{1}{2}$:I. Zeeman levels in the absence of a zero-
field splitting, J. Phys. C., 9:4237.

ACOUSTIC RELAXATION LOSS DUE TO Ni^{3+} IONS IN Al_2O_3

P. J. King and D. J. Monk

Department of Physics,
University of Nottingham, University Park,
Nottingham NG7 2RD, England

If the energy levels of a magnetic ion are modulated by the strain field of an acoustic wave, the dynamic repopulation of the levels can lead to a temperature-dependent acoustic loss, the details of which can be used to study the properties of the ground state and excited states of the ion[1].

Several systems recently investigated reveal acoustic attenuations whose magnitude and dependence on mode are not understood[2,3], and it has been suggested that the influence of static lattice strains may well be responsible.

Recent progress in understanding the Ni^{3+}:Al_2O_3 system[4,5,6], which is known to exhibit acoustic relaxation, suggests that the low-lying levels are now well-understood. We present here a study of the expected relaxation behaviour of this system.

The $3d^7$ Ni^{3+} ion substitutes for aluminium in Al_2O_3 and can be considered as being at an octahedral site with small trigonal distortions. The effect of the dynamic Jahn-Teller effect on the ^2E electronic ground state is to produce ground vibronic ^2E and ^2A states, separated by the *inversion splitting* δ. Residual perturbations, such as strain and the effect of spin-orbit coupling and trigonal field, can be described in zero magnetic field by the Hamiltonian terms

$$\mathcal{H} = 2p\alpha A_z S_z + qV\{Q_\theta U_\theta + Q_\varepsilon U_\varepsilon\}. \tag{1}$$

101

(See, for example, Ham[7] and Salce and de Goer[5].)   Each of the levels is in fact doubly degenerate under spin, but since in zero mag-
netic field no terms couple $\pm|\frac{1}{2}>$ spin states, the system can be
described in terms of three levels.   Under the influence of a stat-
ic lattice strain, described by $Q_\theta + iQ_\epsilon = \rho e^{i\theta}$, the levels are
found at δ and $\pm\{V^2q^2\rho^2 + \alpha^2p^2\}^{\frac{1}{2}}$.   The addition of the strain field
of an ultrasonic wave, given by $Q'_\theta + iQ'_\epsilon = \rho'e^{i\theta'}$, modulates the lower levels by amounts $\pm qV\rho'.\cos(\theta - \theta')[qV\rho/\{(qV\rho)^2 + \alpha^2p^2\}^{\frac{1}{2}}]$.   The
upper ^2A level is unaffected by strain to first order.

    The relaxation loss for such a system is given by
(for $kT >> \alpha p = 0.47$ cm^{-1}):

$$\Gamma = (f'qV)^2 \frac{4.34N}{\rho v^3 kT} . \frac{\omega^2\gamma_+}{\omega^2 + \gamma_+^2} . \cos^2(\theta - \theta')\frac{(\rho qV)^2}{(\rho qV)^2 + \alpha^2 p^2} . \quad (2)$$

Here N is the concentration of Ni^{3+} ions, ρ is the density, v is the
velocity and $\gamma_+ = 2P_{EE} + P_{EA}$, where P_{EE} is the transition probability between the two levels of the doublet and P_{EA} is the transition
probability between either level of the doublet and the singlet.
f is a numerical coupling factor depending on the particular acoustic mode.   Using (1), the probabilities P_{EE} and P_{EA} can be calculated in terms of the parameters of the system.   We find that the relevant matrix elements are independent of the lattice strain and that

$$2P_{EE} = \frac{\Delta^2.kT(qV)^2}{\pi\rho\hbar^4}\{\frac{1}{5v_L^5} + \frac{3}{10v_T^5}\} = AT . \quad (3)$$

Isotropic longitudinal (v_L) and transverse (v_T) phonon velocities
are assumed. Δ is the separation of the doublet levels.

    We also find that

$$P_{EA} = \frac{\delta^3}{2\pi\rho\hbar^4} . \frac{r^2V^2}{(e^{\delta/kT} - 1)} . \{\frac{1}{5v_L^5} + \frac{3}{10v_T^5}\} = \frac{B}{(e^{\delta/kT} - 1)} . \quad (4)$$

Here, r is the Jahn-Teller reduction factor[7].

    Our expression for the attenuation (2) is similar to that of
Sturge[1], but the addition of static lattice strains to the model
produces two reduction factors.   That involving ρqV and αp is of
order $\frac{1}{4}$, but is dependent on the magnitude of the strains and thus

on the sample treatment[5].   The factor $\cos^2(\theta - \theta')$ is equal to 1/2
for a strain distribution random in Q_θ, Q_ϵ space.   A comparison of
the attenuations of different modes (and thus different Q'''s) enables a test of the isotropy to be conducted.   The magnitudes of
the attenuation of the modes measured by Sturge are consistent with
an isotropic distribution.

The relaxation times measured in this laboratory and by Sturge
can be fitted to the expression for γ_+ with $\delta \simeq 60$ cm^{-1},
$A = 1.1 \times 10^7$ s^{-1} K^{-1}, $B = 2 \times 10^{12}$ s^{-1}.   The value of δ is very
sensitive to the precision of the high temperature data.   Abou-
Ghantous et al. obtain $A = 2 \times 10^7$ s^{-1} K^{-1} from linewidth data[4].

Expression (3) is site-dependent since Δ^2 depends on ρ.   The
appropriate average for Δ^2 over all sites does not converge for any
of the strain distributions suggested to date[5], probably due to
their behaviour at high strains.   We can use the experimental values of A, together with (3), to deduce a mean value for $\overline{\Delta^2}$.
$A = 1.1 \times 10^7$ s^{-1} K^{-1} gives $(\overline{\Delta^2})^{\frac{1}{2}} = 1.3$ cm^{-1} and $A = 2 \times 10^7$ s^{-1} K^{-1}
gives $(\overline{\Delta^2})^{\frac{1}{2}} = 1.8$ cm^{-1}.   We have used $V = 30,000$ cm^{-1} and $q = 0.47$.

These values suggest that the strain distribution falls off
extremely rapidly with Δ above ~ 2 cm^{-1} and they can be used as a
test of any future hypotheses concerning strain distribution.

r has recently been estimated from thermal conductivity measure-
ments[5].   Using $\delta = 60$ cm^{-1}, $r = 0.3$, $v_L = 1.08 \times 10^4$ m s^{-1} and
$v_T = 0.64 \times 10^4$ m s^{-1}, we obtain from (4) an estimate for B of
0.6×10^{12} s^{-1}.   This is smaller than the experimental data suggests, but this agreement is probably satisfactory considering our
cavalier treatment of the phonon isotropy and the accuracy of the
data.

REFERENCES

1.   Sturge, M. D., Krause, J. T., Gyorgy, E. M., LeCraw, R. C. and
      Merritt, F. R., Phys. Rev., 155, 218 (1967).

2.   King, P. J., Monk. D. J. and Oates, S. G., J. Phys. C: Solid
      State Phys., 11, 1067 (1978).

3.   Kim, H. and Lange, J., Phys. Rev. B, 18, 1961 (1978).

4.   Abou-Ghantous, M., Bates, C. A., Fletcher, J. R. and Jaussaud,
      P. C., J. Phys. C: Solid State Phys., 8, 3641 (1975).

5.   Salce, B. and de Göer, A.-M., J. Phys. C: Solid State Phys.,
      12, 2081 (1979).

6.   Shen, L. N. and Estle, T. L., J. Phys. C: Solid State Phys.,
      12, 2119 (1979).

7.   Ham, F. S., See, for example, the review article published in
      Electron Paramagnetic Resonance, Ed. S. Geschwind, Plenum
      Press, New York (1972).

PHONON SPECTROSCOPY OF ION PAIRS IN Al_2O_3

F. Hasan, P. J. King, D. T. Murphy and V. W. Rampton

Department of Physics,
University of Nottingham,
University Park, Nottingham NG7 2RD, England

INTRODUCTION

We have studied chromium doped Al_2O_3 and vanadium doped Al_2O_3 using high frequency phonons generated by superconducting tunnel junctions. When the concentration of the dopant ions was low we have found resonance absorptions which can be attributed to single Cr^{2+} ions or V^{3+} ions. More concentrated samples have given evidence of absorptions due to ions coupled together. Cr^{2+} in Al_2O_3 has been extensively studied by a number of methods; the most recent theory of the system is due to Bates (1978) who used a Jahn-Teller theory of the ion coupled to the full lattice modes to obtain a good fit to experimental results. The Bates model predicts a strong phonon absorption at 6.7 cm^{-1} and a weaker absorption at 17.8 cm^{-1} at the temperature of our experiments (~ 1 K). V^{3+} in Al_2O_3 has also been widely investigated and it has been shown that the ground state energy levels consist of a singlet and an excited doublet separated by an energy $D = 8.3$ cm^{-1} (Kinder 1973, Pontnau and Adde 1975).

EXPERIMENTS

Experiments have been made using lead/bismuth alloy tunnel junctions as phonon generators and aluminium tunnel junctions as phonon detectors. This combination gives a spectral range of about 4 cm^{-1} to 24 cm^{-1}. The experiments were made at a temperature of about 1 K.

Nominally pure samples of Al_2O_3 but which contained a few ppm of chromium showed an absorption at 6.7 ± 0.2 cm^{-1}. The line is asymmetric, having a high frequency tail, and has a width at halfheight of ~ 0.5 cm^{-1}. This line we attribute to single Cr^{2+} ions. In samples of Al_2O_3 doped with chromium and containing a higher concentration of Cr^{2+} we find a similar, but very intense, absorption at 6.7 ± 0.2 cm^{-1} and also a weaker broad line at 13.4 ± 0.4 cm^{-1} and a very weak line at 20.2 ± 0.4 cm^{-1}. These extra lines we attribute to clusters of two Cr^{2+} ions and three Cr^{2+} ions respectively.

Samples of Al_2O_3 lightly doped with vanadium show a sharp absorption line at 8.3 ± 0.2 cm^{-1} which we attribute to V^{3+} ions in agreement with the work of Kinder (1973). A sample of Al_2O_3 doped with 1340 ppm of vanadium gave the same absorption line at 8.3 cm^{-1} but now with pronounced shoulders and, in addition, weak but sharp lines at 6.4, 10.6 and 13.4 cm^{-1}. These lines we attribute to pairs of V^{3+} ions.

DISCUSSION

Paramagnetic ions in a dielectric host may be coupled together in various ways. The simplest way of fitting the positions of the satellite lines in the case of V^{3+} in Al_2O_3 is to use an isotropic exchange Hamiltonian $\mathscr{H} = J.S_1.S_2$ with effective spins of one and values of $J = 3.46 \pm 0.10$ cm^{-1} and of $J = 1.86 \pm 0.15$ cm^{-1} which predict the three observed satellites and a fourth which is within the low frequency shoulder of the main line. However, if it is assumed that the spin-phonon coupling is the same for pairs as for single V^{3+} ions then we should expect two things. The satellites above the single ion line would be much more intense than the satellites below the single ion line and also weak lines should appear at a frequency near to twice that of the single ion line. The relative intensities of the satellites can be explained on a model in which nearby point defects different from V^{3+} cause changes in D, but this model will not explain the positions of the satellites which are asymmetric about the single ion line. The coupling between Cr^{2+} ions in Al_2O_3 is probably not exchange interaction because the exchange interaction is much reduced by the strong Jahn-Teller effect (Passeggi and Stevens 1974). Virtual phonon exchange seems a more likely coupling mechanism for Cr^{2+} ions (Baker 1971). The pair lines are centred on twice the frequency of the single ion line whereas it might be expected that the coupling would produce a slight shift in frequency as well as satellites around the single ion line. However the line width observed for Cr^{2+} in Al_2O_3 is more than 2 cm^{-1} so any satellites may be hidden within the line width. It is possible that a range of separations of Cr^{2+} ions occurs and gives a large number of satellites which appear as a very wide line. A second wide line then appears near the double frequency and also consists of several sets of lines. Finally it should be mentioned that the intensity of the double frequency line and the appearance of the triple frequency line due to groups of three ions suggests a tendency of Cr^{2+} ions to cluster together.

REFERENCES

Baker, J. M., 1971, Rep. Prog. Phys., $\underline{34}$, 109-173.

Bates, C. A., 1978, J. Phys. C: Solid State Phys., $\underline{11}$, 3447-59.

Kinder, H., 1973, Z. Physik, $\underline{262}$, 295-314.

Passeggi, M. C. G. and Stevens, K. W. H., 1974, Physica, $\underline{71}$, 141-160.

Pontnau, J. and Adde, R., 1975, J. Phys. Chem. Solids, $\underline{36}$, 1023-4.

****

DISCUSSION

H. Kinder: We have reported satellites of V^{3+} lines in Nottingham, and we saw them in a material which was also vanadium doped, but only to fairly low concentrations. However, it was γ-irradiated, and we didn't see it in the same sample without the γ-irradiation. So I don't think these lines have anything to do with V^{3+} pairs. They must be something related to some other defect.

V. W. Rampton: That may be the case. Our reason for opting for pairs is simply that we found them in the heavily doped sample, but not in the lightly doped one. I think the isotropic exchange parameter we have got is possibly consistent with some of the frequency-crossing parameters that Challis et al have found, but I'm not entirely certain that that's necessarily true. It requires a bit more work to be sure there is an agreement there. I'm not sure that our satellites are quite at the same frequency as yours. I don't think they were quite the same frequency.

OBSERVATION OF HYPERFINE SPLITTING OF THE LEVELS OF V^{3+} IONS AND PAIRS BY PHONON SPECTROSCOPY

L.J. Challis, A.A. Ghazi, D.J. Jefferies and M.N. Wybourne

Department of Physics, University of Nottingham,
University Park, Nottingham NG7 2RD, UK

In principle, frequency crossing spectroscopy[1] has a resolution limited only by the linewidths of the levels involved. In the present work measurements have been made on V^{3+} in Al_2O_3 with a resolution $\gtrsim$ 100 MHz at a crossing frequency of 166 GHz. Measurements of thermal resistivity W have been made for fields along the c-axis near the crossing point at B $\sim$ 3.1T between the $\Delta M = 1$ and $\Delta M = 2$ transition frequencies of V^{3+}: $\nu_1 = D - g_\| \beta B - A_H I_Z$ and $\nu_2 = 2g_\| \beta B + 2A_H I_Z$ where D = 249 GHz, $g_\| = 1.91^2$, A_H = 288 ± 2 MHz[3] and I = 7/2. Both frequencies have 8 hyperfine components which are split by A_H and $2A_H$ respectively. If these components are well resolved signal $\Delta W/W_o$ obtained by crossing them has 21 lines separated by A_H.

Four samples[4] of dimensions $\sim$ 2 x 2 x 10 mm^3 have been investigated. Three were a-axis rods containing 74, 267 and 1360 ppm of V^{3+} and the fourth was a b-axis rod containing 74 ppm of V^{3+}. Measurements of $\Delta W/W_o$ and dW/dB have been made and here we report data on dW/dB obtained by placing the sample between small superconducting coils which produced a modulation field of amplitude 3 mT and frequency 3 Hz[1]. The temperature gradient was measured using 2 carbon resistance thermometers in a 1 kHz inductance bridge with a lock-in amplifier as a detector. A second lock-in amplifier was used to measure the 3 Hz component of the signal from the first amplifier (or 6 Hz component if d^2W/dB^2 was needed). The noise after the first amplifier is dominated by low frequency components - principally vibrational/microphonic noise, 1/f noise from the thermometers and $\sim$ $1/f^2$ noise from temperature fluctuations from the bath. Modulation at 3 Hz results in an improvement of signal to noise ratio of $\sim$ 45 when the modulation amplitude is equal to the linewidth for a time constant of 1 s on the final lock-in amplifier. This improvement should increase by $\sqrt{t}$ for a time constant t.

109

    Fig 1(a) shows data for the a-axis 74 ppm sample.   The 21 lines
are well resolved and the mean value of the hyperfine constant obtained from 4 such traces is 283 ± 5 MHz in good agreement with previous data; the error could be substantially reduced by improvements
in the technique of measuring the small differences in field.   Fig 1
(b) shows the signal computed assuming the scattering to be Lorentz-
ian with $\tau_i^{-1}/\tau_B^{-1} = \Delta_i^2/[(\nu-\nu_i)^2 + \Gamma_i^2]$ where τ_B^{-1} is the scattering by

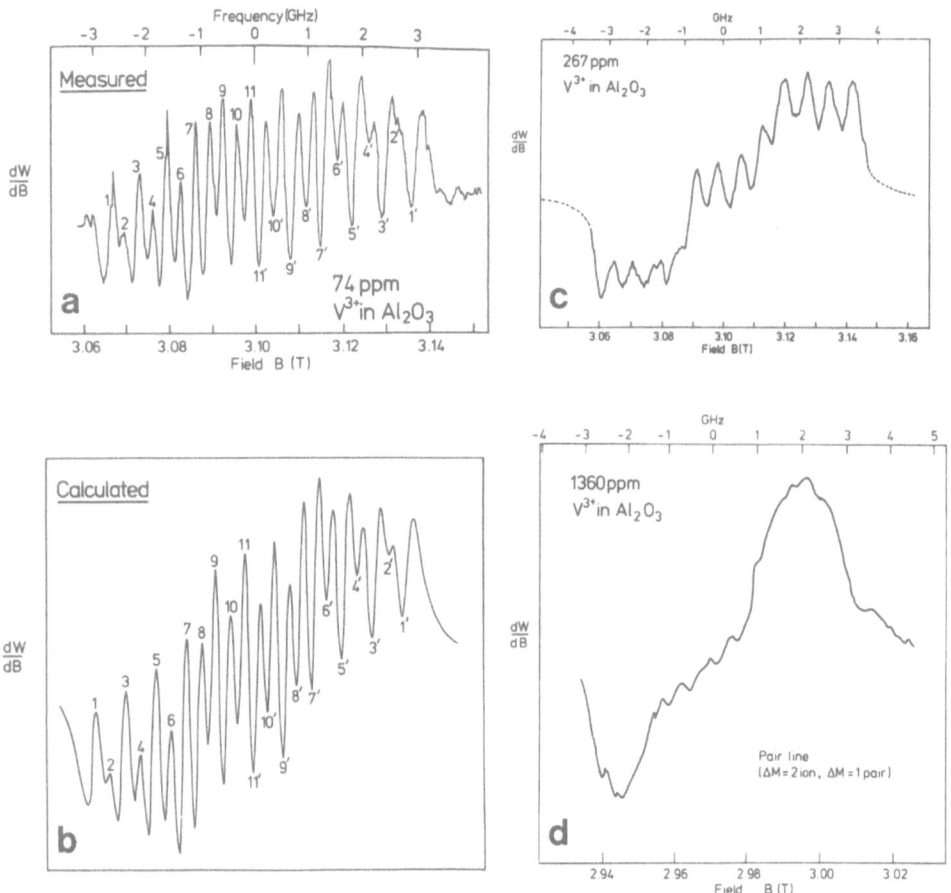

Fig 1: The differential of the thermal resistivity dW/dB for Al_2O_3
samples containing V^{3+} in the following concentrations (a) 74 ppm,
(b) 74 ppm - computed curve, (c) 267 ppm and (d) 1360 ppm; the line
shown here is a V^{3+} pair line.

boundaries and $\Delta_1^2 \propto$ concentration. The parameters used here were $\Delta_1 = 150$, $\Delta_2 = \Gamma_1 = \Gamma_2 = 50$ MHz. The widths Γ_1, Γ_2 are believed to be partly limited by field inhomogeneity; this was demonstrated by an experiment in which the same sample was displaced by 1.5 mm along the axis of the solenoid, where alternate lines were missing from dW/dB[5]. This indicates that the $\Delta M = 1$ lines are strongly overlapping presumably as a result of poorer homogeneity.

However, Fig 1(c) shows data for a sample containing 267 ppm placed at the same position in the solenoid to within < 0.5 mm as the 74 ppm sample was for the data of Fig 1(a) and the alternate lines are again missing. The effective widths in these experiments are not Γ_i but $(\Delta_1^2 + \Gamma_1^2)^{\frac{1}{2}}$ and we presume that in this case the $\Delta M = 1$ lines are overlapping because of the fourfold increase in Δ_1^2. For the sample of concentration 1360 ppm $\Delta_2 \sim 250$ MHz 5 times greater than for the 74 ppm sample and the structure due to the $\Delta M = 2$ transition is just apparent as expected. Finally in 1(d) we show the structure in one of the satellite frequency crossing lines in this sample caused by ~ 1 ppm of V^{3+} exchange coupled pairs with exchange constant J = 8.7 GHz[6]. The line shown is caused by the $\Delta M = 2$ ion line at ν_2 crossing the $\Delta M = 1$ pair line at $\nu_1 - J$.

We conclude that frequency crossing spectroscopy should be capable of quite high resolution ~ 10 MHz or so in fields of high homogeneity. The limitation is presumably the phonon lifetime to boundary scattering of a few MHz. The field modulation technique produces a considerable improvement in signal to noise ratio. The small size of Γ_1, which is clearly appreciably less than A_H (since the $\Delta M = 1$ structure can be seen) is surprising since it appears to imply that the strain broadening due to trigonal strains (or electric fields) is much less than that due to rhombohedral strains which can be measured from the widths of the $\Delta M = 2$ apr spectra. Computer analysis is in progress to determine estimates of the Δ_i and Γ_i parameters for the three samples.

REFERENCES
1. Wybourne M N, Jefferies D J and Challis L J, submitted for publication. This paper includes references to earlier measurements of frequency crossing spectroscopy.
2. Pontnau J and Adde R, J Phys Chem Solids 36, 1023 (1975).
3. Zverev G M and Prokhorov A M, Sov Phys JETP 11, 330 (1960).
4. We are very grateful to Dr de Goër, Professor J Joffrin and Professor H Kinder for providing the samples.
5. These are the data of the present authors shown in Challis L J, Journal de Physique 39-C6, 1553 (1978).
6. Challis L J, Ghazi A A, Jefferies D J, Williams D L and Wybourne M N, Phys Rev Letts 40, 519 (1978).
****

DISCUSSION

A. M. de Goer: May I ask you the final values of the magnetoelastic constant? You have obtained from fitting only one sample?

M. N. Wybourne: Just the one sample, yes. It's not a very good
fitting routine at the moment. It's very sensitive to differences
in height and width of those lines.

THE OBSERVATION OF ANOMALOUS ION SITES IN Al_2O_3 BY PHONON

FREQUENCY CROSSING SPECTROSCOPY

A.A. Ghazi, L.J. Challis, D.L. Williams and J.R. Fletcher

Department of Physics, University of Nottingham,
University Park, Nottingham NG7 2RD, UK

We report measurements on a and b axis samples[1] from a Verneuil grown crystal of Al_2O_3 doped with $\sim$ 74 ppm V^{3+} and containing $\sim$ 1 ppm of Fe^{2+}. The technique used identifies the magnetic fields at which two transition frequencies are equal (cross) from resulting minima in the magnetothermal resistivity. The crystal had been placed in a cyclotron[1] and so could have been γ and n-irradiated.

The spin Hamiltonian describing the ground states of both V^{3+} and Fe^{2+}, if we neglect hyperfine splitting in vanadium, is

$$\mathcal{H} = D\left[S_z^2 - \tfrac{1}{3}S(S+1)\right] + g_{||}\beta B_z S_z + g_{\perp}\beta(B_x S_x + B_y S_y)$$

with S = 1. Fig 1 shows the transition frequencies of these ions for B$||$c-axis (B < 5T) using for V^{3+}: D = 249 GHz, $g_{||}$ = 1.91, $g_{\perp}$ = 1.72[2] and Fe^{2+}: D = 113 GHz[3], $g_{||}$ = $g_{\perp}$ = 3.43[4]. The computed angular dependences of the 8 crossings (A to H) are plotted as solid lines in fig 2. They clearly identify the crossings due to V^{3+} and Fe^{2+} from the 9 or so observed and data from the b-axis sample (b $\perp$ B) are shown in fig 2. Line B was not detected. Lines C and D were also examined in detail for the a-axis sample and agreed with those shown as expected. Line E (Fe^{2+}/Fe^{2+}) is an anticrossing maximum only seen at $\theta \sim 0^{\circ}5$ and lines A (Fe^{2+}/Fe^{2+}) and F (V^{3+}/V^{3+})[5] have also been discussed previously. The agreement with the data could probably be improved by slight adjustment to the parameters. None of the 4 V^{3+}/Fe^{2+} lines (C, D, G and H) has been seen in any of the 11 other Al_2O_3 samples that have been examined (1 'pure' and 10 doped with V (4 samples), Cr (2), Mn (2) and Fe (2)).

Four other crossings O, P, Q and R are readily observable in these 2 present samples. R may also be present in the 'pure' and Cr

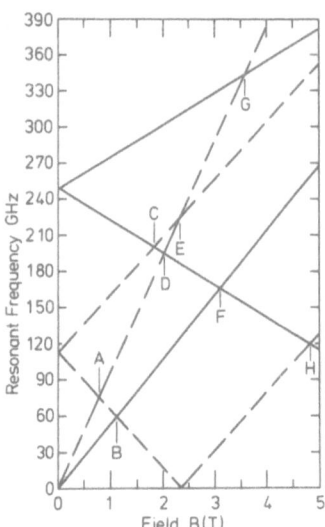

Fig 1: Solid lines - V^{3+} and
broken lines - Fe^{2+}

Fig 2: Points - Experiment
(broken lines are drawn
through the points for O,
P and Q). Solid lines -
Theory.

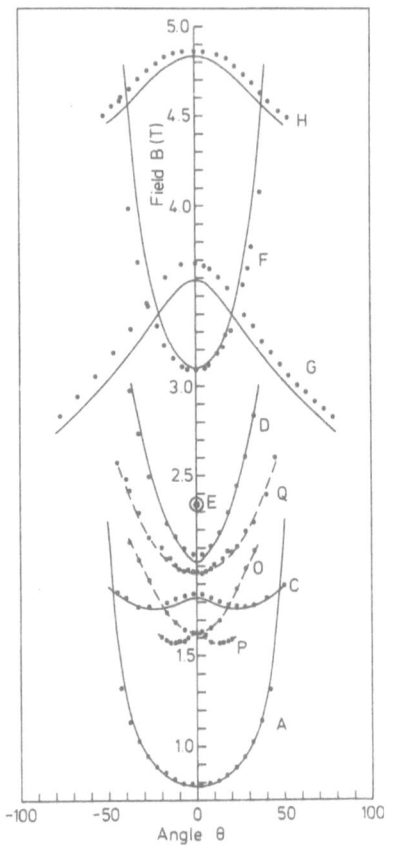

doped samples (its angular dependence has not been studied but there
are lines on the c-axis at about the same field) but lines O, P and Q
are only observed in the present samples. The angular dependences of
P and Q are very similar to those of C and D and it is possible
therefore that they are due to V^{3+} or Fe^{2+} ions associated with
another ion or defect. The data for O, P and Q in fig 2 are for the
a-axis sample, (a ⊥ B) and to look for evidence of a distortion axis
the measurements were repeated for the b-axis sample mounted at
angles $\phi = 0°$, $10°$ and $45°$ to the normal to the field plane while the
c-axis remains in the field plane. However in all cases the data observed were identical to those of fig 2 to within experimental error.

    In an attempt to confirm the association of P and Q with V^{3+}/
Fe^{2+} crossings the a-axis sample was subjected to the sequence of
treatments: oxidation, γ-irradiation and then reduction[6]. Oxidation
apparently converted Fe^{2+} to Fe^{3+} while making little change to the
V^{3+} concentration[7] since line F (V^{3+}/V^{3+}) was unchanged while A, C,
D, E, G and H which involve Fe^{2+} all disappeared. γ-irradiation
restored a small part of the original Fe^{2+} concentration since lines

C, D, E, G and H reappeared (A was not seen but was quite small even
in the as-received state) but were very much weaker and a reducing
heat treatment removed them again. Lines P and Q were removed by
oxidation which is consistent with an association with Fe^{2+} but
(unlike C and D) were not seen after γ-irradiation.

We first consider the possibility that P and Q are caused by
Fe^{2+} frequencies crossing the frequencies of V^{3+} ions associated with
another ion or defect. It seems clear that the resulting distortion
could not be a crystal field effect; this could be represented by a
term

$$\mathscr{H} = D'\left[S_z^2 - \frac{1}{3}S(S+1)\right] + E(S_x^2 - S_y^2) + F(S_x S_y + S_y S_x)$$

and we have shown that this cannot account for the angular dependences of P and Q. (A full test was carried out although in fact the
lack of ϕ dependence rules out these distortions (terms E and F)).
Alternatively the V^{3+} could be paired to another ion. The lack of ϕ
dependence restricts this to isotropic pairing $\mathscr{H} = JS_1.S_2$ and we
have examined $V^{3+}-V^{3+}$ pairs (statistically most probable) and $V^{3+}-$
Fe^{2+} pairs whose concentration would be rather small but would fall
with Fe^{2+} concentration which would explain why P and Q were apparently so much smaller than C and D in the spectra of the final γ-
irradiated state. However neither fits the angular dependence of P
and Q.

An alternative explanation is that P and Q are due to Fe^{2+} ions
in anomalous sites and again it would seem that we can rule out crystal field distortions at least for Q. D is produced by the $\Delta M = 2$
transition of Fe^{2+} with $h\nu = 2g_\| \beta B$ for $B\|C$. For Q this would re-
quire a 5% shift in $g_\|$ which is improbable since this seems very
insensitive to distortion being very nearly the same in cubic (MgO)
and axial (Al₂O₃) crystals.

We conclude that there is some evidence of rather strong phonon
scattering by V^{3+} or possibly Fe^{2+} ions in anomalous sites. We have
been unable to provide a quantitative account of the spectra
observed.

REFERENCES
1. We are very grateful to Professor Kinder for this sample.
2. Pontnau J and Addé R, CR Acad Sci Paris, 280, 301 (1975).
3. Anderson B R and Challis L J, J Phys C 6, L266 (1973).
4. Moore W S, Bates C A and Al-Sharbati T, J Phys C 6, L209 (1973).
5. Anderson B R and Challis L J, J Phys C 7, L440 (1974);
    Challis L J and Williams D L, J Phys C 10, L621 (1977).
6. The details of the treatments are given by Ghazi A A, PhD Thesis
    University of Nottingham 1978.
7. This is known from much more detailed studies. Villedieu M,
    Devismes N and de Goër A M, J Phys Chem Solids 38, 1063 (1977).
****

DISCUSSION

L. J. Challis: Perhaps I could just make a comment. Professor
Kinder mentioned earlier that he has observed V^{3+} in distorted
sites, and perhaps I should say that the initial reason for studying
this was because of those. We didn't find agreement for fields
along the c-axis, but we just cannot get any fit when we try and
reconcile the position of Kinder's lines and our lines when we take
a full analysis and look what happens off the axis.

HARMONIC PHONON ABSORPTION IN A COUPLED SPIN-PHONON SYSTEM

F.W. Sheard

Department of Physics, University of Nottingham,
University Park, Nottingham NG7 2RD, UK

Although interactions between ions due to virtual phonon exchange (VPE) have been extensively discussed[1], experimental evidence for them is meagre. In addition there are difficulties in calculating the magnitude of the VPE coupling. Nevertheless such calculations are desirable for Cr^{2+} ions in Al_2O_3, since VPE interactions can account at least qualitatively for recent experimental results[2] on resonant phonon absorption at frequencies which are multiples of a single-ion transition frequency ω_o, corresponding to a level separation 6.7 cm^{-1}.

We use the simple spin-$\frac{1}{2}$ model previously employed to treat phonon scattering by paramagnetic ions and the analysis of thermal conductivity[3]. However the matrix of the spin-phonon interaction must include both diagonal and off-diagonal elements. For a single spin $\underset{\sim}{S}_j$ at site $\underset{\sim}{R}_j$ the interaction Hamiltonian is thus

$$H_{sp} = N^{-\frac{1}{2}}\sum_{\underset{\sim}{k}}(A^+_{\underset{\sim}{k}}S_{j+} + A^-_{\underset{\sim}{k}}S_{j-} + A^z_{\underset{\sim}{k}}S_{jz})(a_{\underset{\sim}{k}} + a^+_{-\underset{\sim}{k}})e^{i\underset{\sim}{k}\cdot\underset{\sim}{R}_j}$$

where $a_{\underset{\sim}{k}}$, $a^+_{\underset{\sim}{k}}$ are phonon operators and $A^\alpha_{\underset{\sim}{k}}$ ($\alpha = \pm, z$) the coupling parameters. Considering two ions at sites $\underset{\sim}{R}_i$, $\underset{\sim}{R}_j$ we denote the unperturbed pair states by $|-->$, $|+->$, $|-+>$, $|++>$. A transition between $|-->$ and $|++>$ is not allowed in first order since the spin-phonon interaction can only flip one spin at a time. But the effect of VPE interactions is to perturb the states in such a way that a transition is then allowed at the harmonic frequency $2\omega_o$.

To show this, we need the VPE interaction term of the form $J^{+z}_{ij}(S_{i+}S_{jz} + S_{iz}S_{j+})$, where

$$J_{ij}^{+z} = - \frac{2}{N} \sum_k \frac{A_{-k}^+ A_k^z}{\hbar \omega_k} e^{i\underline{k} \cdot (\underline{R}_i - \underline{R}_j)}.$$

This admixes the component $(J_{ij}^{+z}/2\hbar\omega_o)(|+-> + |-+>)$ into the unperturbed ground state $|-->$ and similarly perturbs the upper state $|++>$. The matrix element of the off-diagonal part of H_{sp} between these perturbed pair states is then

$$A_q^+(e^{i\underline{q} \cdot \underline{R}_i} + e^{i\underline{q} \cdot \underline{R}_j})(J_{ij}^{+z}/2\hbar\omega_o)$$

for absorption of a phonon q. The phonon frequency $\omega_q \simeq 2\omega_o$, provided the energy shift $\Delta E \sim J^2/\hbar\omega_o \ll \hbar\omega_o$. There are also VPE interactions of the form $J_{ij}^{++}S_{i+}S_{j+} + J_{ij}^{--}S_{i-}S_{j-}$ which allow second-harmonic transitions induced by the diagonal part of H_{sp}. We note here that an isotropic exchange interaction $\underline{S}_i \cdot \underline{S}_j$ would not admix the states in a manner which permits harmonic absorption.

We now calculate the second-harmonic phonon absorption rate $W_2(\omega_q)$ at low temperatures when the ions are in the ground state. The strain broadening of the levels is described by a Lorentzian lineshape of width Γ_2. Summing over all pairs of ions in the crystal gives the expression

$$W_2(\omega_q) = \sum_{\substack{ij \\ (i \neq j)}} \frac{2\pi}{\hbar^2} \frac{|J_{ij}|^2}{(2\hbar\omega_o)^2} \frac{|A_q|^2}{N}(1+\cos\underline{q} \cdot \underline{R}_{ij}) \frac{\Gamma_2/\pi}{(\omega_q - 2\omega_o)^2 + \Gamma_2^2}.$$

We have assumed here for simplicity that the diagonal and off-diagonal coupling parameters are of similar magnitude A_k. The absorption rate of phonons of frequency $\omega_p \simeq \omega_o$ between strain-broadened single-ion levels is given by

$$W_1(\omega_p) = N_s \frac{2\pi}{\hbar^2} \frac{|A_p|^2}{N} \frac{\Gamma_1/\pi}{(\omega_p - \omega_o)^2 + \Gamma_1^2}$$

where N_s is the total number of paramagnetic ions (spins).

To make numerical estimates we neglect phonon dispersion and employ a spin-phonon matrix element in the form $A_k = \varepsilon \hbar (\omega_o \omega_k/12)^{\frac{1}{2}}$, where ε is a dimensionless parameter. In evaluating the VPE interaction use of a Debye model leads to oscillatory terms. But it has been argued that such terms are unphysical[4] and that the dominant term varies as $1/R_{ij}^3$. Using this approximation gives

$$J_{ij} \simeq - \frac{\varepsilon^2 \hbar \omega_o}{4\pi^2} \left(\frac{a}{R_{ij}}\right)^3,$$

where a^3 is the average volume per atom. In view of this rapid dependence on R_{ij} we need only take account of magnetic ions which are nearest neighbours. For a random distribution of ions with fractional concentration $c = N_S/N$ this gives for the ratio of the peak second-harmonic to peak single-ion absorption

$$\frac{W_2(2\omega_o)}{W_1(\omega_o)} \simeq 2zc \left(\frac{J}{2\hbar\omega_o}\right)^2 \frac{|A_q|^2 \Gamma_1}{|A_p|^2 \Gamma_2} ,$$

where J is the nearest-neighbour interaction and z the number of neighbours. For the transition with $\hbar\omega_o = 6.7$ cm^{-1} of Cr^{2+} in Al_2O_3 we take $\varepsilon \sim 15$ from thermal conductivity data[5] and $R \sim 3a$ from structure data[6]. This gives $J \sim 1.3$ cm^{-1} and the energy shift $\Delta E \sim 0.25$ cm^{-1}, which is much less than the observed inhomogeneous width $\Gamma_2 \sim 3.6$ cm^{-1} of the second-harmonic transition[2]. We also expect $\Gamma_2 \sim 2\Gamma_1$, which is consistent with the experimental results. Since in the investigated sample the estimated Cr^{2+} concentration is 55 ppm ($c = 5 \times 10^{-5}$), we find

$$W_2(2\omega_o)/W_1(\omega_o) \sim 5 \times 10^{-5},$$

which is very much less than the observed absorption ratio[2] of ~ 0.3. The experimental result may be considerably affected by saturation of the absorption line and non-linearity of the phonon detector. Clustering of impurity ions could also enhance the observed ratio but it seems unlikely that these factors could entirely account for the discrepancy between theory and experiment.

REFERENCES
1. J M Baker, Rep Prog Phys 34, 109 (1971).
2. F I Hasan, P J King, D T Murphy and V W Rampton, J Phys C 12, L431 (1979).
3. G A Toombs and F W Sheard, J Phys C 6, 1467 (1973).
4. D H McMahon and R H Silsbee, Phys Rev 135, A91 (1964).
5. J Rivallin, B Salce and A M de Goer, Solid St Comm 19, 9 (1976).
6. N Laurence, E C McIrvine and J Lambe, J Phys Chem Solids 23, 515 (1962).

****

DISCUSSION

P. J. King: Do you expect any angular dependence of the ratio of these two absorption strengths, and if you propagate the phonons in different directions in the crystal would you expect that ratio to change?

F. W. Sheard: What angle are you varying in this?

P. J. King: I'm varying the propagation angle of the phonons.

F. W. Sheard: I'm averaging over all phonons. I don't put any such complications in.

P. J. King: Is there anything in the Hamiltonian which would suggest that there is an angular dependence?

F. W. Sheard: Yes. It could well vary with angle because if you take the more precise expression for the virtual phonon exchange, then it depends on the relative positions of the ions relative to the crystal axes, for instance.

V. W. Rampton: How much can one increase the value of J without changing the ΔE so much that one pushes the harmonic line out of our measured linewidth?

F. W. Sheard: By much less than an order of magnitude. If you increase J by a factor of 10, then the ΔE goes like a factor of 100.

PHONON SCATTERING BY CHROMIUM IONS IN GaAs

L.J. Challis and A. Ramdane

Department of Physics, University of Nottingham,
University Park, Nottingham NG7 2RD, U.K.

Chromium in GaAs can exist in 3 ionic states, Cr^{1+}, Cr^{2+} and Cr^{3+} of which there is particular discussion about the ground state of Cr^{2+}. In the present work we have studied the thermal conductivity of the 6 samples of GaAs(Cr) described in table 1 from 1 to 20K and in magnetic fields up to 13T.

The zero field conductivities are compared with those of a pure sample[3] of Casimir length 4.3 mm in fig 1. From the data for GA735

TABLE 1: Details of 6 GaAs samples kindly supplied by the Plessey Co (GA) and Dr Ishiguro (TI). The valence states (MAJ = majority, MIN = minority) are deduced from the model of Brozel et al[1]. SI = semi-insulating, NA = not available. TI#4 and #5 are neighbouring slices to those used in ref 2. The GA samples were doubly doped with Cr/Si (GA735, 785) or Cr/Zn (GA 781).

SAMPLES	DOPANT (cm^{-3})			VALENCE STATES		AXIS	MM DIMENSIONS
	Cr	Si	n_{el}	MAJ	MIN		
GA735(e)	$2x10^{17}$	$2x10^{17}$	SI	Cr^{2+}	–	<110>	2.9x3.0x15
GA735(064)	NA	NA	$\sim 10^{15}$	Cr^{1+}	Cr^{2+}	<001>	3.0x3.0x10
GA785	$6x10^{16}$	$>>6x10^{16}$	$\sim 10^{18}$	Cr^{1+}	–	<110>	4.0x4.2x20
GA781	NA	$3x10^{16}$	$p\sim 10^{14}$	Cr^{3+}	Cr^{2+}	<001>	3.0x3.0x10
TI#4	$6x10^{15}$	NA	SI	Cr^{1+}	Cr^{2+}	<110>	2.7x5.0x24
TI#5	$7x10^{16}$	$1.1x10^{16}$	SI	Cr^{3+}	Cr^{2+}	<110>	4.1x5.3x21

(e) and 785 it seems clear as expected that the phonon scattering bv Cr^{1+} (^6S in T_d) is small compared with that by Cr^{2+}, and from Ga 781 and TI #5 it seems that Cr^{3+} also scatters phonons strongly: any Cr^{4+} present in 781 is weakly coupled ($^3A_2$ in T_d). The dip in K(T) at ∿ 8K in some samples was earlier observed by Chaudhuri et al[4] in an SI sample and suggests resonant scattering at ∿ 600 GHz. The dip occurs at a temperature slightly higher than the minimum at 5 ± 1K of the attenuation in heat pulses recently observed by Narayanamurti on 4 SI samples[5]. Other earlier thermal conductivity measurements on GaAs(Cr) were on highly doped samples where the scattering is believed to be by structural defects[6]. There have also been extended ultrasonic measurements by Tokumoto and Ishiguro[2]; we note that the ultrasonic absorption of TI#5 like the thermal resistivity is bigger than that in TI#4.

The photoresistivity W(t) of GA735(064) was measured after ill-umination by a flash of light from a torch bulb. $W(t)/W_o$ is shown in fig 2 starting 15s after the flash when the sample had returned to thermal equilibrium with the helium bath. In the dark it contains largely Cr^{1+} and this is photoionized to form Cr^{2+} and Cr^{3+} and an increase in n_{el}[7]. Neutral Cr^{3+} ions should trap electrons rapidly (<<15s) to form Cr^{2+} and the long decay time of ∿ 54 minutes seems

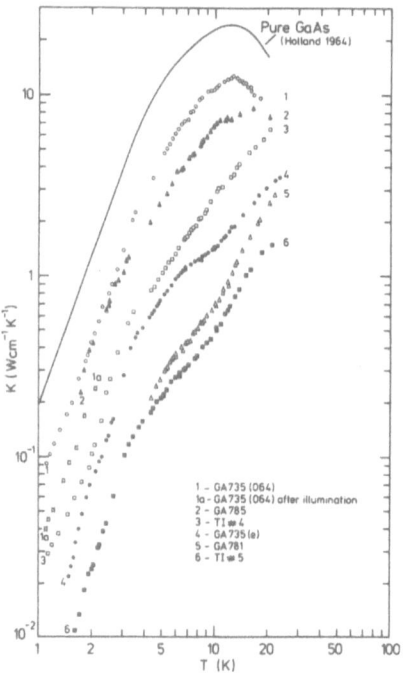

Fig 1: Thermal conductivity K(T).

consistent with slow trapping by the negative Cr^{2+} ion and a very
similar decay constant has been observed in the photoconductivity
(Eaves and Williams, to be published). Of particular interest is
that a large amount of the photo-induced thermal resistivity and
photoconductivity remains after many hours - thermal conductivity
data in this state is shown in fig 1 - although can be removed by
warming to room temperature. It is not known however whether these
stable phonon scattering centres are Cr^{2+} ions in different sites or
other photoionized defects.

Finally we report measurements on thermal magnetoresistance
which have been made on 5 samples from 1-4K for fields in the (110)
plane. Negligible magnetoresistance was observed in GA785 and 735
(064) up to 5.5T consistent with the small phonon scattering by chromium seen in fig 1. Large effects were observed in GA735(e), T1#4
and #5 and in GA735(e) the measurements were extended to 13T where
the phonon scattering appeared to be quenched. For a sample of int-

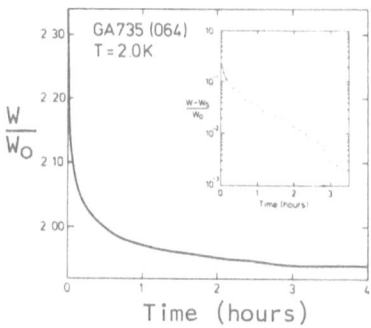

Fig 2: Photoinduced thermal resistivity $W(t)/W_o$ of GA735(064);
       W_o is the value before illumination.

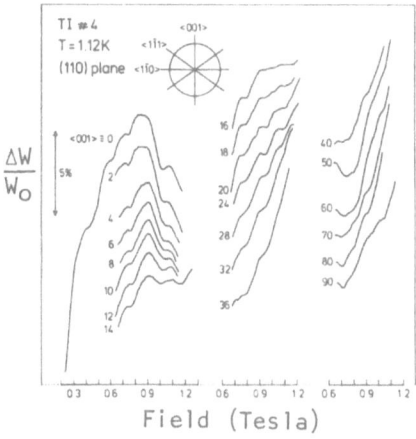

Fig 3: Thermal magnetoresistance of TI#4..

ermediate conductivity T1#4 where the Cr scattering can reasonably be attributed to Cr^{2+} temperature independent minima were observed superimposed on a broad overall change (Fig 3). It is thought that these occur at fields where 2 transition frequencies are equal (frequency crossings) so that it is of interest that their positions do not change as the field is rotated. Similar isotropy was observed in crossings seen in a sample of SI GaP(Cr).

The data so far do not seem consistent with the spin-Hamiltonians deduced for either Cr^{3+} or Cr^{2+} from epr data[8] and we note that in the Cr^{2+} case there are also problems in reconciling high resolution optical work[9] and some of the ultrasonic work[2] with the epr data, although this is selfconsistent at 2 very widely differing frequencies (10 and 300 GHz)[8]. The resonant frequency of $\sim$ 600 GHz is comparable to the span of the 5 levels of the optical work but twice that of the epr level schemes. The isotropy of the crossing spectrum is also inconsistent with the epr scheme for Cr^{2+}. There is clearly much more to learn about this intriguing system.

We are very grateful to Drs M J Cardwell, L Eaves, T Ishiguro, P R Jay and H Tokumoto for help in various ways.

REFERENCES
1. Brozel M R, Butler J, Newman R C, Ritson A, Stirland D J and Whitehead C, J Phys C 11, 1857 (1978).
2. Tokumoto H and Ishiguro T, J Phys Soc Japan 46, 84 (1979).
3. Holland M G, Phys Rev 134, A471 (1964).
4. Chaudhuri N, Wadhwa R S, Toku Phoola and Sreedhar A K, Phys Rev B 8, 4668 (1973).
5. Narayanamurti V, Chin M A and Logan R A, Appl Phys Lett 33, 481 (1978).
6. Vuillermoz P L, Langier A and Mai C, J Appl Phys 46, 4623 (1975).
7. Work on photo-induced charge conversion in SI GaAs(Cr) is reported by Stauss G H and Krebs J J in Proc 6th Int Symp on GaAs and Rel Compounds (Edinburgh 1976) IOP Conf Ser 33A, 84.
8. Krebs J J and Stauss G H, Phys Rev B 15, 17 (1977) and 16, 971. Wagner R J and White A M, to be published.
9. Lightowlers E C, Henry M O and Penchina C M, in 'Physics of Semiconductors 1978' (ed Wilson B L H) p 307. The Institute of Physics, Bristol and London (1979).

TWO-FLUID EQUATIONS FOR PHONONS WITH AND WITHOUT MOMENTUM

A. Thellung

Institut für Theoretische Physik
der Universität Zürich
8001 Zürich, Switzerland

In the customary microscopic derivation of the two-fluid equations of liquid He II explicit use is made of the assumption that an elementary excitation of wave vector $\underline{k}$ carries a momentum $\hbar\underline{k}$[1]. We show that: (i) there are two natural ways of defining phonons in a liquid, one kind (Eulerian phonons) carrying a momentum $\hbar\underline{k}$, the other kind (Lagrangian phonons) having no momentum, and (ii) the resulting two-fluid equations are the same in both cases.

EULERIAN AND LAGRANGIAN PHONONS

We consider liquid He II below $0.6°$ K, where the thermal excitations consist entirely of phonons. Since their wave length is large compared to the interatomic distance they can be obtained from a continuum theory, i.e. from the quantum theory of the irrotational motion of a non-viscous fluid[2].

Hydrodynamic motion can be described either in the formulation of Euler (local coordinates $\underline{x}$) or of Lagrange (material coordinates $\underline{a}$)[3]. In either case at any instant of time the motion can be decomposed in a spatial Fourier series, leading to modes that are sinusoidal in the coordinates used (longitudinal for irrotational motion) plus a uniform translation at velocity $\underline{U}$ (mode with wave vector $\underline{k}=0$).

Quantization with periodic boundary conditions in a cube of volume V yields in Euler's formulation[2] the following expressions for energy, total momentum and average velocity

$$H_E = \frac{1}{2}\rho_0 V\underline{U}_E^2 + \sum_{\underline{k}} n_{\underline{k}}(\varepsilon_{\underline{k}} + \hbar\underline{k}\underline{U}_E) + \text{anharmonic terms} \qquad (1)$$

$$\underline{P}_E = \int_V d^3x \; \rho\underline{v}_E = \rho_o V\underline{U}_E + \sum_{\underline{k}} \hbar\underline{k} \; n_{\underline{k}} \tag{2}$$

$$\underline{u}_{sE} : = (1/V)\int_V d^3x \; \underline{v}_E = \underline{U}_E \tag{3}$$

Here ρ means the density, ρ_o its average or equilibrium value, $\underline{U}_E$ the translational velocity, $n_{\underline{k}}$ the number of phonons of wave vector $\underline{k}$, $\varepsilon_{\underline{k}} = \hbar\omega_{\underline{k}} = \hbar c|\underline{k}|$ their energies in the harmonic approximation (c = sound velocity), $\underline{v}_E = -\underline{\nabla}\phi + \underline{U}_E$ the hydrodynamic velocity field. Expressions (2) and (3) are exact. The anharmonic terms in (1) lead (a) to a modification of the phonon energies $\varepsilon_{\underline{k}}$ (dispersion) and (b) to transitions between the unperturbed phonon states.

In Lagrange's formulation the quantities corresponding to (1), (2), (3) are

$$H_L = \tfrac{1}{2}\rho_o V\underline{U}_L^2 + \sum_{\underline{k}} n_{\underline{k}}\varepsilon_{\underline{k}} + \text{(different) anharmonic terms} \tag{4}$$

$$\underline{P}_L = \int_V d^3a\rho_o\underline{v}_L = \rho_o V\underline{U}_L \tag{5}$$

$$\underline{u}_{sL} : = (1/V)\int_V d^3x \; \underline{v}_L = (1/V)\int_V d^3a(\rho_o/\rho)\underline{v}_L$$

$$= \underline{U}_L - (1/\rho_o V)\sum_{\underline{k}} \hbar\underline{k} \; n_{\underline{k}} + \text{terms of higher order in } \frac{\rho-\rho_o}{\rho_o} \tag{6}$$

Note that the spatial average in (6) is taken in local, i.e. Eulerian, coordinates. Expression (5) is exact.

Eqs. (2), (3), (5), (6) show that Eulerian phonons carry a momentum $\hbar\underline{k}$, but do not contribute to the average velocity, whereas Lagrangian phonons possess no momentum, but give a contribution $-(1/\rho_o V)\hbar\underline{k}$ to the average velocity. An analogous result is true for classical sound waves that (at a given moment) are purely sinusoidal either in Eulerian or Lagrangian coordinates.

TWO-FLUID EQUATIONS

If a phonon with energy $\varepsilon_{\underline{k}}$ in the background medium at rest is considered in the laboratory system, with respect to which the bearer fluid is moving with a velocity $\underline{u}_s$, the wave vector is the same, $\underline{k}'=\underline{k}$, whereas the phonon energy in the latter frame is $\varepsilon'_{\underline{k}} = \varepsilon_{\underline{k}} + \hbar\underline{k}\underline{u}_s$. This follows from the formula for the Doppler shift, $\omega'_{\underline{k}} = \omega_{\underline{k}} + \underline{k}\underline{u}_s$.

The crucial point now is that $\underline{u}_s$ is to be identified with the translational velocity $\underline{U}$ of the background medium plus the average velocity due to the presence of other phonons (if they give a contribution), i.e. with $\underline{u}_{sE}$ (3) for the Eulerian and $\underline{u}_{sL}$ (6) for the Lagrangian case.

Since $\sum_k n_k \varepsilon_k'$ and, in a liquid, $\sum_k n_k \underline{k}$ are constants of the motion, the mean phonon number in thermal equilibrium is given by $\langle n_{\underline{k}} \rangle = [\exp(\varepsilon_{\underline{k}} + \underline{u}_s \hbar \underline{k} - \underline{u}_n \hbar \underline{k})/k_B T) - 1]^{-1}$, where $\underline{u}_n$ is interpreted as the drift velocity of the phonon gas. The mean wave vector density is

$$(1/V)\sum_k \underline{k} \langle n_{\underline{k}} \rangle = (2\pi)^{-3} \int d^3k \; \underline{k} \langle n_{\underline{k}} \rangle = :(1/\hbar)\rho_n(\underline{u}_n - \underline{u}_s), \qquad (7)$$

which defines ρ_n, the so-called density of the normal component in the two-fluid model.

The two-fluid equations are just the balance equations (always written in local coordinates) for the following conserved quantities: (i) energy of the phonons, (ii) wave vector of the phonons, (iii) momentum of the liquid as a whole, (iv) mass.

In order to obtain the balance equations for (i) and (ii) one multiplies the Boltzmann equation by ε_k or $\underline{k}$ and integrates over $\underline{k}^1$. No difference arises between Eulerian and Lagrangian phonons, and the resulting two-fluid equations are the same.

To deal with (iii) the expression for the momentum density is needed. For Euler's case it is according to (2), (3), (7)

$$\rho \underline{U}_E + \rho_n(\underline{u}_n - \underline{u}_s) = \rho \underline{u}_s + \rho_n(\underline{u}_n - \underline{u}_s) = \rho_n \underline{u}_n + \rho_s \underline{u}_s, \qquad (8)$$

where $\rho_s: = \rho - \rho_n$ is the so-called density of the superfluid component. For Lagrange's case the momentum density is according to (5), (6), (7) given by $\rho \underline{U}_L = \rho \underline{u}_s + \rho_n(\underline{u}_n - \underline{u}_s) = \rho_n \underline{u}_n + \rho_s \underline{u}_s$, i.e. again expression (8). This leads in both cases to the equation (P = pressure)

$$\partial(\rho_n \underline{u}_n + \rho_s \underline{u}_s)/\partial t = - \underline{\nabla}P. \qquad (9)$$

Finally, the conservation of mass (iv) is described by the continuity equation. For a system of massive particles, mass current density and momentum density are the same thing, so that in either case the continuity equation reads

$$\partial \rho/\partial t = - \text{div} (\rho_n \underline{u}_n + \rho_s \underline{u}_s). \qquad (10)$$

Thus the whole set of two-fluid equations is shown to be identical for Eulerian and Lagrangian phonons.

REFERENCES
1. I.M. Khalatnikov, "Introduction to the Theory of Super-fluidity", Benjamin, New York, 1965
2. R. Kronig and A. Thellung, Physica 18, 749 (1952)
3. H. Lamb, "Hydrodynamics", Cambridge, University Press, 1957
****

DISCUSSION

G. P. Srivastava: In the case of solids do you still have two types of phonons - Lagrangian and Eulerian phonons.

A. Thellung: In the case of solids the only natural way of defining phonons is in Lagrangian coordinates because there you label the atoms. You could do it with Eulerian coordinates, but it's much more complicated.

G. P. Srivastava: But in any case you argue that there is no meaning of associating a momentum with a phonon.

A. Thellung: It's not necessary. You can.

G. P. Srivastava: Some books say there is. Some books disagree with that, so it's a matter perhaps of definition.

TWO-FLUID THEORY OF PHONON SCATTERING IN DIELECTRIC CRYSTALS

Baxter H. Armstrong

Department of Aeronautics and Astronautics
Stanford University, and
IBM Scientific Center
Palo Alto, CA 94304

INTRODUCTION

In order to describe the transport characteristics of thermal phonons, we propose to divide them into two categories. The first consists of relatively nondispersive phonons with interactions dominated by anharmonic normal (N) processes. This category defines a propagation group that transports energy. The second category is dominated by resistive processes and constitutes a group that extracts momentum and energy from the propagation group, and passes it on to an external heat bath. Thus, this second group acts as a thermal reservoir, thermal resistance appearing through the scattering of propagating phonons into these reservoir modes. Towards high temperature (T), the resistive processes that determine the reservoir are principally umklapp (U) processes.

Towards low T, extrinsic resistive processes (isotope, impurity scattering, etc.) provide the reservoir resistivity, and the distinction between propagation and reservoir modes becomes dependent on T. This approach is modeled after acoustic theory, where the attenuation of a wave is calculated from the N-processes that couple it to a reservoir of thermal phonons.[1,2] In addition to utilizing this distinction between propagating and reservoir phonons as suggested by the acoustic theory, we use the average transition rate expression of acoustic theory to provide the N-process rate for thermal phonons.

THEORY AND RESULTS

Phonon wavenumber $(\vec{q})$ space is divided into two regions according to whether the N-process transition rate $\tau_{NR}^{-1}(\vec{q})$ from modes in the first region (#1) to modes in the second (#2) exceeds the total resistive process rate $\tau_R^{-1}(\vec{q})$ out of the given mode or not. In isotropic approximation, this condition can be expressed in terms of the reservoir, or region, boundary frequency ω_R defined by

$$\tau_{NR}^{-1}(\omega_R) = \tau_R^{-1}(\omega_R) \tag{1}$$

such that $\omega < \omega_R$ implies region 1 and $\omega > \omega_R$, region 2. $\tau_{NR}^{-1}(\omega)$ is obtained from Eq. 4.15 of Ref. 1 for attenuation in the regime $\omega\tau > 1$. Because τ_{NR}^{-1} includes interaction only with a subset of thermal modes, namely, the reservoir modes, the Woodruff-Ehrenreich result is modified to yield

$$\tau_{NR}^{-1}(\omega) = \gamma^2 T \Delta C_v \omega \tan^{-1}(2\omega\bar{\tau})/[2 \rho c^2] \tag{2}$$

for a mean phonon (averaged over polarizations) of velocity c, γ is a Grüneisen constant, ρ is density, $\bar{\tau}$ the average reservoir relaxation time, and ΔC_v is the specific heat of the reservoir phonons (which lie between ω_R and the maximum spectral frequency). We note that (2) can also be obtained from Landau-Rumer theory extended to account for finite thermal phonon lifetimes. The transition rate for processes T + L → L and T + T → T, or L + L → L and L + T → T given in Ref. 2 (Eqs. 240 and 241 or 238 and 243) can be summed to (2) in the mean phonon approximation.[3]

Toward high T, τ_R^{-1} will become dominated by the transition rate τ_U^{-1} for U-processes. It is a good approximation to assume that U-processes occur in dispersive modes near the Brillouin zone boundary which do not contribute to propagation, and, further, that $\tau_U^{-1} \geq \tau_{NR}^{-1}$ at the U-process threshold so that dispersive modes above this point go into the reservoir. One or more polarization branches often continue nondispersively above the lowest U-process frequency. Hence, the highest propagation mode will usually lie above the reservoir threshold, and propagating and reservoir phonons coexist here. To accommodate this situation, the propagation modes are assumed to extend to a maximum nondispersive frequency ω_P generally greater than ω_R. Towards low T, isotope/impurity scattering according to the rate[4]

$$\tau_I^{-1} = A(\hbar/k)^4 \omega^4 \tag{3}$$

will usually become dominant. When this occurs, the reservoir region descends in frequency to include nondispersive acoustic modes. Thus, while only nondispersive phonons act in the propagation role,

both nondispersive and dispersive phonons can contribute to the
reservoir. At sufficiently low T, only nondispersive acoustic modes
will contribute to ΔC_V, and $\omega\bar{\tau} \gg 1$ for important frequencies. In
this approximation, Eqs. (1)-(3) combine to yield

$$x_R^3 = Q\,T\,H(x_R)A^{-1} \tag{4}$$

for determination of x_R. The constant $Q \equiv 9\pi\gamma^2\beta R(k/\hbar)/(4wc^2 T_D^3)$
and $H(x_R)$ is the complementary Debye integral $\int_{x_R}^{\infty} t^4 \exp(t)dt/$
$[\exp(t) - 1]^2$, β is number of atoms/molecule, R the gas constant, w
the molecular weight, and T_D, Debye temperature. The N-process rate
to the reservoir follows as

$$\tau_{NR}^{-1}(x) = Q\,H(x_R)T^5\,x \tag{5}$$

where $x \equiv \hbar\omega/kT$. Equation (4) is only weakly dependent on T, so x_R
is roughly independent of T. In this approximation, (5) has a form
similar to that obtained from Callaway analyses of thermal conductivities.[5] Performing the customary average $\langle\ \rangle_{av}$ over
$x^4 \exp(x)/[\exp(x) - 1]^2$, we obtain $\langle\tau_{NR}^{-1}(x)\rangle_{av} = 4.7903\,Q\,H(x_R)T^5$. For
LiF, $Q \cong 0.16$ sec^{-1} deg^{-5}, using $\gamma = 1.63$. This result reproduces
the value $\langle\tau_N^{-1}\rangle_{av} = 0.5\,T^5$ sec^{-1} of Ackerman and Guyer[5] if $H(x_R) = 0.6$,
$x_R \cong 10$, corresponding to a reservoir in the far tail of the thermal
phonon frequency distribution. This result, however, depends on the
crystal purity through A, in contrast to Callaway theory[6] which is
presumed to yield an intrinsic constant. The relationship between
the two-fluid and Callaway approaches, obtained[3] to clarify this
difference, indicates that the relaxation time approximation is not
valid for N-processes ("N N collisions") that begin and end in region 1. If we assume these collisions do not contribute to the heat
flux loss from the region 1 modes, a condition is obtained analogous
to Callaway's imposition of conservation of momentum on N-processes,
and yields a thermal conductivity (n_q = occupation number):

$$K = \sum_q \hbar\omega_q\,c_q^2\,(dn_q/dT)\tau_R'\,\cos^2\theta \tag{6}$$

of the kinetic-theory type, where the two-fluid mean free time
$(\tau_R')^{-1} = \tau_{NR}^{-1} + \tau_R^{-1}$ is independent of the NN collisions. Equation
(8) has been used to calculate K for LiF for three levels of iso-
topic purity for comparison with Thacher's[4] measurements. The
U-process threshold frequency inferred from fitting this two-fluid
K to Thacher's data is much closer to the lowest zone-boundary inter-

section frequency of neutron-scattering dispersion curves than is the
value obtained by Callaway analysis.

REFERENCES

1. T. O. Woodruff and H. Ehrenreich, Phys. Rev. $\underline{123}$, 1553 (1961).
2. Humphrey J. Maris in Physical Acoustics, ed. by W. P. Mason and
    R. N. Thurston (Academic Press, Inc., N.Y., 1971) Vol. VIII.
3. Baxter H. Armstrong, submitted for publication, Phys. Rev.
4. Philip D. Thacher, Phys. Rev. $\underline{156}$, 975 (1967).
5. C. C. Ackerman and R. A. Guyer, Ann. Phys. (N.Y.) $\underline{50}$, 128 (1968).
6. Joseph Callaway, Phys. Rev. $\underline{113}$, 1046 (1959).

ANALYSIS OF THE MACROSCOPIC EQUATIONS

FOR SECOND SOUND IN SOLIDS*

W. C. Overton, Jr.

Los Alamos Scientific Laboratory

Los Alamos, N.M. 87545

The microscopic theories of second sound in solids can be expressed in macroscopic form (Guyer and Krumhansl, 1966; Ackerman and Guyer, 1968) when the normal and resistive process phonon relaxation times $\tau_N(\omega,T)$ and $\tau_R(\omega,T)$ have been averaged appropriately over all ω's. Thus, for given temperature T, $\tau_N(T)$ and $\tau_R(T)$ can be regarded as fixed parameters. The macroscopic equations given in Ref. 2 for the heat current in 3-dimensions can be expressed in one dimension according to

$$(1 + \tau_o \frac{\partial}{\partial t}) \frac{\partial^2 Q}{\partial x^2} - \frac{1}{v^2 \tau_R} \frac{\partial Q}{\partial t} - \frac{1}{v^2} \frac{\partial^2 Q}{\partial t^2} = - F_Q(x_o,\omega) \, \delta(x-x_o) \, e^{-i\omega t}, \quad (1)$$

where $v = c_D/\sqrt{3}$ = second sound velocity, c_D = mean Debye velocity, $\tau_o = 9\tau_N/5$, and we have here inserted an impulse driving function over the plane $x = x_o$. Eq. (1) represents the experimental situation in which the single crystal solid is excited by a heat pulse applied on one face at $x = 0$. Both the face at $x = 0$ and the opposite face at $x = L$ serve as reflectors of the heat current Q. Sensitive thermometers on the plane at $x = L$ measure the second sound phenomenon.

Our purpose is to show a method for solving (1) in closed form and for calculating the temperature excursion $\Delta T(L,t)$ for comparison with experiment. Letting $Q(x,t) = \Psi(x,\omega)\exp(-i\omega t)$, we obtain

$$d^2\Psi/dx^2 + k^2\Psi = - F_Q(x_o,\omega) \, \delta(x-x_o)/(1 - i\omega\tau_o) \quad , \quad (2)$$

$$k^2 = (\omega^2 + i\omega/\tau_R)/[\, v^2(1 - i\omega\tau_o)] \, . \quad (3)$$

---

*Work performed under the auspices of the U.S.D.O.E.

In (1), (2) the driving force is confined to the plane $x=x_0$ by the delta function, but this is but a special case of the more general function $F(\xi,\omega)$; $0< \xi< L$. For the more general case the particular integral is

$$\Psi_p = -[k(1 - i\omega\tau_0)]\int_0^x F(\xi,\omega) \sin k(x-\xi)\, d\xi , \qquad (4)$$

and the general solution is $\Psi = A \sin kx + B \cos kx + \Psi_p$. Using boundary conditions for total reflection of the heat current at $x = 0$, $x = L$, we obtain $B = 0$ and, using (4), we can evaluate A. The solution can be put in the final form,

$$\Psi(x,\omega) = [k(1 - i\omega\tau_0)]^{-1}[\int_x^L d\xi\, F(\xi,\omega)\sin k(L-\xi) \sin kx/\sin kL$$

$$+ \int_0^x d\xi\, F(\xi,\omega) \sin k(L-x) \sin k\xi/\sin kL] . \qquad (5)$$

The response to a forcing function of unit amplitude and frequency $\omega/2\pi$, concentrated at x_0, is then obtained from (5) as the Green's function $\Psi(x,\omega)/F(x_0,\omega)$, in the form

$$G(x,x_0,\omega) = \begin{cases} \dfrac{\sin k(L-x)(1- \cos kx_0)}{k^2(1 - i\omega\tau_0) \sin kL} ; & 0< x< x_0 , \\[4mm] \dfrac{\sin kx[1 - \cos k(L-x_0)]}{k^2(1 - i\omega\tau_0) \sin kL} ; & x_0< x< L . \end{cases} \qquad (6)$$

A sharp heat pulse applied at $x = 0$ can be represented by $f(t) = (Ka/2)\exp(-|t|a)$, the Fourier transform of which is $F(0,\omega)=(Ka^2)/(\omega^2+ a^2)$. Thus, we obtain

$$Q(x,t) = (K/2\pi) \int_{-\infty}^{\infty} d\omega\, G(x,x_0,\omega) \exp(-i\omega t)/(1 + \omega^2/a^2) . \qquad (7)$$

This line integral can be evaluated for $t \geq 0$ by summing residues of the poles of the $-i\omega$ half plane. This number is finite because the number of allowed values of k is finite, although large. The poles at $- i/\tau_R$, i/τ_0, and 0 have zero residues.

For the plane $x = L$, where thermometers of the experiment are located, the poles associated with $kL = n\pi$ ($n = \pm1, \pm2,\ldots,N_{max}$) are determined by solving the quadratic

$$k^2 L^2 = (L^2/v^2)(\omega^2+ i\omega/\tau_R)/(1 - i\omega\tau_0) = n^2\pi^2 . \qquad (8)$$

We find

$$\omega_n = \pm \Omega_n - i\zeta_n/2 ; \quad \Omega_n = (n^2\pi^2v^2/L^2 - \zeta_n^2/4)^{\frac{1}{2}} ; \qquad (9)$$

$$\zeta_n = n^2\pi^2v^2\tau_0/L^2 + 1/\tau_R ; \quad W_n= i\Omega_n ; \quad \Omega_n \text{ real for } M<n<N .$$

A second set falls on the $- i\omega$ axis at $\omega_n = \pm iW_n - i\zeta_n/2$ for all $n > N$. In view of (8), and since n occurs only as n^2 in (9), no two poles ever fall at the same point. Thus, it is only necessary to sum residues on the first Riemann sheet. In order to determine N, let $\pm \eta$ be the solution of the quatratic obtained from (9) by setting $\Omega_n = 0$, then N is in the range $\eta-1 < N < \eta+1$. Using τ_N and τ_R for solid 3He at T < 1 K, we find N varies from 4 to about 20 and M usually = 1. With $x_0 = 0$ in (7) we can obtain the residues and determine $Q(x,t)$. The derivative $\partial Q/\partial x$ then gives $C_v \partial T/\partial t$, which, when evaluated at x = L, can be expressed in the form

$$C_v \Delta T(L,t) = KL^3 \{ \sum_{n=1}^{N} (n^4 \pi^4 D_n v^2)^{-1}(1-\cos\pi n)[-\Omega_n^2 \cos\Omega_n t-(\Omega_n \zeta_n/2)\sin\Omega_n t]$$

$$\times e^{-\zeta_n t/2} + \sum_{N+1} (n^4 D_n \pi^4 v^2)(1-\cos\pi n)[W_n^2 \cosh W_n t+(W\zeta/2)\sinh W_n t]e^{-\zeta_n t/2}\} \quad (10)$$

$$D_n = 1 - (n\pi v \tau_0/L + L/n\pi v \tau_R)^2 . \quad (11)$$

Instead of using τ_N, τ_R values obtained from theory, we select chosen   arbitrary parameters to calculate the waveforms described by (10). The computational procedure is to adjust values until agreement with the experimental second sound waveshape is obtained. The adjusted τ_N, τ_R are then used separately to calculate the thermal conductivity which is then compared with static experimental data. Although the agreement is not quantitative, the result obtained for solid 3He at 0.4 K is 60% of the static value. Thus, the two approaches are at least qualitatively consistent.

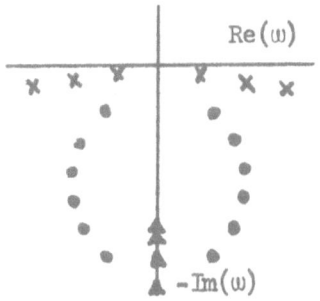

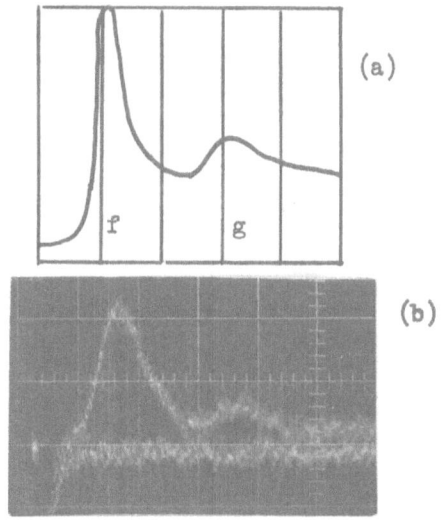

(a)

(b)

Fig. 1. ● - poles for M< n< N;
▲ - for n > N; ✗ - poles for
incorrect transmission line
analog of Ref. 2.

Fig. 2. (a) Calculated waveform
with τ_N= 0.09 μs, τ_R= 4.4 μs. (b)
Observed for solid 3He at 0.5 K;
f- first arrival; g- first reflection; 10 μs per division. Ref. 3.

REFERENCES

1. R.A. Guyer and J.A. Krumhansl, Phys. Rev 148 (1966) 766.

2. C.C. Ackerman and R.A. Guyer, Ann. Phys.(N.Y.) 50 (1968) 128.

3. C.C. Ackerman and W.C. Overton, Phys. Rev. Letters 22(1969)764.
                        ****

DISCUSSION

K. Weiss: You report an approach using the macroscopic equations
for heat flux and temperature involving two constant relaxation
times. Now there is also the possibility which was done about ten
years ago to take a Boltzmann equation with two constant relaxation
times to analyze second sound involving dissipative processes on the
one hand and normal collisions on the other hand. What is the
connection between the two?

W. C. Overton: Do you have results for that calculation?

K. Weiss: I can give you the reference for the paper.

W. C. Overton: I'd appreciate that because these results when you
calculate it out show a wave shape which is very similar to the observed phenomenon, and most of the theoretical results do not agree
on that.

S. J. Rogers: Is there any possibility in your model of putting in
a variable second sound velocity which would take one from the
regime of proper second sound into the ballistic regime?

W. C. Overton: In this regard the second sound velocity is found
not to agree with the Debye velocity divided by $\sqrt{3}$. It's always
considerably lower than that. You can interpret this as meaning
that it's only the transverse modes that are contributing to second
sound and the longitudinal model hasn't very much to do with it.
That's one interpretation. I think Dynes and Narayanamurti observed
the same effect in their work, and they also interpreted it in a
manner like that in which primarily the transverse modes contribute.

HIGH TEMPERATURE CRYSTALS: LIMITATIONS ON THE PHONON DESCRIPTION

Rosemary A. MacDonald and Raymond D. Mountain

National Bureau of Standards
Washington, D. C. 20234

INTRODUCTION

An anomalously large increase in specific heat is observed in many solids that melt at temperatures, T_m, much greater than their Debye temperature, Θ. These high values have been attributed to vacancies and anharmonicity[1], therefore it is important to know the full anharmonic contribution to the specific heat. We have carried out such a calculation by the Monte Carlo method for a model of rubidium ($T_m \sim 6\Theta$)[2] and obtain higher values than those given by lowest order perturbation theory. In this paper we summarize our results and discuss the implications for real solids.

RESULTS OF MONTE CARLO CALCULATIONS FOR RUBIDIUM

The calculations were carried out for 250 atoms, initially on bcc sites in a cube, interacting via a pair potential appropriate for rubidium[3]. The pressure, P, the potential part of the internal energy U_i, and the specific heat at constant volume, C_v, were obtained by standard procedures[2]. The values of P and U_i obtained after 10^6 trials were stable to 1 part in 10^4. Atomic migration was monitored throughout the calculations by recording the distribution of particles, $\rho(d)$, where d is their distance from their initial positions.

Table I gives the values obtained for C_v at V=56.914 cm^3/mole, together with C_v^{anh}, calculated by lowest order perturbation theory, for comparison. To estimate the error in C_v, at several temperatures the calculation was repeated with a different starting point in the random number sequence. The end result was quite sensitive to the amount of atom migration that had occurred. We have denoted by *

Table I. The specific heat at constant volume, C_v, and at constant pressure, C_p, from Monte Carlo calculations; C_v^{anh} from lowest order perturbation theory; in dimensionless units. V=56.914 cm^3/mole. * denotes moderate, ** denotes extensive displacement of atoms from original sites, (see text).

T	C_V/R	C_p/R	C_v^{anh}/R	T	C_V/R	C_p/R	C_v^{anh}/R
260	3.43	3.58	3.082	380	3.42	3.63	3.12
					4.02*	4.37*	
290	3.32	3.46	3.092				
	3.19	3.31		400	4.05*	4.39*	3.127
320	3.48	3.66	3.101	410	3.90*	4.26*	3.130
					5.08**	5.79**	
350	3.33	3.49	3.111				
	3.52*	3.71*		415	4.67**	5.30**	3.131

those cases where the atomic displacements were moderate ($\sim$ 1 lattice spacing) and by ** those cases where the atomic displacements were extensive (2 or more lattice spacings). No appreciable displacement occurred in the remaining cases. Clearly, there is a marked increase in specific heat associated with increased atomic migration.

In Fig. 1 we show a typical histogram of $\rho(d)$ obtained after 7.5×10^6 trials at 400 K. Subsidiary structure has developed at the first, second, and third neighbor positions, the second neighbor peak overlapping the first. If the atoms had stayed near their original lattice sites, $\rho(d)$ would have had a much stronger peak near the origin and no other features. If the lattice structure had become unstable, the structure in $\rho(d)$ would have disappeared, leaving a fairly uniform distribution over a wide range of lattice spacings. The particular run from which the histogram shown here was computed was taken to 10 million trials instead of the usual 1 million, which is already a long Monte Carlo calculation for a system of this size[4]. Although P and U_i were acceptably stable at the end of 10^6 trials, there was evidence of considerable strain in the early stages of this run and we felt it would be instructive to see if additional large strain events occurred. We found that the lattice became unstable after 8.0×10^6 trials. We see that while migration leads to an enhanced specific heat, too much migration leads to instability.

DISCUSSION

The evidence of substantial atomic displacement in this model system and the marked effect this has on the pressure, internal

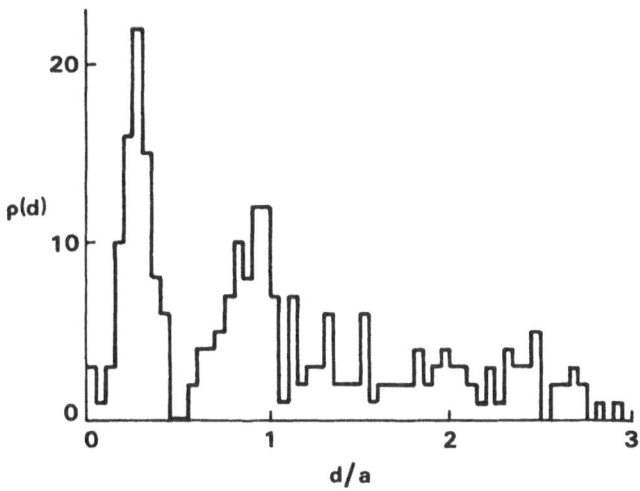

Fig. 1.   Histogram showing $\rho(d)$, the distribution of particles
           according to their distance, d, from their initial positions.
           a is lattice spacing.

energy, and specific heat of the system, give us some indication of
the processes that must be accounted for in a theoretical treatment
of solids at high temperatures. Clearly, a phonon description of the
solid cannot encompass the large scale motions that occur in our
model system. There is no way of dealing with atom migration from
one site to another in the perturbation theoretic approach. Whether
this migration is the whole answer to our original problem of the
anomalous rise in specific heat is another matter. Many other factors
such as boundaries and point defects (in particular, vacancies
which lower the barrier to migration) are involved in the melting
transition of a real solid. However, in solids with $T_m >> \Theta$, it is
reasonable to expect anomalous behavior since atomic migration will
become increasingly important as the temperature rises relative to Θ.

REFERENCES

1.   J. D. Filby and D. L. Martin, Proc. Roy. Soc. <u>A284</u>, 83 (1965).

2.   R. A. MacDonald, R. D. Mountain, and R. C. Shukla (to be published).

3.   D. L. Price, K. S. Singwi, and M. P. Tosi, Phys. Rev. B2, 2983 (1970).

4.   S. S. Cohen, M. L. Klein, M. S. Duesbery, and R. Taylor, J. Phys. F: Metal Physics 6, 337 and L271 (1976).

PHONON RESONANT ABSORPTION PROBED BY ACOUSTIC-OPTIC INTERACTION IN SOME IONIC MOLECULAR CRYSTALS

F. Michard, B. Perrin

Département de Recherches Physiques, Tour 22
Université P. et M. Curie
75230 PARIS Cedex 05

The large ultrasonic absorption which occurs in several mole-
cular crystals |1| |2| is believed to result from a coupling bet-
ween lattice and intramolecular vibrations. This hypothesis was
first suggested by LIEBERMANN |3| who predicted an excess of acous-
tic absorption for weakly bonded crystals such as molecular crystals.
This large ultrasonic absorption is attributed to the weak coupling
between lattice and intramolecular vibrations, with a relatively
slow exchange of energy taking place between coupled lattice modes
and internal modes that are nearly in resonance. Although the lattice vibrational frequencies are usually lower than those of the
internal vibrations, the LIEBERMANN'S theory predicts the possible
coupling between the higher lattice modes and the lowest vibrational motions via multiphonon process.

Difficulties encountered studying resonance absorption may
be enhanced in molecular crystals by other possible damping mechanisms than the relaxational absorption due to molecular resonance
such that due to dislocations or excessive mosaïcity which curen-
tly occur in these organic crystals. So it may be advantageous to
study resonance absorption in ionic molecular crystals, crystalline
imperfections being not so spread in these crystals. Furthermore
they generally exhibit higher symmetrical structure and their phonon dispersion curves are often better studied. The isomorphous
cubic crystals $Sr(NO_3)_2$, $Ba(NO_3)_2$ and $Pb(NO_3)_2$ belong to this kind
of compounds.

Acousto-optic studies were previously developed to obtain
ultrasonic absorption versus frequency in the frequency range
100 MHz - 1 GHz by means of BRAGG diffraction and at about 20 GHz
by means of BRILLOUIN scattering. The more important results are

1. The well-known square law dependence of the ultrasonic attenuation α versus frequency ν typical of AKHIESER regime is not observed in these crystals for frequencies of the order of a few
   hundred megahertz. Actually we observe a large ultrasonic absorp-
   tion with a ν^p dependance (p <2) |4|.

2. In the case of barium and lead nitrates ultrasonic absorption
   measured at about 20 GHz is higher than the extrapolated curve
   obtained from absorption measurements in the frequency range
   100 MHz - 1 GHz (figure 1).

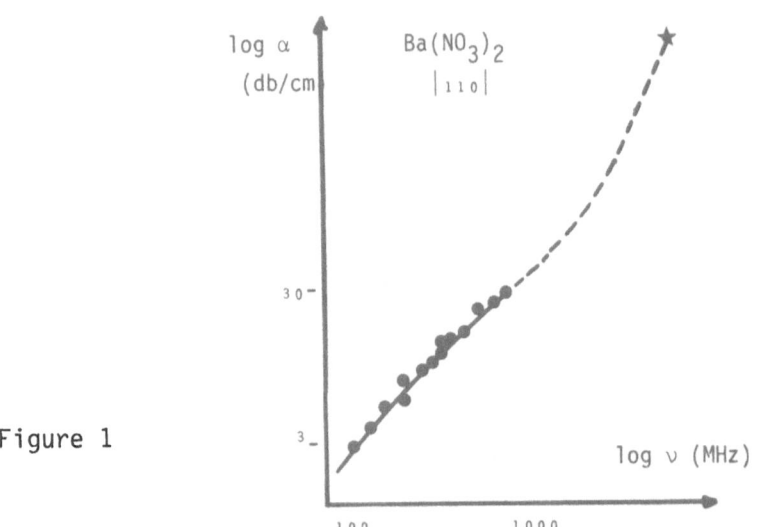

Figure 1

        These results may be interpreted in terms of a superposition
of two damping processes : AKHIESER and LIEBERMANN absorption
$\alpha = \alpha_A + \alpha_L$. At low frequencies (ν <1GHz), AKHIESER damping is negli-
geable with regard to resonance absorption and attenuation may
be approximated by:

$$\alpha_L = \mu k \nu^2 / (\nu^2 + k^2) \qquad (1)$$

        The values of the exchange frequency parameter k obtained
from the curves $\alpha = f(\nu)$ are:

$$k_{Sr} \simeq 630 \text{ MHz}, \quad k_{Ba} \simeq 330 \text{ MHz}, \quad k_{Pb} \simeq 300 \text{ MHz}$$

These results are completed by attenuation measurements versus
temperature in the range 300K-4K for several ultrasonic frequencies
obtained by means of BRAGG diffraction|5| . Typical curves $\alpha = f(T)$
are given in figure 2.

We can observe:

1) The behaviour of α from room temperature towards low temperature exhibits a T^5 law dependence. In the range 140K-60K, the
attenuation does not depend on T, as expected in AKHIESER regime.
Below about 60K, the behaviour of α is characteristic of the transition between AKHIESER and LANDAU-RUMER regime.

2) The curve related to $Sr(NO_3)_2$ attenuation behaviour allows
us to separate the contribution of k and μ in formula (1) in the
location of the observed maximum of α versus T.

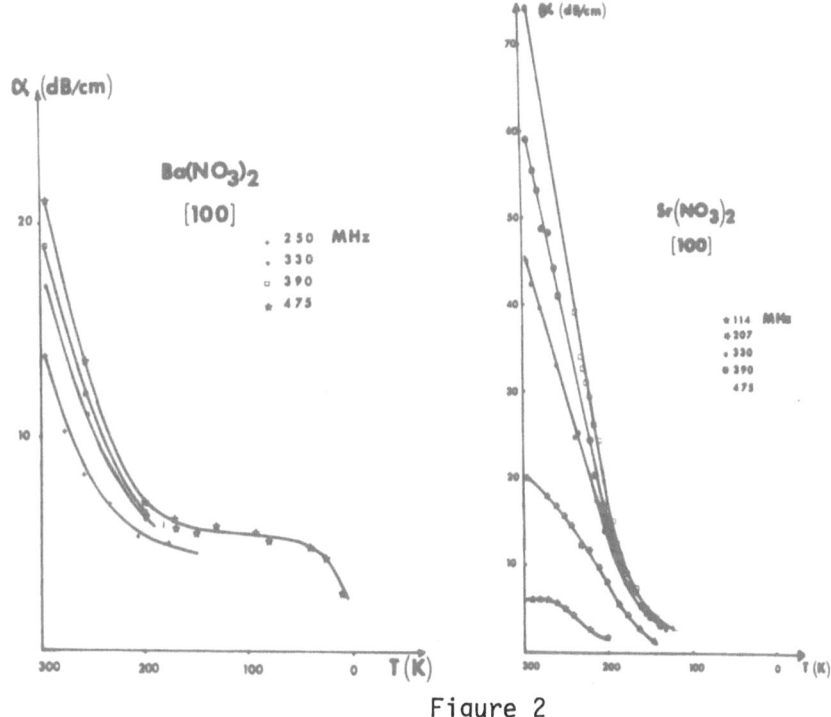

Figure 2

The observation of phonons resonant absorption in the isomorphous
Sr, Ba and Pb nitrates is certainly correlated with the existence
in these compounds of very low frequency intramolecular vibrations|6|

Acknowledgements: We wish to thank Professor A.ZAREMBOWITCH for
many helpful discussions and suggestions on this work.

References:
|1|S.S.YUN,R.T.BEYER- J.Chem.Phys.,40,2538-2546,(1964)
|2|J.D.WILSON,S.S.YUN-J.Ac.Soc.Am.,50,164-171,(1971)
|3|L.N.LIEBERMANN- Physical Acoustics,Mason ed., IVA,183-193,(1966).
|4|F.MICHARD et al, Int.Conf.Phonon scattering in solids,87-89,(1976)
|5|F.PLICQUE et al., to be published
|6|M.H.BROOKER,J.B.BATES- Spectr.Acta,29A,439,(1973)
                            ****

DISCUSSION

R. Colella: You mention two techniques - Bragg diffraction of light
and Brillouin scattering. In waht respect are they different?

F. Michard: In the case of Bragg diffraction we diffract light on
the ultrasonic beam we generate in the sample, but in the case of
Brillouin scattering we use thermal waves to diffract light.

DECAY OF OPTICAL INTO ACOUSTICAL PHONONS

D. Huet and J. P. Maneval

Group de Physique des Solides
E.N.S., 24 rue Lhomond
74231 PARIS Cedex 05, France

One channel for the relaxation of photoexcited (P.E.) hot carriers in semiconductors is, at T=0, spontaneous emission of highly energetic phonons. The purpose of the present study is to trace back the complicated cascade mechanisms leading to thermalization.

The direct-gap semiconductor InSb offers a relatively simple picture[1]. Because of the small effective mass ratio between electrons and holes, most of the kinetic energy resulting from the absorption of a photon is imparted to the electron system; it is subsequently released to the lattice via $q \approx 0$ longitudinal optic (L O) phonons, due to polar coupling, a process which takes place on the picosecond time scale[2].

Consider the problem in one dimension (Fig. 1). Zone centered LO can give rise to either $L + L$ or $L + T_1$ in a first step obeying momentum and energy conservation laws. The next step will in addition involve T_2 phonons. Further decays will take place for L and T_1 phonons (but no escape for T_2!).

Time-of-flight experiments resolve the eventual populations belonging to each acoustic branch. We compare (Fig. 2) P.E. phonons resulting from pulse irradiation (20 nsec; Argon laser) of the InSb surface[3] and conventional heat pulses. Immediately apparent are (i) the broadening of P.E. phonon peaks (scattering plus dispersion) and (ii) the weakness of the T_1 signal due to the efficient decay mechanism $T_1 \rightarrow T_2 + L$.

The large ballistic T_2 peak in the P.E. case at least confirms the long lifetime expected for these modes. How they were created, while it is forbidden by the symmetry along [100] of the third-order interaction potential, is another question. Four-phonon processes may be very important, as testified by the thermal dilatation coefficient of InSb at low temperatures[4].

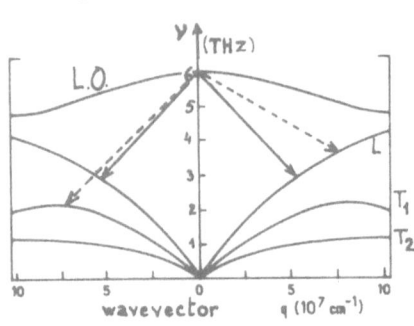

Fig. 1. Dispersion relations for InSb along [110]. Only the first steps of LO decay by 3-phonon processes are shown.

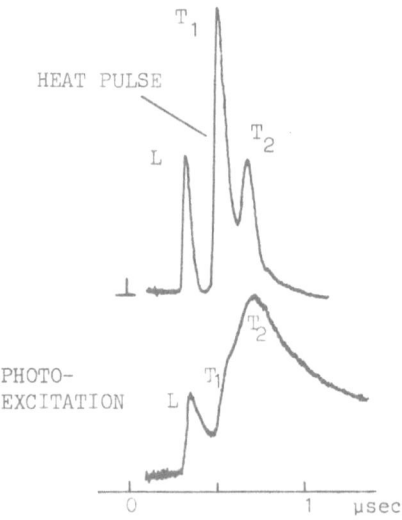

Fig. 2. Bolometric signals from heat pulses (upper trace) and photoexcitation (lower trace). Sample length: 1.04 mm. T = 1.3 K. Note the accumulation in the T_2 peak.

We have also analyzed P.E. phonons by absorption spectroscopy with the Fermi sphere[5] or "$2k_F$ spectroscopy." For a propagation length of 7 mm, the L spectrum is as shown in Fig. 3, where it is compared to a heat-pulse spectrum. Significant shift of the maximum, and poorer content in low-frequency modes show that in an average time of $\sim$ 2 μsec, thermalization of P.E. longitudinal phonons is not complete.

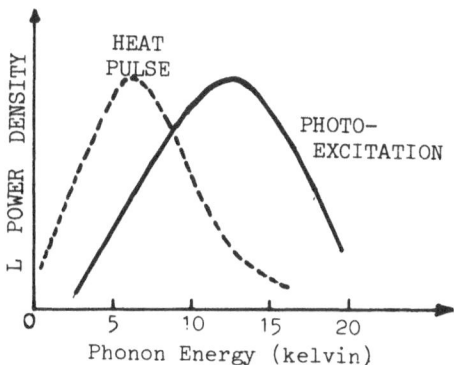

Fig. 3. $2k_F$ spectroscopy of the longitudinal peak after 7 mm of propagation along [111].

The spectrum cannot be interpreted as a blackbody spectrum even with intrinsic scattering[5] included.

Finer analysis of times of flight in Fig. 2 reveals that the T_2 pulse is early in the photoexcitation case by about 12% compared to the velocity of sound (1.67 km/sec). Thus, before decaying into T_2 modes, the energy has travelled as faster acoustic waves over about 0.2 mm. From this figure, one can put an estimate of $\sim$ 70 nsec on the lifetime in higher branches (L and T_1) before decay into the T_2 branch.

REFERENCES

1. Semiconductors and Semimetals, Vol. I, ed. by R. K. Willardson and A. C. Beer, Academic Press (1966).
2. D. Von der Linde and R. Lambrick, Phys. Rev. Letters, 42, 1090, (1979).
3. Charge carriers are not expected to take a significant part in heat transport.
4. D. F. Gibbons, Phys. Rev. 136, 112 (1958).
5. D. Huet et al., J. de Physique (Paris), 37, 521 (1976).

DISCUSSION

V. Narayanamurti: What is the laser excitation? Have you chosen the
laser frequency appropriately?

J. P. Maneval: No, the gap is not equal to the laser frequency.
It's an argon laser. It may be 30 times the gap energy.

R. Bray: You made the statement that the creation of T_2 should be
forbidden. But they are obviously created.

J. P. Maneval: T_2 is forbidden in 3-phonon processes because of the
symmetry of the third order elastic constants, but it's not forbidden
by 4-phonon processes. This means that 4-phonon processes may be
present. There may be other explanations too.

ACOUSTIC-OPTICAL PHONON INTERACTIONS IN SOLIDS

G. P. Srivastava

Physics Department,
New University of Ulster,
Coleraine, N. Ireland, BT52 1SA

We present here a scheme of studying the role of optical phonons in heat transfer and in generating heat resistance via various three-phonon acoustic-optical N and U processes. Using the Debye dispersion law in the reduced zone scheme we write for acoustic modes $\omega = cq$ and for optical modes $\omega_o = \omega_D + \omega_g + c_o(q_D - q)$, where ω_D is the Debye frequency for the corresponding acoustic polarization mode, q_D is the Debye radius, c and c_o are, respectively, acoustic and optical phonon velocities, and ω_g is the frequency gap at the zone boundary. Following Klemens[1] we introduce a factor $r = 2/\sqrt{3}(\alpha-\beta)/(\alpha+\beta)$ to reduce the cubic anharmonic Hamiltonian for optical phonons. Here α and β are two effective force constants such that at the zone boundary $\alpha/\beta = (\omega_o/\omega_D)^2$.

Using the standard procedure we obtain in our earlier notation the following expressions for acoustic-optical interactions of the types $\vec{q}_s + \vec{q}'_{s'} \rightarrow \vec{q}''_{s''}$ and $\vec{q}_s \rightarrow \vec{q}'_{s'} + \vec{q}''_{s''}$:

$$\tau^{-1}(ac+ac\rightarrow op) = B \sum_{\substack{s's'' \\ \varepsilon}} \frac{r''}{c_o''} \int dx' \frac{x'}{x} WW'W_o''$$

$$(1+\varepsilon y'') \frac{\bar{N}'(\bar{N}_o''+1)}{(\bar{N}+1)} \tag{1a}$$

$$\tau^{-1}(ac+op\rightarrow op) = B \sum_{\substack{s's'' \\ \varepsilon}} \frac{r'r''}{c''_o} \int dx' \frac{x'}{x} WW_o'W_o''$$

$$(1+\varepsilon y'') \frac{\bar{N}_o'(\bar{N}_o''+1)}{(\bar{N}+1)} \qquad (1b)$$

$$\tau^{-1}(op\rightarrow ac+ac) = \frac{Br}{2} \sum_{\substack{s's'' \\ \varepsilon}} \frac{1}{c''} \int dx' \frac{x'}{x} W_o W'W''$$

$$(1-\varepsilon+\varepsilon z'') \frac{\bar{N}'\bar{N}''}{\bar{N}_o} \qquad (1c)$$

$$\tau^{-1}(op\rightarrow op+ac) = Br \sum_{\substack{s's'' \\ \varepsilon}} \frac{r'}{c''} \int dx' \frac{x'}{x} W_o W_o'W''$$

$$(1-\varepsilon+\varepsilon z'') \frac{\bar{N}_o'\bar{N}''}{\bar{N}_o} \qquad (1d)$$

$$\tau^{-1}(op+ac\rightarrow op) = Br \sum_{\substack{s's'' \\ \varepsilon}} \frac{r''}{c''_o} \int dx' \frac{x'}{x} W_o W'W_o''$$

$$(1+\varepsilon y'') \frac{\bar{N}'(\bar{N}_o''+1)}{(\bar{N}_o+1)} \qquad (1e)$$

Here $x = |\vec{q}|/q_D$, $W = q/q_D$, $W_g = \omega_g/q_D$, etc., with similar quantities with a single prime and a double prime corresponding to polarizations s' and s'', respectively. Also in the above equations $z'' = W''/c''$, $y'' = (c_o'' + W'' - W'')/c''$, $B = \hbar q_D^5 \gamma^2/4\pi\rho c_L^2$, where γ = Grüneisen constant, c_L = longitudinal acoustic phonon velocity, ρ = mass density of the solid, and $\varepsilon = + 1/-1$ for N/U processes. The sums are over polarizations, and the integral limits are decided by consideration of conservation of energy and wave-vector.

Table 1 gives our numerical results in the single-mode relaxation time approximation for the natural Ge sample of Geballe and Hull[2] for which the following parameters have been used: boundary length L = 0.212cm, mass-defect constant A = 3.5×10^{-44} sec^3, and γ = 0.67. The phonon dispersion curves of Cochran[3] were approximated for our calculations.

Table 1   Role of optical phonons in heat transfer and scattering in Ge

T(°K)	Resistivity due to ac-op scattering (%)	Conductivity due to optical phonons (%)	Total conductivity (Watt/cm/°K)
5	0	0	3.8
17	0	0	11.4
100	16.3	0.05	2.04
300	33.9	0.5	0.63
900	40.2	1.5	0.23

We conclude that in the sample studied, optical phonons do not carry any significant heat. This is because the group velocities associated with these phonons are small and also since optical phonons are characterized by high frequencies they are strongly Rayleigh scattered by the mass defects present in the solid. On the other hand the various acoustic-optical phonon scattering processes reduce significantly the high-temperature thermal conductivity due to acoustic phonons. Although this helps to increase the temperature dependence of the conductivity at high temperatures, it can not completely explain the experimentally observed variation T^{-n}, n>1. The role of four-phonon processes might be of help in this regard. Our findings are to some extent in accord with the conclusions of Logachev and Yur'ev[4] and Ecsedy and Klemens[5].

References

1. P. G. Klemens, Phys. Rev. 148: 845 (1966).
2. G. P. Srivastava, Phil. Mag. 34: 795 (1976).
3. W. Cochran, Proc. Roy. Soc. (London) A253: 260 (1959).

4.   Yu. A. Logachev and M. S. Yur'ev, Soviet Phys. -
     Solid State 14: 2826 (1973).
5.   D. J. Ecsedy and P. G. Klemens, Phys. Rev. B15:
     5957 (1977).

EFFECTS OF ^{3}He IMPURITIES ON THE ACOUSTIC PROPERTIES OF hcp ^{4}He

Izumi Iwasa and Hideji Suzuki

Department of Physics, Faculty of Science
University of Tokyo
Hongo, Bunkyo-ku, Tokyo 113, Japan

INTRODUCTION

The acoustic properties of pure hcp ^{4}He crystals have recently been investigated by the present authors.[1] It is concluded that the resonance vibration of dislocations causes very large anomalies in the velocity and attenuation in the MHz-region below 1 K. The purpose of the present study is to examine the effects of ^{3}He impurities on the acoustic properties of ^{4}He crystals.

EXPERIMENTAL

The experimental set-up is the same as before[1] except that the sample gas was pressurized by means of a Toepler pump system.[2] Measurements have been made on samples with ^{3}He concentrations, x_3, of 30 ppm, 300 ppm and 1 % between 0.12 and 1.8 K at 10, 30 and 50 MHz with various amplitudes and repetition frequencies of the input pulse. All the crystals were grown at a nearly constant pressure of 39.0 atm. The pulse-echo method with two X-cut quartz transducers was employed. The input RF pulse had a width of about 1 μsec and its amplitude was varied by means of a step attenuator. The peak power of the ultrasonic wave excited in the specimen was estimated to be about 0.1 erg/sec at 10 MHz and at the smallest input (denoted by 0 dB), corresponding to the stress amplitude of about 100 dyn/cm^2. The signal was amplified by 40 dB and displayed on the oscilloscope. Several echoes were observed but only the first traversed signal was used for the measurement.

RESULTS

The temperature dependence of the velocity and attenuation was measured for a crystal of $x_3 = 1\%$ at 10, 30 and 50 MHz. At each frequency the velocity was represented by $v/v_o = -4.2 \times 10^{-4} \, T^4$ with $v_o = 557$ m/sec and no anomaly was observed. The attenuation is shown in Fig.1. The value at 0.3 K is supposed to be zero and a relation, $\alpha = 1.9 \times 10^{-8} \Omega T^4$ (dB/cm), fits the data below 1 K quite well, where $\Omega/2\pi$ is the sound frequency and T is the temperature. The results of the 300 ppm sample are almost the same as those of the 1% sample except for a slight power dependence at 10 MHz.

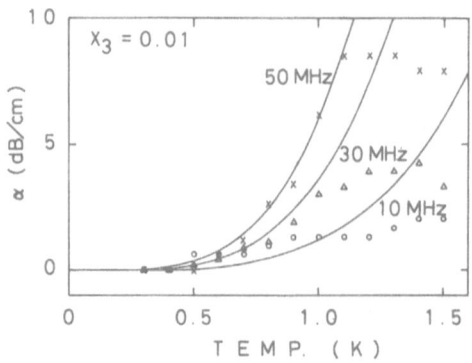

Fig.1. Temperature dependence of the attenuation in a 1% crystal at 10, 30 and 50 MHz. The curves are proportional to ΩT^4.

Most interesting data were obtained for crystals of $x_3 = 30$ ppm. Figure 2 shows the temperature dependence of the sound velocity at various input amplitudes. A marked power dependence is evident at

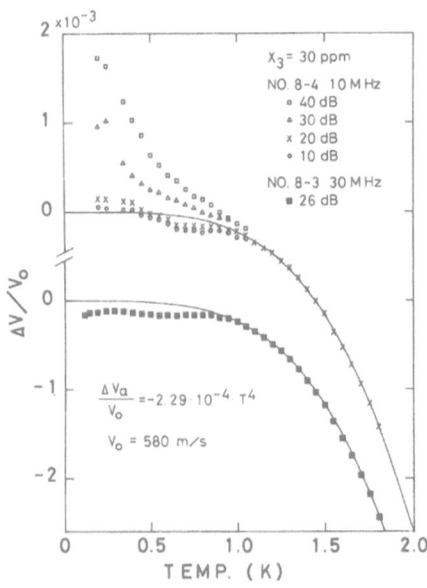

Fig.2. Temperature dependence of the velocity at various frequencies and input amplitudes in a 30 ppm crystal.

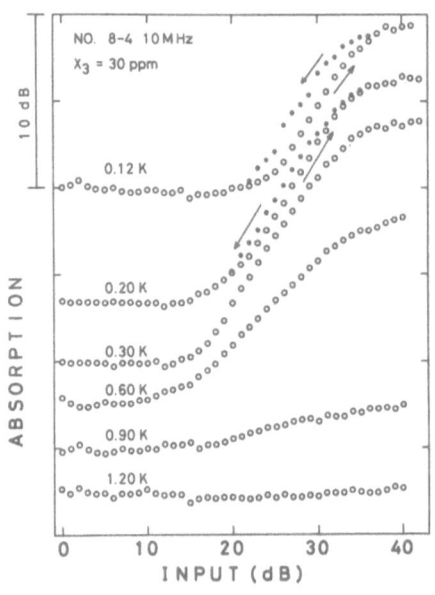

Fig.3. Attenuation in a 30 ppm crystal at 10 MHz as a function of the input amplitude at various temperatures.

10 MHz below 1 K. The effect at 30 MHz (not shown) is about 1/4 of
that at 10 MHz. Above 1 K the data can be fitted by a T^4-curve with
the same coefficient at both frequencies. It seems that the velocity
at 10 MHz at sufficiently low power and temperature approaches the
continuation of the T^4 curve fitted at higher temperatures. In Fig.
3 the power dependence of the attenuation is plotted at various
temperatures. The size of the power dependence is negligible above
1.2 K but it becomes substantial below 0.3 K. There are two plateaus
(below $\sim$ 20 dB and above $\sim$ 36 dB) and a transition region between
them. Below 0.2 K a hysteresis is observed. An increase of 3 dB
in the attenuation was observed at 10 MHz at 0.12 K in the transition
region (power = 30 dB) by changing the repetition frequency from 5
to 90 Hz.

DISCUSSION

     We have observed that in contrast to pure ^{4}He crystals there
are no anomalies in the velocity and attenuation in the crystals
with 1% ^{3}He. We consider that the ^{3}He atoms pin down the dislocations so that their resonance vibration is completely suppressed.
In this case the attenuation is solely due to three-phonon processes
and proportional to ΩT^4 when $\Omega \tau > 1$, where τ is the average lifetime
of thermal phonons. The condition is met below 1 K for $10 \sim 50$ MHz.
Berberich et al.[3] have reported the lifetime of hypersonic phonons
of 1.5 GHz in pure hcp ^{4}He to be nearly proportional to T^4. When
their value is scaled to our frequencies, we obtain $\alpha = 2.5 \times 10^{-8} \Omega T^4$
(dB/cm), which is consistent with our experimental value.

     The input dependence observed for the 30 ppm crystals clearly
shows that a breakaway process of the dislocations from the ^{3}He
atoms occurs in the transition region. The binding energy between
^{3}He atom and dislocation is estimated to be $\sim$0.2 K from a 1/T plot
of the input amplitude at which the transition region begins. It is
somewhat lower than the calculated values based on the dislocation
theory (0.6 K for a perfect edge dislocation and 0.3 K for a 60°
partial dislocation). In the high-input plateau the dislocations
are totally broken away from the atmosphere of ^{3}He atoms. We obtain
from the velocity data at 40 dB that $R\Lambda \simeq 2 \times 10^4$ cm^{-2}, where Λ is the
dislocation density and R is an orientation factor of about 0.1.
Finally we note that the existence of the transition region together
with the repetition-frequency dependence and the hysteresis in it
is closely related to the mobility of ^{3}He atoms. A detailed calculation is still in progress.

REFERENCES

1. I. Iwasa, K. Araki and H. Suzuki, J.Phys.Soc.Jpn.46,1119(1979).
2. A. Sakai, Y. Nishioka and H. Suzuki, J.Phys.Soc.Jpn.46,881(1979).
3. P. Berberich, P. Leiderer and S. Hunklinger, J.Low Temp.Phys.22,
   61(1976).

DISCUSSION

H. J. Maris: You didn't see any transition between first sound and zero sound. Is that right?

I. Iwasa: Yes, I expect a transition with 1% impurity, but I have not observed it.

H. J. Maris: Do you have any idea why?

I. Iwasa: I don't know how big the effect is.

Phonon-Quasiparticle Coupling in Dilute He^3-He^4 Mixtures[*]

F. Guillon and J.P. Harrison

Department of Physics, Queen's University

Kingston, Ontario, Canada

Abstract. Below 10 mK the dominant thermal resistance between the quasiparticles in dilute mixtures and a heat exchanger should be the phonon-quasiparticle resistance in the mixture. It is demonstrated that empirically this resistance (R) is given by $RVT^3/d \sim 10^{-2} m^2 K^4 W^{-1}$ where V is the volume of the mixture and d is the size of the pores containing the mixture. This empirical relation is remarkably similar to the bulk liquid theoretical result $RVT^3/\ell_{q-p} \sim 2 \times 10^{-2} m^2 K^4 W^{-1}$ for He^4 + 5% He^3 where ℓ_{q-p} is the mean free path for quasiparticle-quasiparticle scattering.

· · · · · ·

There is continued interest in cooling dilute He^3 in He^4 mixtures to 1 mK and below. Heat transfer from the He^3 quasiparticles (q-ps) to a refrigerator should proceed via the He^4 phonons, the phonons in the heat exchanger material and the electron gas in this material. Of the four resistances in series (He^3 q-p to He^4 phonon resistance (3-4), He^4 phonon to solid phonon (thermal boundary resistance), phonon to electron resistance and bulk resistance of the metal heat exchanger) the 3-4 resistance should dominate below 20 mK. Energy is transferred from He^3 q-ps to He^4 phonons because He^3 q-ps carried along with the acoustic wave motion collide with other q-ps[1]. This transfer is proportional to the number of phonons and the probability of q-p - q-p scattering[2] (i.e., $T^3 \times T^2$). Numerically, $RVT^5 \sim 10^{-11} m^3 K^6 W^{-1}$ for a 5% mixture and $\sim 6 \times 10^{-11} m^3 K^6 W^{-1}$ for 1.3% mixture[3]. The coupling constants required for these evaluations have been confirmed by thermal conductivity and acoustic attenuation measurements.

There have been several measurements of heat flow between He^3 q-ps in dilute mixtures and heat exchangers below 20 mK. In no case has the T^{-5} resistance been observed and in all cases the total resistance was always less than the expected 3-4 resistance. On

157

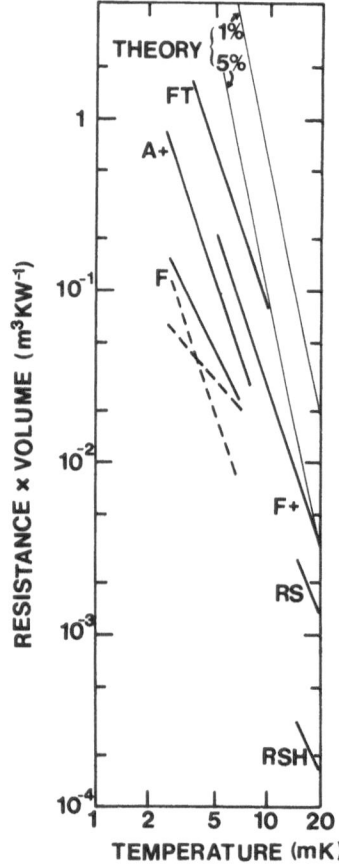

Figure 1. Thermal resistance x volume of mixture within sinter or powder for: RS - 6% He[3] and 2 μm silver[4]; RSH - 6% He[3] and 0.15 μm silver[5]; A+ - 1.6% He[3] and 0.2 μm thick copper flakes[6]; FT - 6% He[3] and nominal 0.5 μm silver ; F+ - 1.1% He[3] and nominal <70 μm CMN[8]; F - 6% He[3] and 0.27 μm silver[9].

the other hand the measured resistances were significantly larger than those between q-ps in <u>pure</u> He[3] and heat exchangers, particularly below 10 mK. The low resistance between pure He[3] and solids has often been attributed to magnetic coupling but a recent survey[3] of the experiments has shown that the small size of the solid particles is a more likely explanation. If this is so there is no reason for the thermal resistance between He[4] phonons and finely divided solids to behave differently (note that in pure He[3] q-ps and phonons are strongly coupled). The measured dilute mixture resistances can therefore be attributed to the 3-4 resistance. They are shown multiplied by the volume of helium within the sinter or powder in figure 1, together with the

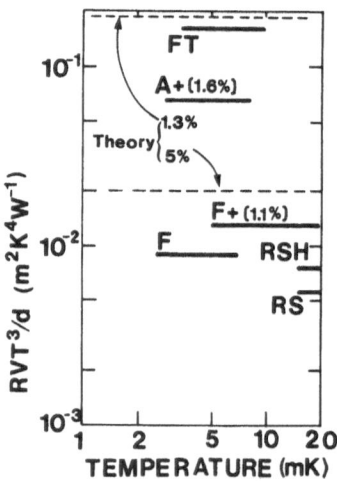

Figure 2. Plot of RVT^3/d. The dashed lines are the theoretical values of RVT^3/ℓ_{q-p}.

theoretical 3-4 resistances for 1.3% and 5% He^3 in liquid He^4. The
CMN/1.1% mixture results (F+) have had a temperature independent
phonon bottleneck term subtracted. The 40 nm silver results (F)
have been decomposed into a T^{-3} term and T^{-1} term, the latter being
attributed to one of the other thermal resistances. In figure 2 the
experimental results are replotted as RVT^3/d to show more clearly
the dependence upon temperature and pore size, d. There is considerable uncertainty in the chosen value of d. For F, RS and RSH,
d was chosen to be the sinter particle diameter deduced from surface
area measurements. For A+, d was equated to the thickness of the
flakes since electron microscope pictures show that the flakes pack
parallel to each other. Electron microscopy showed the pore sizes
in the CMN (F+) to be $\sim$ 2 µm. No area measurements or microscopy
are available for FT and so the nominal particle size (0.5 µm) was
chosen, even though for other silver powders it is known that a 200°C
sinter coalesces the particles.

From figure 2 it seems that empirically the 3-4 resistance is
given by
$$RVT^3/d \sim 10^{-2}m^2K^4W^{-1}$$

First of all this means that with 5 cm^3 of mixture in 50 nm pores,
$R \sim 10^5$ K/W at 1 mK which is low enough to permit sub-mK cooling
of dilute mixtures. However of even greater interest is that the
above empirical formula is very close to the Baym-Ebner theoretical
result,

$$RVT^3/\ell_{q-p} \sim \begin{cases} 2 \times 10^{-2}m^2K^4W^{-1} \ (5\%) \\ 20 \times 10^{-2}m^2K^4W^{-1} \ (1.3\%) \end{cases}$$

where ℓ_{q-p} is the mean free path of a quasiparticle scattered by
other quasiparticles. These equations are shown as dashed lines on
figure 2. The significance of this similarity is that phonons are
attenuated in the dilute mixture by q-p - boundary scattering with
the same coupling constant as by q-p - q-p scattering.

. . . . .
 Research supported by N.S.E.R.C.

REFERENCES
1. G. Baym and C. Ebner, Phys. Rev. 164, 235 (1967).
2. J.C. Wheatley, O.E. Vilches and W.R. Abel, Physics 4, 1 (1968).
3. J.P. Harrison, J. Low Temp. Phys. (to be published).
4. R. Radebaugh and J.D. Siegwarth, Proc. LT 13 1, 398 (1974).
5. R. Radebaugh, J.D. Siegwarth and J.C. Holste, Proc. ICEC-5,
    242 (1974).
6. A.I. Ahonen, P.M. Berglund, M.T. Haikala, M. Krusius,
    O.V. Lounasmaa and M. Paalanen, Cryogenics 16, 521 (1976).
7. G. Frossati and D. Thoulouze, Proc. ICEC-6, 116 (1976).
8. G. Frossati, H. Godfrin, B. Hebral, G. Schumacher and
    D. Thoulouze, Proc. ULT, 306 (1977).
9. G. Frossati, Proc. LT15, 1578 (1978).
****

DISCUSSION

H. J. Maris: I think you were hinting at a rather nice analogy then.
A free electron cannot absorb a single photon. The lowest order
process is Compton scattering because of conservation of energy and
momentum. But the electron can absorb the photon if there is another
particle nearby to absorb the momentum - and that's why an atom can
absorb a photon - it's really the electron does the absorbing and
the nucleus takes up the momentum. I guess one can view the Baym
theory in that context - the phonon is sort of absorbed while two
quasi-particles are making a collision. Your physical argument
fits in with this picture because now we have the phonon being absorbed by the quasi-particle while it's near the wall, so that the
wall picks up the momentum. So this seems a very nice consistent
picture, and of course when you do have a size effect like this
where maybe it makes more sense to think of the phonon as a standing
wave inside some cavity rather than a freely propagating wave the
phonon becomes a sine standing wave and you do lose, as you say,
the selection rules and the process becomes allowed.

J. P. Harrison: Yes. That's a nice way of looking at it.

K. Fossheim: At which temperature do you expect the size effect to
appear?

J. P. Harrison: It depends on the concentration of ^{3}He in the ^{4}He,
but basically 20 to 30 mK is when these size effects come in. The
quasi-particle mean-free-path becomes very long in dilute mixtures
at low temperatures, much bigger than the size of the pores that
anybody is using to contain them. The whole thing is dominated by
confined geometry in any experimental cell, whether it's a dilution
refrigerator or a nuclear refrigerator.

VELOCITY OF ROTON SECOND SOUND

Richard W. Cline[†] and Humphrey J. Maris

Department of Physics
Brown University
Providence, RI 02912

Roton second sound is a collective excitation similar to
ordinary second sound, but involving only excitations from the
roton part of the spectrum. If number is conserved in collisions
between rotons, the velocity is predicted to be[1]

$$c_N = \frac{3kT}{p_o} \tag{1}$$

where p_o is the momentum at the roton minimum. If number is not
conserved, the velocity should be[2,3] c_o, which is smaller than c_N
by a factor of $\sqrt{3}$. Roton scattering is believed to be almost
entirely of the type

$$R + R' \rightarrow R'' + R''' \tag{2}$$

Thus, roton number is expected to be conserved. Measurement of the
velocity (1) would therefore provide confirmation of current views
on roton-roton interactions.

Using a heat pulse technique, Dynes, Narayanamurti, and
Andres[4-6] demonstrated the existence of roton second sound, but
because of strong amplitude dependence of the velocity, they were
unable to determine whether the velocity was c_N or c_o. Recent
experiments by Castaing[7,8] have suffered from the same difficulty.
Through the use of generators and detectors of large areas, we have
been able to observe roton second sound at significantly lower
amplitude than was possible previously. For powers less than about
5 mW mm^{-2} the velocity becomes independent of amplitude. Results
are shown in Fig. 1. These data are obtained from the arrival time
of the _peak_ of the received pulse. At the shortest path length,
0.23 mm, the velocity is the slowest and approaches c_o. For longer

path lengths, the velocity is greater and lies between c_o and c_N. The changes of the velocity with path length are larger than the uncertainties in the measurement.

Weiss[10] investigated the effects of finite number-conserving and number non-conserving collision times on the velocity. By considering the relaxation of the number density he found that at low frequencies roton second sound should have the velocity c_o, while at high frequencies the velocity would tend to c_N. In a heat pulse experiment the effective frequency of second sound decreases as the path length increases. This is because the higher frequency components usually have higher attenuation. Thus, the number density relaxation process gives a pulse velocity which decreases with distance, in disagreement with our observations.

There is another effect which may be important even if there are no number non-conserving collisions. In a medium in which heat conduction occurs, the velocity <u>decreases</u> from the adiabatic velocity (equal to c_N) to an isothermal velocity c_T as the frequency is increased. For a roton gas the midpoint of the transition is

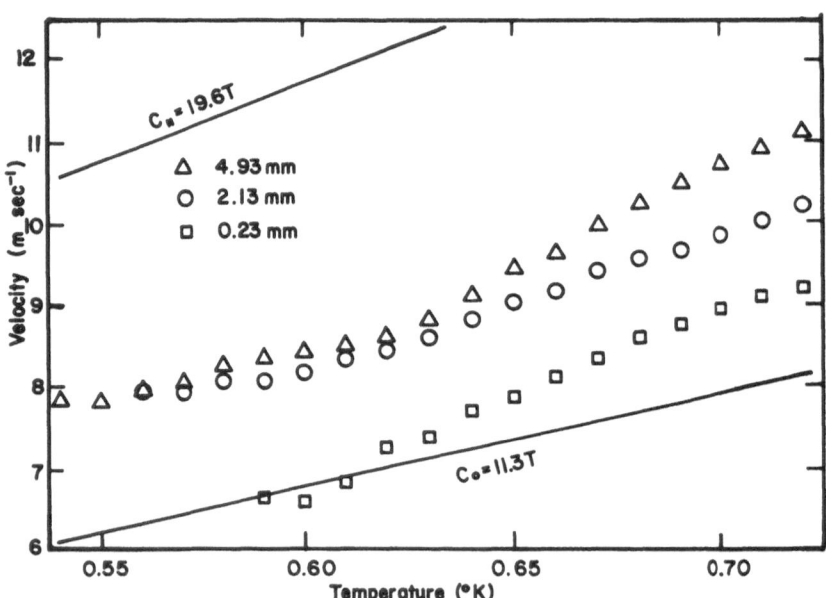

Fig. 1.   Velocity of roton second sound as a function of temperature for 3 different path lengths. The velocity corresponds to the arrival time of the peak of the pulse.

when Ω is of the order of $27 \, \mu kT/p_o^2 \tau_{RR}$, where μ is the roton effective mass and τ_{RR} is the roton-roton scattering time. Khalatnikov and Chernikova's estimate of τ_{RR} implies that at $0.6^o K$ roton second sound should propagate isothermally for $\Omega > 10^6 \, sec^{-1}$. One can show that the isothermal velocity is

$$c_T = \sqrt{3} \, \frac{kT}{p_o} = c_o \qquad\qquad (3)$$

This effect causes a pulse to travel at the isothermal velocity for short path lengths, but to travel with the adiabatic velocity once the high-frequency components have been attenuated. This is the path-length dependence we observe, and indicates that we might possibly be seeing the adiabatic-isothermal transition in a roton gas. An additional point is that isothermal second sound cannot propagate unless the dominant roton interactions are number-conserving. Thus our observation of this mode would be consistent with this theoretical picture. However, a difficulty with this idea (K. Weiss, private communication) is that at high frequency the effect of viscosity will also be important, and this tends to increase the velocity and cause attenuation. Thus further theoretical work is required to determine if weakly-damped isothermal roton second sound can in fact occur.

We have also studied the propagation of roton second sound at higher heater powers for which the velocity is amplitude dependent. Our observations can be explained in terms of the propagation of a large amplitude density wave with shock-like characteristics in a roton gas with number conservation. Further details of this work will be published elsewhere.

This research was supported in part by the National Science Foundation under Grant DMR 77-12249 and through the Materials Research Laboratory of Brown University.

References

[†] Present address: Department of Physics, Massachusetts Institute of Technology, Cambridge, MA 02139

1. H. J. Maris, Phys. Rev. Lett. 36, 907 (1976).
2. I. M. Khalatnikov and D. M. Chernikova, Zh. Eksp. Teor. Fiz. 49, 1972 (1965)(Sov. Phys. JETP 22, 1336 (1966)).
3. I. M. Khalatnikov and D. M. Chernikova, Zh. Eksp. Teor. Fiz. 50, 411 (1965)(Sov. Phys. JETP 23, 274 (1966)).
4. R. C. Dynes, V. Narayanamurti, and K. Andres, Phys. Rev. Lett. 30, 1129 (1973).
5. V. Narayanamurti, R. C. Dynes, and K. Andres, Phys. Rev. B11, 2500 (1975).

6. V. Narayanamurti and R. C. Dynes, Phys. Rev. B13, 2898 (1976).
7. B. Castaing, Phys. Rev. B13, 3854 (1976).
8. B. Castaing and A. Libchaber, J. Low Temp. Phys. 31, (1978).
9. R. W. Guernsey, Jr., K. Luszczynski, and W. C. Mitchell, Cryo-
    genics 7, 110 (1967).
10. K. Weiss, Phys. Rev. B15, 4227 (1977).
                            ****

DISCUSSION

D. V. Osborne: Could you tie any of these loose ends together by
carrying out a single-frequency at resonance experiment?

R. W. Cline: We did try to do this, and the problem you run into
is that you always generate phonons, and the phonons are running
back and forth and there is some scattering between rotons and
phonons. The phonons pass through the rotons 20 times because their
velocity is more than an order of magnitude larger than the rotons.
They cause so much scattering that you can't do cavity-type experiments. If you could get rid of the phonons, it would be a nice
thing to do.

H. J. Maris: Another effect that does happen which is very interesting is that if you send a pulse of phonons or rotons at an interface you certainly get some of the other sort coming back. There's
a little bit of conversion goes on which makes it hard to set up a
standing wave made up of purely phonons or rotons.

D. V. Osborne: I was going to suggest an annular standing wave.

INTERFACE EFFECTS AND THE PROPAGATION OF WELL DEVELOPED SECOND SOUND

PULSE TRAINS IN THIN SLABS OF SOLID ^{4}He

G.P. Dance and S.J. Rogers

Physics Laboratory,
University of Kent at Canterbury,
Kent, CT2 7NR,
England

For a given density the temperature range for second sound in solid ^{4}He depends upon the sample thickness [1]. The present work extends the range of observation to higher temperatures by studying second sound propagation in slabs with thicknesses from 0.1 to 0.025 cm. Crystals were grown at constant pressure around a LiF single crystal (see inset to Fig. 1) at pressures in the range of 25 to 100 atm. at rates from 2mm/hr to 2mm/min. The second sound pulses propagate in the thin slab between this crystal and the glass plate P. A, B, and C are InSn films which serve as heaters or bolometers [2] as required. The second sound could be generated either directly (by pulsing B or C) or indirectly (by pulsing A) via the propagation of a ballistic phonon pulse through the LiF. In this latter arrangement the phonon focussing properties of the LiF [3] were exploited to look for differences in the effects of incident longitudinal and transverse phonons.

All the crystals studied showed well developed second sound, and in the thinnest slabs at the highest pressure the transition from diffusive to second sound propagation was observed at nearly 1.5K; for such samples a number of echoes could be seen at 1.2K. The heater and detector areas were sufficiently large for edge effects to be negligible, and the pulses generated both directly and indirectly were accordingly Gaussian rather than derivative in shape [4].

For many of the crystals several second sound echoes were seen over a considerable temperature range. Even in the absence of internal dissipative processes reflection losses will give rise to an exponential decay in the amplitude of successive echoes, but the change in the decay constant with temperature is a measure of the

165

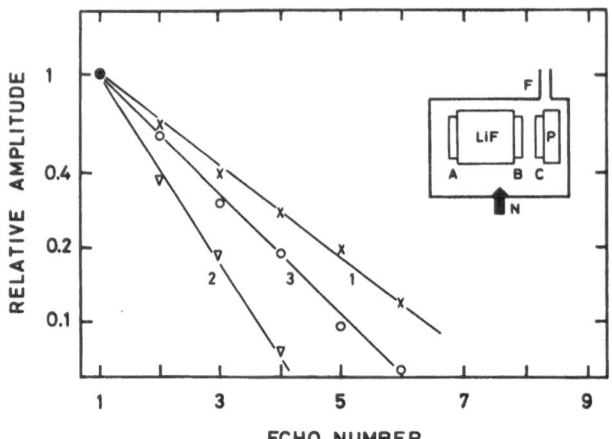

Figure 1   Amplitude of successive second sound echoes relative to
            the first. The solid was grown at 55 atm. and the decay
            lines 1, 2, and 3 are respectively for 0.77K, 0.82K and
            0.6K.

            Sample Chamber (Inset): InSn films (A,B,C); fill line (F);
            nucleating stud (N); glass plate (P).

variation of the second sound attentuation. Three decay curves in
the neighbourhood of the attenuation minimum are shown in Fig. 1.
The additional losses at higher and lower temperatures seen in curves
2 and 3 are due to increased resistive scattering and decreased N
process scattering respectively. The latter gives rise to damping
via second viscosity [5] with an associated increase in pulse width.
The damping varies considerably from crystal to crystal at a given
pressure and there are corresponding variations in the temperature
range for the second sound propagation. For the worst crystals the
onset of second sound is depressed by as much as 0.3K as compared
with that for the best crystals formed at the same pressure. In
almost all such cases a considerable improvement was effected by
annealing the crystal within a few mK of the melting curve for about
15 minutes.

     The effect of crystal quality on the second sound attenuation
is reflected in the velocity data. Fig. 2 shows measurements for
both directly and indirectly generated second sound pulses in a
number of crystals grown at a pressure of 55 atm. To within a 5%
scatter the points fall on a universal curve with branches which
represent differences in crystal quality and/or thickness. It will

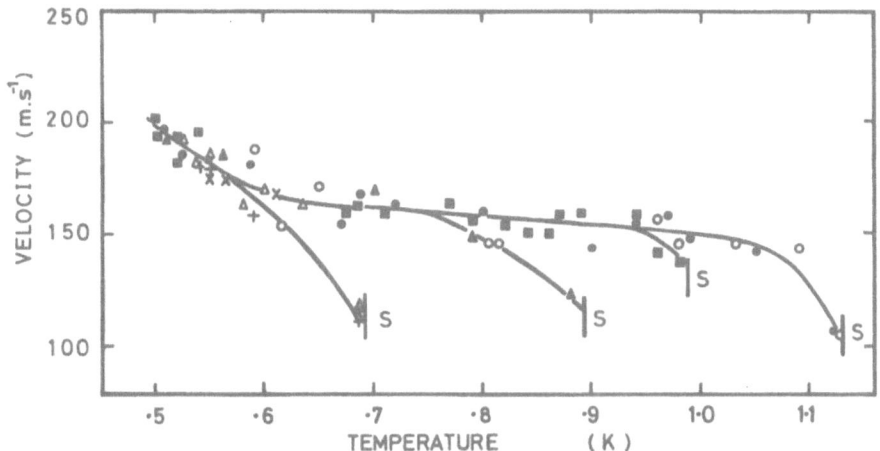

Figure 2   Samples: 0.24mm: direct ●;   indirect o
              0.64mm: direct ■, Δ→after anneal ▲, X→after anneal +.

be seen that decreasing the crystal thickness extends the range of
second sound propagation to significantly higher temperatures.
Allowing for sample variations, the results for the indirectly
generated second sound do not appear to differ significantly from the
other data. It may be, however, that the distribution of phonons as
between the various modes in the incident pulse is changed in the
polished layer at the LiF/helium interface [6]. It would be of
considerable interest to use a freshly cleaved surface but this is not
possible for the [111] direction of propagation.

    We had hoped that the transition to ballistic propagation that we
begin to see at low temperatures would provide information concerning
the crystallographic orientation of the various samples. This has not
proved to be the case but, if it is assumed that the results for
various crystals at a given pressure correspond to a range of crystallographic orientations, the universal curves for velocity, as for
example in Fig. 2, suggest that the second sound is not very anisotropic. It has proved possible to determine directly the effect on
the second sound propagation of changing the solid density for a
sample of a given orientation. The crystal quality is not drastically affected by an expansion to a lower pressure, and a family of
velocity curves can be obtained in a series of such expansions.

REFERENCES
[1] C.C. Ackermann and R.A. Guyer, Ann. Phys. 50, 128 (1968)

[2]   R.M. Kimber and S.J. Rogers, Cryogenics, 13, 350 (1973)
[3]   B. Taylor, H.J. Maris and C. Elbaum, Phys. Rev. B3 1462 (1971)
[4]   V. Narayanamurti, R.C. Dynes and K. Andres, Phys. Rev. B11,
      2500 (1975)
[5]   S.J. Rogers, Phys. Rev. B3 1440 (1971)
[6]   J. Weber, W. Sandmann, W. Dietsche and H. Kinder, J. Phys.(Paris)
      39, Colloq. C6-242 (1978).

MEASUREMENTS OF THE SPECIFIC HEAT OF LIQUID AND SOLID ^{4}He IN A CONFINED GEOMETRY AT LOW TEMPERATURES

B. Hébral, G. Frossati, H. Godfrin, D. Thoulouze, and A.S. Greenberg[*]

Centre de Recherches sur les Très Basses Températures, C.N.R.S., BP 166 X, 38042 Grenoble-Cedex, France

We present the results of specific heat measurements in liquid (0.43 and 21 bars) and solid hcp (20.9 cm^3/mole) ^{4}He in the temperature range 100 mK to $\approx$ 2 K. For the case of the liquid, we were interested in studying the effects of a confined geometry and, in particular, if there is a modification of the anomalous phonon dispersion curve.[1,2] For the case of the solid we were interested in extending C_V measurements to lower temperatures than have been reported[3] in order to examine more carefully the persistent question of a "specific heat anomaly" in quantum crystals.[4,5] We also searched for possible signatures of a supersolid state which has been studied theoretically in some detail.[6]

Our experimental cell was constructed using araldite epoxy, and contained 21.85 g of powdered CMN (pores 1-5 μ) with a sample volume of 6.25 cm^3. For each data taking run the CMN was demagnetized initially from 1.1 kG at 15-20 mK to zero field reaching a typical minimum magnetic temperature, $T^* = 2$ mK. Details on the experimental procedure are given elsewhere.[7]

In fig. 1 we present the total heat capacity data with liquid ^{4}He at 21 bars from 3 mK to 1.5 K. The dashed line represents the ^{4}He contribution taken from other experiments[2] and the difference, which gives the background heat capacity, is the mixed line. At high temperature, T > 300 mK, we measure unambiguously the heat capacity of the ^{4}He. At low temperatures, T < 70 mK, we measure primarily the heat capacity of the paramagnetic cerium spins. We shall discuss here the results above 100 mK where the ^{4}He contribution becomes appreciable compared to the background. There are, nevertheless,

[*]On leave from Colorado State University, Fort Collins, Colorado, USA.

interesting features below 10 mK depending on the helium isotope in the cell, as observed previously.[7,8]

The main contribution to the liquid ^{4}He specific heat for $T \lesssim 500$ mK is due to the phonons, and can be expressed as $C_v = AT^3 + CT^5 + \ldots$. We find C is negative, $C \approx -100$ mJ/mole/K^6 at zero pressure which implies in $\varepsilon = c_o p(1 - \alpha_2(p/\hbar)^2 \ldots)$, the energy momentum relation, a negative α_2. This means that three phonon processes are allowed in phonon scattering.[9] At 21 bars : $C \approx 0$ and the variation of C with pressure is similar to that reported elsewhere.[1,2]

The agreement between our results at 0.43 and 21 bars with those obtained for bulk samples[1,2] is very good. Thus the phonon dispersion remains essentially unchanged due to the confined geometry.

The Debye temperature of hcp ^{4}He (20.9 cm^3/mole) is presented on fig. 2 from 100 mK to 1.4 K. Below 600 mK, θ_D is constant : $\theta_D = 26.1 \pm 0.1$ K and the heat capacity of solid ^{4}He is well described by the classical Debye model. Below 100 mK the total heat capacity is the same as measured with liquid ^{4}He in the cell. No change in C_v due to a possible supersolid transition is observed. However the heat capacity of solid ^{4}He is about 0.5 % of the background for these temperatures and only very large effects (≈ 1 erg/mK) could have

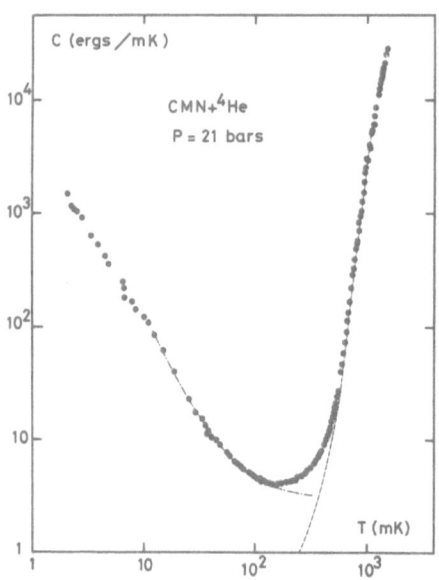

Fig. 1 Total heat capacity of the cell filled with liquid ^{4}He at 21 bars.

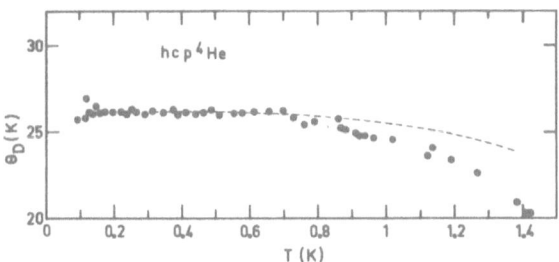

Fig. 2 Debye temperature of hcp ^{4}He.

been detected. Our results together with recent thermal conductivity data[10], indicate that the transition, if it exists, is a relatively small effect, or is to be found for T $\lesssim$ 25 mK.

Between 300 mK and 1 K these results are in excellent agreement with previous measurements[3] on bulk sample represented by the dashed line on fig. 2. In the high temperature regime, T > 1 K, we measure a larger heat capacity than previously reported.[3] An anomaly due to the hcp-bcc or hcp-liquid transitions has been observed[3,11] but in our case the anomaly appears near 1 K, which seems somewhat too low for a similar explanation. To account for this excess we suppose a metastable liquid phase is formed near the heater during the heating pulses. The liquid probably remains in a supercooled state after the temperature has become uniform thus giving rise to a larger measured heat capacity.

The precisely measured background allows us to reanalyze our measurements with bcc ^{3}He. The data are now well described from 5 mK to 300 mK as the sum of an exchange term ($|J|/k_B$=.65 mK) and a phonon term (θ_D=19.6 K) for a molar volume of 24 $\pm$ 0.1 cm^3/mole. This agrees within 3 %, with the results obtained in bulk samples.[12] The large C_V excess reported around 100 mK in solid ^{3}He confined in 100 μ channels[4] is not observed here. In our restricted geometry (1-5μ)we find no "specific heat anomaly" for either bcc ^{3}He or hcp ^{4}He.

In summary it appears from these heat capacity measurements that in our confined geometry there is a general agreement with the results for open geometries. Specific heat behavior of quantum crystals or liquids of micron size is well represented by their bulk properties down to 100 mK.

1. N.E. Phillips, C.G. Waterfield, J.K. Hoffer, Phys. Rev. Lett. 25, 1260 (1970).
2. D.S. Greywall, Phys. Rev. B18, 2127 (1978).
3. W.R. Gardner,J.K. Hoffer,N.E. Phillips,Phys.Rev. A7,1029 (1973).
4. S.H. Castles, E.D. Adams, Jour. of Low Temp. Phys. 19, 397 (1975).

5. B. Hébral, G. Frossati, H. Godfrin, G. Schumacher, and
   D. Thoulouze, J. de Phys. Lett. 40, L-41 (1979).
6. For a review on this point see C.M. Varma, N.R. Werthamer in
   The Physics of Liquid and Solid Helium, p. 503, edited by
   K.H. Bennemann and J.B. Ketterson (John Wiley, 1976).
7. B. Hébral, G. Frossati, H. Godfrin, G. Schumacher, D. Thoulouze,
   Rev. de Phys. Appl. 13, 533 (1978).
8. J.C. Wheatley, Ann. Acad. Sci. Fennicae A210, 15 (1966).
9. See,for example,H.J. Maris, Rev. Mod. Phys. 49, 341 (1977).
10. G. Armstrong, A.A. Helmy, A.S. Greenberg, to be published in
    Phys. Rev. B.
11. J.K. Hoffer, W.R. Gardner, C.G. Waterfield, N.E. Phillips, Jour.
    of Low Temp. Phys. 23, 63 (1976).
12. D.S. Greywall, Phys. Rev. B15, 2604 (1977).

KAPITZA RESISTANCE STUDIES USING PHONON PULSE

REFLECTION

H. Kinder, J. Weber, and W. Dietsche

Physik-Department E 10, TU München

8046 Garching, West Germany

The transport of heat across interfaces between solids and liquid helium is generally described by a thermal boundary resistance which is named after Kapitza.[1] Two microscopic transport mechanisms have been distinguished. One is the transmission of phonons under the conservation of energy as first proposed by Khalatnikov[2] and later refined by many authors.[3] This mechanism is very ineffective because of the large acoustic mismatch between helium and all solids. A far more effective second transport mechanism was invoked to account for the many experimental facts which were mostly obtained by phonon pulse experiments.[4-8] This mechanism was found on all surfaces studied, if the phonon frequencies were above $\sim$50 GHz or, correspondingly, T>1K. Therefore, it was considered to be an intrinsic effect and some models were put forward which were all based on the assumption of an ideally plane surface of the solid.[9,10] None of these models was able to account at least for the magnitude of the heat transport without lumping the problem into adjustable parameters. Thus, the second heat transfer mechanism was called "anomalous", and was considered to be - if not miraculous - a tantalizing problem at least.[11]

In what follows we first show that there is no miracle, rather that the assumption of an ideally plane surface was not justified. To work with a truly clean surface, it is necessary to keep it under ultra high vacuum conditions. This can be easily done in a chamber surrounded by liquid helium below $\sim$3K when H_2 is frozen out. To obtain fresh surfaces, we cleaved crystals of LiF or NaF remotely by an apparatus shown schematically in the inset of Fig.1.[12] Phonon pulses were reflected from the freshly cleaved surfaces by using superconducting tunnel junctions on the far face of the crystal as generators and detectors of monochromatic phonons of 290 GHz.[13]

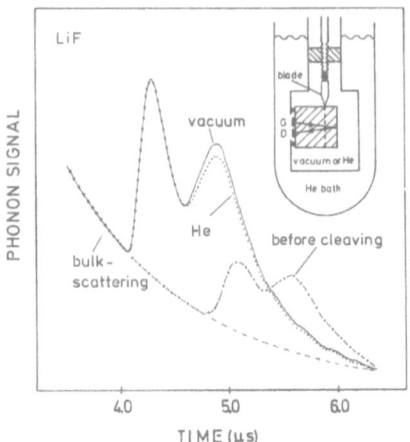

Fig. 1: Transverse phonon pulses reflected from the "normal"
(dash-dotted) and the freshly cleaved surface with vacuum (solid)
or liquid helium (dotted) in the vacuum chamber (inset).

Fig. 1 shows the phonon intensity as a function of time for LiF.
The dash-dotted trace shows the echo pattern before cleaving, the
solid line after cleaving in vacuum. The time of flight is
shortened because of the reduced thickness of the crystal. The
dotted trace was taken after the chamber was filled with helium.
The first pulse, a transverse mode with polarization parallel to
the cleaved surface, shows no effect of the helium. This is in
marked contrast to previous experiments on "normal" surfaces which
showed a decrease in echo height by ~50%. Still there is a small
effect on the second pulse, an oblique transverse mode, whose polarization has a finite angle with the surface. Thus, one might
think that the absence of the anomalous transport mechanism is a
peculiarity of the pure transverse mode which did not show up in
previous experiments because of the rougher surfaces.[4-8] Therefore,
we repeated the experiment with NaF where phonon focusing also allows
one to study longitudinal (L) phonons.

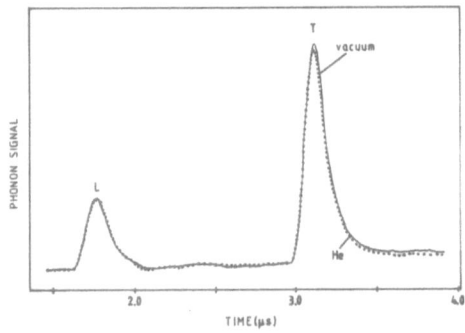

Fig. 2: Longitudinal (L) and transverse (T) echoes in NaF without
and with helium, at normal incidence on a fresh surface.

Fig. 2 shows that there is no helium effect on the L phonons
either, and there is only a very small effect on the transverse (T)
pulse. The latter may in fact come from the tail of diffusely
scattered phonons as will be discussed below. Still, we may have
observed here a peculiarity of phonons at nearly normal incidence.
To check this, we looked at 5 different angles up to 33° away from
normal incidence.[14] This was achieved by using a "multidetector"
consisting of 5 tunnel junctions which were simultaneously evaporated in series connection. Each junction had a different impedance
so that a particular junction could be selected by the setting of
the operation point of the device. The result for two different
angles at the same surface is shown in Fig.3.

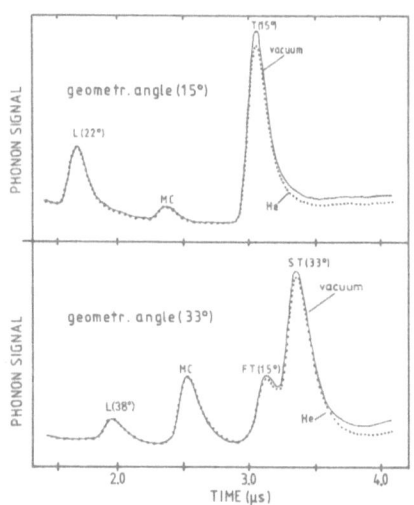

Fig. 3: Phonon pulses at 15° and 33° incidence away from normal in
the (001) plane. Note that the mode conversion (MC) pulse shows no
helium effect.

Besides the L and T modes, a mode converted (MC) echo also emerges
as expected for non-normal incidence. In previous experiments the
MC echo showed a particularly strong decrease with helium.[5,6] This is
not the case here, where the mode conversion arises from specular
reflection rather than from diffuse scattering. Only on the tail of
the T modes, we observe again a very small helium effect which we
attribute again to diffuse reflection from the surface.

All measurements presented so far were taken at 290 GHz. We
have also varied the frequency[13] as shown in Fig. 4 where the height
of the FT echo at normal incidence in LiF is plotted versus frequency, without and with helium. No noticeable helium effect was observed in the frequency range between 80 GHz and 280 GHz.

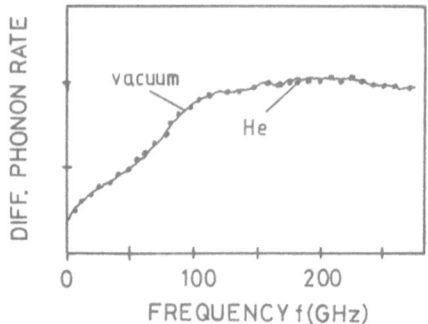

<u>Fig. 4:</u> Height of the fast transverse echo in LiF as a function
of frequency.

        Thus we are led to extrapolate that on ideally flat surfaces
which would not show any diffuse scattering, the anomalous transmission will be absent for all angles of incidence for all polarizations, and for all frequencies up to 290 GHz at least.

        Consequently, we can now understand why the models based on
smooth surfaces appeared to fail. Only with more complicated surface structures can we hope to explain the anomalous transport
mechanism. Since the structure of "normal" surfaces is not known
in sufficient detail, one way to proceed is to vary the surface in
a controlled manner.

        At high frequencies, only a small variation is necessary.
Fig. 5 shows an experiment which was done exactly in the same way
as that of Fig.2, except that the cleaved surface was not quite as
smooth. (Some small steps and craters were visible with an electron
microscope.)

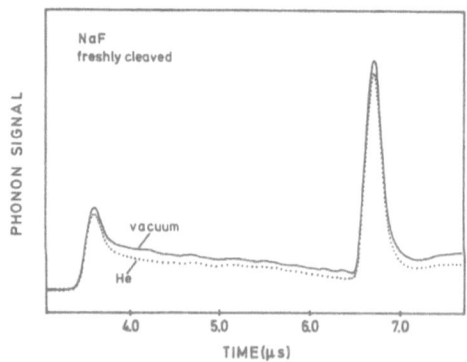

<u>Fig. 5:</u> Echoes in NaF as in Fig. 2. The surface is freshly
cleaved but less smooth than that of Fig. 2.

In contrast to Fig.2, the pulses have long tails which correspond to phonons with longer time of flight, i.e. to non-specularly reflected ones. With helium, a large relative decrease is observed in the tails, but only a small one on the pulses which in fact should be also assigned to a contribution of diffusely scattered phonons in view of Fig. 2.

Larger variations of the surface, e.g. produced by warming to room temperature and letting air into the vacuum chamber, caused stronger diffuse scattering and stronger helium effects.[12,14]

Thus, diffuse scattering and the helium effect appear to be strongly correlated. Either roughness or the presence of impurities leads to both diffuse scattering and anomalous transmission of the diffusely scattered phonons.

At low frequencies, more drastic variations of the surface are required for the anomalous transport to appear.[15] Schubert, Leiderer and Kinder have carried out an experiment at 25 GHz, by using the method of stimulated Brillouin scattering for generation, and spontaneous Brillouin scattering for detection of longitudinal coherent phonon pulses. Thin quartz platelets were used so that many successive reflections were observed. Details will be given elsewhere.[16]

With surfaces cleaned with pure acetone, no helium effect on the echo pattern was observed. Therefore, the surface were "varied" by coating them with thin layers of paraffin oil ($\sim$50 nm). This coating did not change the reflection coefficient under vacuum conditions. However, in helium a 5% reduction of the echo height per reflection was observed at a low intensity of the phonons, see Fig.6.

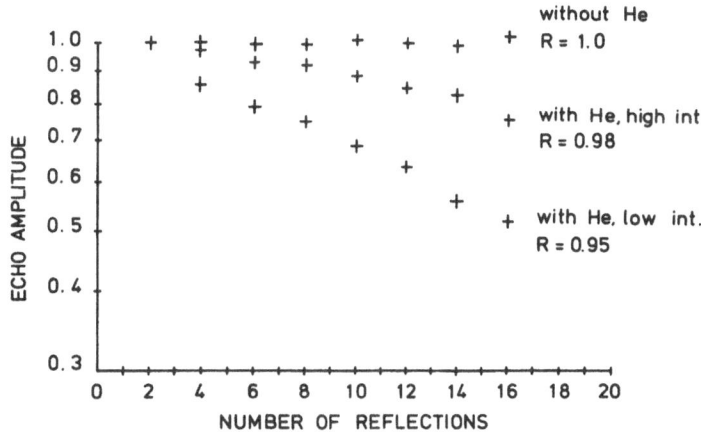

Fig. 6: The decay of 25 GHz longitudinal phonon pulse trains (generated and detected by Brillouin scattering) in quartz. High intensity: $\sim$10 W/mm^2; low intensity: $\sim$2 W/mm^2. Ref. 16.

With the setup used, the intensity of the generated phonons was variable. At the highest intensities reached, the helium effect was remarkably diminished. Careful checks have ruled out all trival effects, so that we are led to assume that we have seen saturation of the anomalous transmission.[16] Thus, for this model system at least, we found that saturable excitations are able to mediate the anomalous Kapitza conduction. In the field of glassy materials, such saturable excitations are commonly termed "two level systems".[17] However, any sufficiently anharmonic oscillator is saturable, because of the unequal level spacing. So it is interesting to look for additional anharmonic properties of the interface.

In fact, anharmonic effects at high frequencies have already been observed, using "normal" surfaces.[5] Fig. 7 shows an experiment on Si where the echo height with thin helium films was measured with two different detectors, a Sn junction and a bolometer. The generator was also a Sn junction. Hence, the Sn detector was only sensitive to the reflected phonons when their frequencies were not decreased by the helium film. On the other hand, the bolometer was

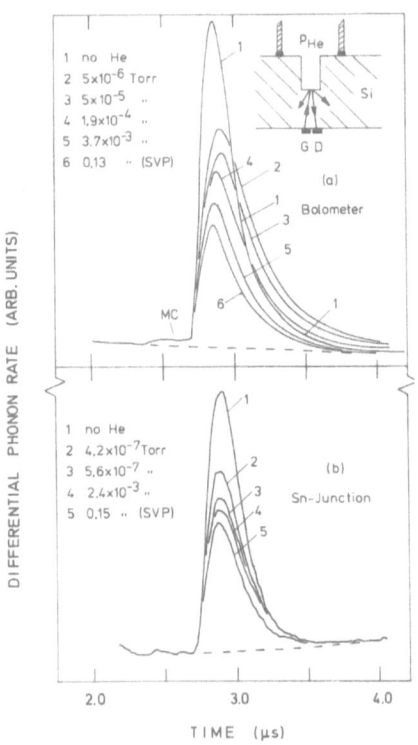

Fig. 7: Transverse 290 GHz phonons reflected from a "normal" surface detected by (a) bolometer, and (b) Sn junction. The difference indicates that the delayed phonons in (a) were shifted to lower frequencies by the helium film.

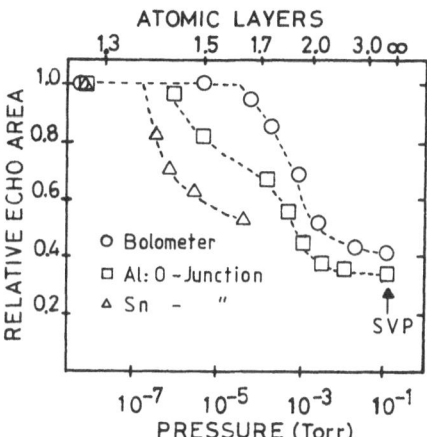

Fig. 8: Areas of pulses of Fig.7 and of a similar set taken with
a granular Al junction.

sensitive to all phonons. With thin films of helium, e.g. trace 2,
the signal height of both junction and bolometer is decreased.
However, a long tail of the pulse is detected with the bolometer which
outweighs the loss in signal height. Qualitatively, this shows that
some of the phonons are down-converted in frequency and delayed in
time by the helium film.

In Fig.8, the pulse areas are plotted as a function of helium
pressure or film thickness. Also, the data obtained with a granular
Al junction detector (140 GHz threshold) have been added. For a film
thickness around 1,5 atomic layers, there is a marked difference
between the Al and the Sn dectector, indicating that an appreciable
number of reflected phonons is only slightl shifted such that their
frequencies are in the range 140-290 GHz. With thicker films, the
thermalization becomes more and more complete and eventually conduction to the gas dominates at higher pressures, as shown by the bolometer curve. Similar frequency conversion effects have been observed
recently in a transmission experiment by Wyatt and Crisp.[18]

To summarize, we have shown that a "normal" surface in contact
with He is much more anharmonic than a bare surface alone on one
hand and bulk helium on the other for both low and high frequencies.
Similar anharmonicities have been found in glasses at low frequencies[16] and more recently at high frequencies.[19] In addition, we have
seen from the experiments on freshly cleaved surfaces that the
"active" parts of the surface must be those which scatter the phonons
diffusely. Such parts may have a grossly damaged and disordered
structure and may be even broken up into little cracks, slits, and
pores. The "normal" surfaces probably also have some porous structure due to condensed gases, oxides (Si, Ge), steps,and cracks.

Thus, our present thinking is that such grossly disordered or glassy
structures may provide the better matching required for the anomalous
transport.[10] The longer wavelength phonons then would require
thicker disordered layers. That the observed anharmonicity plays the
role of breaking the bottleneck has been found for one type of interface[16] at low frequencies, but it is still an open question in
general.

1. P. L. Kapitza, Zh. Eksp. Teor. Fiz. 11:1 (1941).
2. I. M. Khalatnikov, Zh. Eksp. Teor. Fiz. 22:687 (1952).
3. L. J. Challis, K. Dransfeld, and J. Wilks, Proc. Roy. Soc. (London)
      A 260:31 (1961); I. N. Adamenko and I. M. Fuks, Zh. Eksp.
      Teor. Fiz. 59:2071 (1970); R. E. Peterson and A. C. Anderson,
      Phys. Lett. 40A:317 (1972); H. Haug and K. Weiss, Phys. Lett.
      40A:19 (1972); J. L. Opsal and G. L. Pollack, Phys. Rev. A
      9:2227 (1974).
4. C.-J. Guo and H. J. Maris, Phys. Rev. Lett. 29:855 (1972).
5. H. Kinder and W. Dietsche, Phys. Rev. Letters 33:578 (1974);
      W. Dietsche and H. Kinder, J. Low Temp. Phys. 23:27 (1976);
      J. Weber, W. Dietsche, and H. Kinder, Phys. Lett. 64A:202(1977).
6. A. R. Long, R. A. Sherlock, and A. F. G. Wyatt, J. Low Temp. Phys.
      15:523 (1974).
7. R. A. Sherlock, N. G. Mills, and A. F. G. Wyatt, J. Phys. C
      8:300 (1975); A. F. G. Wyatt, G. J. Page, and R. A. Sherlock,
      Phys. Rev. Lett. 36:1184 (1976).
8. T. J. B. Swanenburg and J. Wolter, Phys. Rev. Lett. 31:693 (1973).
9. A. R. Long, J. Low Temp. Phys. 17:7 (1974); T. Nakayama, J. Phys.
      C 10:3273 (1977); W. M. Saslow and M. E. Lumpkin, Solid State
      Comm. 29:395 (1979).
10. H. J. Maris, Phys. Rev. B 19:1443 (1979).
11. A. C. Anderson in: "Phonon Scattering in Solids", L. J. Challis,
      V. W. Rampton, and A. F. G. Wyatt eds., Plenum, New York
      (1976), p. 1.
12. J. Weber, W. Sandmann, W. Dietsche, and H. Kinder, Phys. Rev.
      Lett. 40:1469 (1978).
13. for a review of the method, see H. Kinder in: "Proc. of the 14th
      Int. Conf. on Low Temp. Phys. LT14", M. Krusius and M. Vuorio
      eds., North Holland/Elsevier, Amsterdam (1975) Vol.5, p. 287.
14. J. Weber, W. Dietsche, and H. Kinder, Verhandl. DPG(VI) 14:409
      (1979), and to be publ.
15. A. C. Anderson and W. L. Johnson, J. Low Temp. Phys. 7:1 (1972);
      E. S. Sabisky and C. H. Anderson, Solid State Comm. 17:1095
      (1975).
16. H. Schubert, P. Leiderer, and H. Kinder, to be publ.
17. S. Hunklinger and W. Arnold in "Physical Acoustics"
      R. N. Thurston and W. P. Mason eds., Academic, New York
      (1976), Vol. 12, p. 155.
18. A. F. G. Wyatt and G. N. Crisp, Journal de Physique 39:
      Colloque C-6, 244 (1978).
19. W. Dietsche and H. Kinder, this issue, and to be publ.

CHARACTERISTICS OF KAPITZA CONDUCTANCE

A.F.G. Wyatt

Department of Physics
University of Exeter
Stocker Road, Exeter EX4 4QL U.K.

The Kapitza conductance problem is still with us. It almost went away last year in some people's view when Weber et al[1] reported that the reflectivity of LiF and NaF crystals hardly changed when they were coated with ^{4}He if the crystal was cleaved at 1K. However, we can now report[2] that the two channels for conductance that we have described previously still apply to NaF surfaces cleaved under ^{4}He liquid. The second, non classical channel is therefore a property to be explained for a real clean surface with no factors other than the solid, the helium and possible physical defects in the solid.

In this limited review I shall concentrate on the properties of interfaces between well defined solid surfaces and liquid ^{4}He. Metal surfaces have been recently improved[3,4] but many more detailed experiments can be done on insulators. Therefore the extensive thermal resistance work on metals will not be considered here as these data do not yet give many clues to the microscopic processes which increase the conductance. The reader is also referred to the reviews[5,6,7] and the references contained in them for details of the boundary resistance between solids and liquid ^{3}He and the solid Heliums. Points worth noting here are that the conductance is similarly anomalous to solid ^{4}He as it is to liquid ^{4}He [8] and that the situation with ^{3}He is still confused by the lack of clear evidence for or against a magnetic coupling to the ^{3}He nuclei.

The experiments that have been most helpful in recent years are reflection and transmission measurements with solids which are transparent to phonons. These, coupled with the frequency selectivity of superconducting tunnel junction generators and detectors and liquid ^{4}He at dilution refrigerator temperatures, have enabled differentia-

tion between phonons of different polarisation, different propagation
directions and different frequencies. Such experiments have identified both the classical and the non-classical transmission channels,
they have shown that both L and T polarisations couple to ^{4}He phonons
and that it is the first two to three atomic layers of ^{4}He which are
important. We have also been able to show the correspondence of these
detailed measurements with the interfacial thermal conductance and
hence that the non-classical channel can plausibly account for the
anomalously high boundary conductance and its temperature dependence.

The Classical Channel

The classical model of transmission of phonons through an interface depends on the displacement of the interface by incident phonons
from the body of the material on one side of the interface[9,10]. The
usual boundary conditions enable the transmission coefficients and
the propagation directions to be determined. An estimate of the size
of the transmission coefficient, α is given by the normal incidence
value $\alpha_o = 4Z_1Z_2/(Z_1+Z_2)^2$ where $Z_i = \rho_i c_i$; ρ is density and c is
phase velocity of the phonon. For NaF and ^{4}He at SVP $\alpha_o \approx 0.01$ for
longitudinal phonons. This transmission coefficient is the same for
both directions through the interface. However, for angles, other
than normal incidence, the value of α does not change much in the
solid out to 90° to the normal whereas in the ^{4}He liquid it drops
to zero at a small angle of incidence $\sim 5^o$ to the normal. The cone
defined by this angle is the so called critical cone and phonons
incident from the ^{4}He liquid outside of this cone are in the
classical limit totally reflected.

This classical treatment of the transmission also gives the
angles of transmission from the condition that the parallel component of the wave vector (q_{11}) is conserved. This is simply Snell's
law which can be written $c_2 \sin\theta_1 = c_1 \sin\theta_2$. As a direct result of
this it can be seen that as the velocity of phonons in the solid is
always higher than in the liquid ^{4}He, phonons incident from the solid
have a non zero probability of going into the ^{4}He (with the obvious
exception of waves with displacement vectors parallel to the interface) and they all are transmitted into the critical cone in the ^{4}He.
Conversely, only phonons incident within the critical cone from the
^{4}He side are transmitted into the solid. These two results are found
experimentally and are shown in figure 1 traces a and b. In these
figures we see the well defined peak centred at normal incidence
superimposed on a broad background.

Detailed confirmation that the central peak corresponds to the
classical channel comes from the close agreement between the
measured and calculated values of the critical cone angle for
solids with quite different velocities: MgO[11], NaF[12,13,17] and
KCl[11]. The measured angles are necessarily broadened by the finite
size of crystal sample and detector but when these are taken into

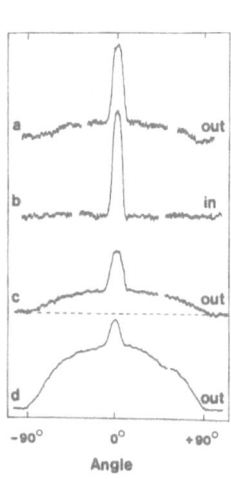

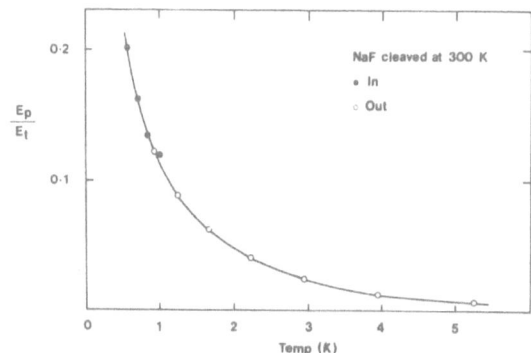

Fig. 2. Ratio of energy in classical peak to the total energy as function of heater temperature for NaF cleaved at 300K. ref:13.

Fig. 1. Angular phonon distributions between NaF and liquid ^{4}He, a and b: emission and acceptance at low heater power, c: emission from NaF cleaved under liquid ^{4}He, d: as c but after room temperature exposure.

account the agreement is to within the experimental accuracy ±1° except for a slight rounding at the foot of the peak.

The clear identification of this channel is important as it implies that at least some area of the interface is ideal in the sense that it has translational symmetry. It also shows that the anomalous conductance is not due to a simple broadening of this critical cone, although it must be said that it is not impossible that other areas of the surface might have a very broadened cone. It is very clear however that there is a negligible amount of 'intermediate' broadening. It is also difficult to see how the cone broadening proposals[14,15] resulting from short lifetime phonons can account for broadening out to 90° as the mean free paths would have to be unphysically short.

The q_{11} conserving property of the radiation into the central peak is independent of phonon frequency, which is consistent with the classical model. However, it appears that the fraction of incident energy transmitted by this channel does diminish as the frequency is increased in the range 10–100 GHz. This is due to competition between the two channels[13]. At lower frequencies the relative importance of the classical channel increases and although we cannot reach a low enough frequency for this channel to be seen to be dominant it is clear that this is the trend, figure 2. From ultrasonic work at normal incidence at a frequency of 1 GHz[16] the magnitude of the transmission probability has been shown to agree with the classical model and so we make the assumption that the

magnitude of the peak in the angular distribution at low frequencies
is given by the classical transmission probabilities. This is
necessary for calibrating the total heat flux transmitted.

To summarise the position on the classical channel, we can say
it clearly exists and q_{11} is well conserved in all cases studied.
However the energy conductivity of this channel is between 1 and 2
orders of magnitude too small to account for the measured conductance at high frequencies. Until the discovery of the background
channel[17] this was a mystery.

The non-Classical Channel

It now seems obvious, with the clarity of vision that hindsight affords, that the high interfacial conductivity between a
solid and liquid must come from the enormous amount of phase space
which lies outside of the critical cone on the ^{4}He side. The
angular distribution of phonons emitted from a plane interface
shows that there is a considerable energy flux outside of the
critical cone angle and in fact, there is emission out to 90°
to the normal[13,17].

This result has now been found to be qualitatively unchanged
when all classical contamination of the surface is avoided by
cleaving the crystal face under liquid ^{4}He at 1K (Fig. 1, trace c).
This is even more foolproof than cleaving in vacuum at 1K and
gives transmission results which may be directly compared to the
reflection results of Weber et al[1].

The phonons injected into the NaF crystal and which are then
incident on the interface with the ^{4}He can be varied in frequency
by changing the temperature of the heater on the back of the NaF
crystal. The energy transmitted in the background channel (E_b)
relative to the peak (q_{11}) channel (E_p) changes rapidly with
frequency as can be seen in figure 2 which is for a 300K cleaved
NaF crystal[13]. The transmission probability for the background
channel increases with phonon frequency in the range 10-100 GHz
and this channel carries $10-10^2$ times more energy than the
classical channel.

From a simple model which takes into account the competition
between the two channels the effective mean phonon transmission
coefficient $\bar{\alpha}$ is readily calculated[13]. The values of $\bar{\alpha}$ for NaF
cleaved in air[13] are shown in figure 3 plotted against heater
temperature, which because of the classical transmission between
solids is a good measure of the phonon frequencies in NaF. This
increase in $\bar{\alpha}$ with phonon temperature has been seen many times by
direct thermal conductance as Challis[6] has emphasised. This correspondence between the two measurements in both absolute value and
frequency dependence reinforces the notion that the background

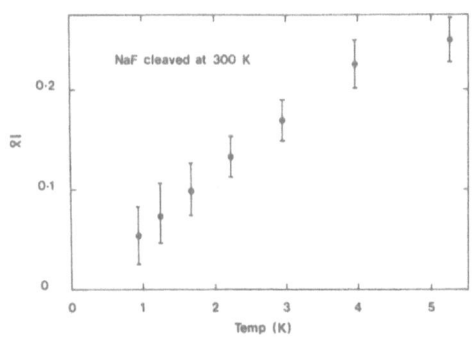

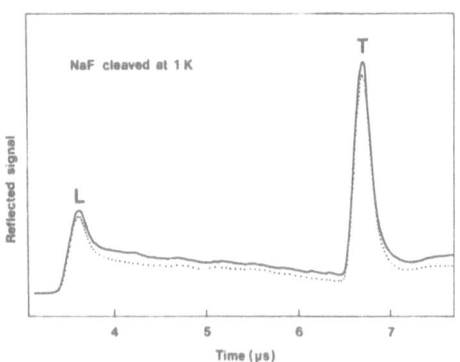

Fig. 3. Values of the average
transmission coefficient as a
function of phonon temperature
calculated from data in fig. 2;
ref:13.

Fig. 4. Reflected signals from
an interface between NaF cleaved
at 1K and vacuum (solid line)
and bulk ^{4}He (dotted line)
Weber et al[1]. Note the large
change in delayed signal after
L peak.

channel is the source of the anomalous conductance.

Emission frequencies

The classical model implicitly assumes that ω is conserved in
the transmission process. As the transmission probability in this
model is independent of ω then we assume that the spectrum of
frequencies in the q_{11} peak is identical with the spectrum incident
on the interface in the solid. The background channel is different:
as there is no q_{11} conservation there is no a priori reason for ω
conservation.

A knowledge of whether ω is conserved or not is important for
the modelling of microscopic processes to account for the background. It is possible to answer this question with frequency
selective superconducting tunnel junction detectors. The angular
distribution has been measured with both a bolometer, which is
sensitive to all frequencies, and with an Al tunnel junction which
only detects phonons with $\hbar\omega/k$ > 4K. If the spectra in the peak
and background are the same then the two detectors will measure the
same angular distribution. The measured results[18] show that the
background is smaller by a factor of ~0.3 when measured by the
tunnel junction. This clearly indicates that the background
contains lower frequencies than the peak and so we should search
for inelastic processes which scatter into the background.

There is also a considerable amount of indirect supporting
evidence for low frequencies in the background. The clearest comes
from the analysis of spectra from thin metal film phonon generators.

A metal film is usually rough compared to a cleaved surface and so
there is no spatial separation of the peak and background. This
means that the peak phonons are emitted at all angles in the ^{4}He
and so for all directions the spectrum is dominated by the background in the ratio E_b/E_p. The spectrum from a heater has been
measured directly by the dispersion in the liquid ^{4}He at 24 bar[19].
It shows the effective temperature of the phonons in the ^{4}He is
considerably lower than the temperature of heater. For example for
a heater temperature of 3K the phonon temperature is 1.4K.

This result enables the transmission from liquid He to a solid
(called the In direction) to be explained in a satisfactory manner
which is consistent with the transmission in the opposite direction
(Out). The In results show[13] two channels but the background channel
is much smaller than in the Out direction with the same heater power.
However we have seen that in the Out case the phonons incident on
the interface correspond to the heater temperature while the
phonons injected into the liquid ^{4}He are much lower. These low
frequency phonons are incident on the interface for the In
measurements. If the measured phonon temperatures[19] are used for
the In case then the values of E_p/E_t are consistent in the two
directions. This is a necessary result for thermodynamic equilibrium
between the solid and the Helium at the same temperature.

L and T polarisations

The phonons in the solid can be L and T polarised while only L
modes are propagated in the He. Salemink et al[20] has shown that
both L and T modes in polished sapphire give rise to phonons in
^{4}He liquid. The important question here is whether phonons with
displacement vectors parallel to the interface can couple to the
liquid He. Unfortunately Salemink's result does not answer this
unambiguously because the mechanically polished crystal surface
would not be sufficiently flat for the displacement to be parallel
over the whole of the area. It would be most interesting if this
measurement were done on a cleaved surface as Weber et al[1] have
shown that the reflected FT mode in LiF, which is polarised
parallel to the surface, does not change with ^{4}He coverage.

Reflection of phonons

A number of groups have made reflectivity measurement on the
solid side as a function of ^{4}He coverage[21-27]. To derive information
about transmission from these studies requires the assumption that
what is not reflected must be transmitted. The technique is that a
short phonon pulse generated on one face of a crystal is reflected
from the opposite side and detected on the same face as the generator. The detected signal shows peaks due to the specular reflection
of the different polarisations and intermediate peaks due to mode
changes at the reflecting surface. There is usually a substantial

signal arriving after each peak which is considered to be due to
phonons of the same mode as the peak preceeding it but which are
delayed. Delays could be caused by interface states and non-specular
reflections. The reflected signals are measured with a vacuum and
then He on the reflecting surface. The results on both room temperature cleaved and polished surfaces show that all polarisations
are reduced by ^{4}He. However the delayed signals generally show a
greater reduction than the specular signals. The change in reflectivity can be measured as a function of He film thickness. This
shows the dramatic result that the reflectivity finishes changing
when only 2-3 atomic layers of ^{4}He are present; especially
surprising as the first 1-2 layers are quasi solid.

Weber et al[26] have shown that energy from the transmitted pulse
can be stored in ^{4}He layers ~1.6 atoms thick. This energy is somewhat thermalised and so is reemitted at a slightly later time of
~250ns and at a lower frequency than the incident pulse. In a
helium film so thin it is not clear in what form the energy is
stored; it would seem to be too thin for phonons but it could be
stored by localised or propagating states. This phenomenon does
suggest that the first few atomic layers of Helium do act as an
intermediate state between phonons in the solid and background
phonons in the ^{4}He. It is interesting to note that these time delays
are absent when no He is present, even on chemically dirty surfaces.
So if these ^{4}He surface states, which cause the delay, are also
causing the emission into the background, this is consistent with
the fact that the background is an intrinsic effect on ^{4}He on
clean surfaces.

The reflection measurements on crystals cleaved at 1K, by Weber
et al[1], show a smaller change with ^{4}He coverage than the same
crystals exposed to air. The effect of ^{4}He on LiF was less than on
NaF but, except for the FT mode in LiF which showed no change,
the changes are still much larger than the classical mismatch model.
The changes seen in NaF with ^{4}He shown in figure 4 is 3-4% for the
specular L peak, ~20% for the delayed L signal and only 3% for the
specular T peak if the shift in base line is taken into account.
The large change in the delayed signal might be reasonably associated with transmission in the background seen on NaF crystals
cleaved under ^{4}He. However it should be noted that the phonons used
by Weber et al[1] are not typical thermal phonons as $2\Delta_{Sn}$ is almost
coincident with the maxon peak of the ^{4}He dispersion curve. Only
frequency down-converted phonons would therefore be able to
propagate away from the surface into the ^{4}He.

Some of the crystals which were cleaved under ^{4}He and
measured, were warmed to room temperature and then cooled again
and remeasured. This treatment causes an absolute increase in the
background channel and decrease in the classical channel. This
can be seen in figure 1 curve d which may be directly compared with

curve c as it is for the same crystal, heater and bolometer but
measured after the room temperature cycle. The value of $\bar{\alpha}$ at a
temperature of 3.4K is increased from 0.14 to 0.41. A similarly
large change, in the effect of ^{4}He coverage on phonon reflectivity,
is seen with room temperature exposure of the surface[1]. The major
change is again for the time delayed phonons which reinforces the
association of the time delayed signal with the background channel.

Conclusion

From the experiments described above it seems clear that the
anomalously high Kapitza conductance is a feature of clean
surfaces. This is consistent with the older result of Johnson[28].
The transmission and reflection measurements seem to be in accord
and both indicate that the classical channel is completely
inadequate to account for the results. The angular distribution
measurements show that the high conductance is due to a second
(background) channel which emits a lowered frequency spectrum
into all angles in the ^{4}He. It is very frequency dependent: it
contributes negligibly to the conductance below 1GHz but is
completely dominant at 100 GHz. Theories must now account for
this considerable body of measured detail.

References

1. J.Weber, W.Sandmann, W.Dietsche, H.Kinder, Phys. Rev. Lett.
   40:1469 (1978).
2. A.F.G.Wyatt and S.T.Lehni, to be published.
3. N.S.Snyder, J.Low Temp. Phys. 22:257 (1976).
4. A.W.Pattullo and J.C.A.van der Sluijs, preprint (1979) and
   this volume.
5. N.S.Snyder, Cryogenics 10:89 (1970).
6. L.J. Challis, "The Helium Liquids" ed. J.M.G.Armitage and I.E.
   Farquer (N.Y. Academic Press) ch. 10 (1975).
7. A.C. Anderson, "Phonon Scattering in Solids" ed. L.J.Challis,
   V.W. Rampton and A.F.G. Wyatt (N.Y. and London:Plenum)1 (1975).
8. J.T.Folinsbee and A.C.Anderson, Phys. Rev. Lett. 31:1580 (1973).
9. W.A. Little, Can. J. Phys. 37:334 (1959).
10. I.M.Khalatnikov, "Introduction to the theory of superfluidity)
    (N.Y. Benjamin) (1965).
11. G.J.Page, R.A.Sherlock, A.F.G.Wyatt, K.R.A.Ziebec, "Phonon
    scattering in Solids", ed. L.J.Challis, V.W.Rampton and
    A.F.G.Wyatt (N.Y. and London:Plenun) 18 (1975).
12. R.A.Sherlock, A.F.G.Wyatt, N.F.Mills and N.A.Lockerbie, Phys.
    Rev. Lett. 29:1299 (1972).
13. A.F.G.Wyatt and G.J.Page, J.Phys. C. Solid State Physics 11:
    4927 (1978).
14. H.Haug and K.Weiss, Phys. Lett. 40A:19 (1972).
15. R.E.Peterson and A.C.Anderson,Phys. Lett. 40A:317 (1972).
16. B.E.Keen, P.W.Matthews and J.Wilks, Proc. Roy. Soc. A284:125.

(1965).

17. R.A.Sherlock, N.G.Mills and A.F.G.Wyatt, J.Phys. C. Solid State
    Phys. 8:300 (1975).

18. A.F.G.Wyatt and G.N.Crisp, J.Physique Colloque C6 39 suppl.
    8:244 (1978).

19. R.A.Sherlock, A.F.G.Wyatt and N.A. Lockerbie, J.Phys. C. Solid
    State Phys. 10:2567 (1977).

20. H.W.M.Salemink, H.Van Kempen and P.Wyder, Phys. Rev. Lett.
    41:1733 (1978).

21. H.Kinder and W.Dietsche, Phys. Rev. Lett. 33:578 (1974).

22. C.J. Guo and H.J.Maris, Phys. Rev. Lett. 29:855 (1972)
    Phys. Rev. A10:960 (1974).

23. A.R.Long, R.A.Sherlock and A.F.G.Wyatt, J. Low Temp. Phys.
    15:523 (1974).

24. J.S.Buechner and H.J.Maris, Phys. Rev. Lett. 34:316 (1975).

25. W. Dietsche and H.Kinder, J. Low Temp. Phys. 23:27 (1976).

26. J.Weber, W.Dietsche and H.Kinder, preprint.

27. J.T.Folinsbee and J.P.Harrison, J. Low Temp. Phys. 32:469 (1978).

28. R.C. Johnson, Bull. Am. Phys. Soc. 9:713 (1964).

****

DISCUSSION

N. Shiren: You mentioned experiments that had been done in ^{3}He as
well as in ^{4}He.

A. F. G. Wyatt: These are reflection experiments. Two groups have
now done reflection experiments in ^{3}He and ^{4}He. There's the Kinder
group and Harrison's group.

H. Kinder: May I just briefly point out why we never published this
result. We just put it in the rerprint. Our helium wasn't pure
enough, and we did't realize this right away. When we realized it,
we decided not to publish this result because we might have had a
^{4}He film at the solid surface and the pressure that we measured
would be that of the ^{3}He.

H. Kinder: Have you ever looked at these surfaces where you showed
the last experiment with an electron microscope? Could you see
whether they are rough under the electron microscope?

A. F. G. Wyatt: There is no point in looking at the surfaces
cleaved under helium in the electron microscope because you may as
well look at one that you have cleaved at room temperature, and in
fact you can get them into the electron microscope a lot faster if
you cleave them and put them in. Now we've done the same thing that
you have done and we find it's very hard to detect any damage there
with the electron microscope - this is with a resolution of something like 100 Å - a scanning electron microscope.

H. Kinder: With the very same surfaces where you did these experiments?

A. F. G. Wyatt: They were with the same sodium fluoride sample. They obviously weren't with the same surface because that surface was necessarily exposed to air a lot longer. So what we would do was take a sample of sodium fluoride, cleave it, put it in the electron microscope and look at it. The point is that we found it very difficult to find damaged areas. You can find damaged areas, but on the scale of 100 $\overset{\circ}{A}$ there are large areas of surface which apparently are defect-free, which I think is the same as you found.

H. Kinder: Sure, but you have no direct knowledge of the low temperature cleaved surfaces. That is what is essential when you say we have a discrepancy here.

A. F. G. Wyatt: I don't think we have a discrepancy because before I knew of your previous results then I think I agreed entirely with your results of last year, and there was no difference. I would say that you have showed that the anomalous reflection has disappeared. I am saying that the anomalous transmission may still be there.

C. Elbaum: I would like to make a comment about the matter of the cleaving. There is something to be said about potential differences in the semi-microscopic flatness of the surface when cleaved at 1°K versus 300°K. I wouldn't take it as a foregone conclusion that they need be the same.

R. C. Dynes: In view of the fact we know that in some focusing directions in solids there is enormous difference in flux, have you looked at other directions in your crystals?

A. F. G. Wyatt: No. You'll appreciate that one needs an atomically flat surface for these experiments and therefore you've got to use cleavage planes.

R. C. Dynes: So all this is done for the same set of planes.

A. F. G. Wyatt: 100 - planes.

KAPITZA CONDUCTION VIA NON-LOCALIZED INTERFACE MODES

F.W. Sheard and J.F. Fletcher

Department of Physics, University of Nottingham
University Park, Nottingham NG 2RD, UK

The origin of the anomalous phonon Kapitza conductance has been shown experimentally[1] to lie in the emission of energy into the helium at angles outside the acoustic critical cone. The power emitted per unit solid angle followed a $\cos\theta$ law which cannot be explained by the acoustic-mismatch theory modified to allow for phonon damping in the solid[2], or by theories based on localized tunnelling states in the helium[3]. Recently Guyer has suggested that non-localized states such as vacancy waves[4] may be a general feature of the partially disordered helium layers adjacent to a solid surface. He also proposed that vacancy waves may participate as intermediate states in phonon transmission processes calculated in second order perturbation theory, but this again does not give the observed angular distribution.

Here we propose a model for the Kapitza heat flow which involves direct first-order absorption and emission of phonons by non-localized interface modes, which are thereby thermally excited. We do not specify these modes more closely except to remark that they arise from the quantum tunnelling of defects in a similar way to the formation of excitons from localized electronic excitations. On a real surface there are inevitably imperfections or impurity atoms which introduce a randomness in the defect sites and energies. In view of this the interaction of a bulk phonon with an interface mode will contain coherent and incoherent components (as in neutron scattering). We therefore model the composite solid-interface-liquid system by means of the Hamiltonian

$$H = H_S + H_L + H_I + H_{SI} + H_{LI}$$

where $H_S = \Sigma_q \hbar\omega_q a_q^+ a_q$, $H_L = \Sigma_k \hbar\omega_k b_k^+ b_k$ refer to the phonons in the solid and liquid respectively and

$$H_I = \hbar\omega_0 \sum_{\underline{p}} c_{\underline{p}}^+ c_{\underline{p}}$$

refers to the interface modes which are annihilated and created by Bose operators $c_{\underline{p}}$, $c_{\underline{p}}^+$. The wave vector $\underline{p}$ is parallel to the interface. By taking ω_0 constant we have neglected the bandwidth of the modes but we assume that due to disorder in the helium layers there will be a broad distribution of band energies $\hbar\omega_0$ as for localized two-level defects in glasses. For the coupling terms we consider only the incoherent part

$$H_{SI} = \sum_{\underline{q},\underline{p}} A_{\underline{q}\underline{p}} a_{\underline{q}} c_{\underline{p}}^+ + \text{h.c.}$$

of the solid-interface coupling but take the coherent part, which conserves the parallel component of wave vector, of the liquid-interface coupling

$$H_{LI} = \sum_{\underline{k}\underline{p}} B_{\underline{k}\underline{p}} \delta_{\underline{k}_{||},\underline{p}} b_{\underline{k}} c_{\underline{p}}^+ + \text{h.c.}$$

We consider the steady flow of heat when the solid is at temperature $T + \Delta T$ and the liquid helium at T. The thermal excitation of the interface modes is determined by a balance equation and corresponds to an excess population

$$\Delta n_{\underline{p}} = n_{\underline{p}} - n^0(T) = \frac{R_S}{R_S + R_L} \left(\frac{\partial n^0}{\partial T}\right) \Delta T,$$

where R_S and R_L are transition rates for direct phonon emission into the solid and liquid respectively. More explicitly

$$R_L = \sum_{\underline{k}} R_{\underline{k}\underline{p}} \delta_{\underline{k}_{||},\underline{p}}, \qquad R_{\underline{k}\underline{p}} = (2\pi/\hbar^2)|B_{\underline{k}\underline{p}}|^2 \delta(\omega_{\underline{k}} - \omega_0)$$

and R_S is similarly defined. The power emitted into solid angle $d\Omega$ in the helium (at frequency $\omega_k = \omega_o$) may thus be written

$$P(\theta)d\Omega = (2\pi)^{-3}\left(\int \hbar\omega_k R_{kp}\Delta n_p k^2 dk\right)d\Omega$$

where $p = k_{||}$ is implied. If we now assume that the interaction of the interface modes with the bulk liquid phonons is much greater than with the phonons in the solid, ie $R_L \gg R_S$, then $\Delta n_p = (R_S/R_L)$ $(\partial n^o/\partial T)\Delta T$ and the LI matrix element cancels from the expression for $P(\theta)$. This is because the temperature drop ΔT then occurs mainly between the solid and the interface modes so that the magnitude of the heat flow is limited only by the SI coupling. Indeed evaluation gives $R_L = 2|B_{kp}|^2/\hbar^2|d\omega_k/dk_z|$, where k_z is the component of $\underline{k}$ normal to the interface, and hence

$$P(\theta)d\Omega \propto \hbar\omega_k k|k_z|R_S(\partial n^o/\partial T)\Delta T d\Omega.$$

Since the rate R_S involves only an angular average over phonon wave vectors in the solid we see that the emitted power $P(\theta) \propto |k_z| \propto \cos\theta$ as found experimentally.

An important feature of the calculation is that $P(\theta)$ is independent of the detailed structure of the matrix elements and of the distribution of the energies of the interface modes, although these determine the magnitude and temperature dependence of the Kapitza conductance. Also, in agreement with experiment, the model allows for the appearance of a time delay $\tau_S = 1/R_S$ in the reflection of a phonon pulse[5] inside the solid, for the situation when the surface is covered by a few helium monolayers so that relaxation of the interface modes to the bulk liquid cannot occur. However no provision has been made in the model for frequency non-conserving processes which have been detected experimentally[6].

REFERENCES

1. R A Sherlock, N G Mills and A F G Wyatt, J Phys C 8, 300 (1975).
2. M Vuorio, J Low Temp Phys 10, 781 (1973).
3. T Nakayama, J Phys C 10, 3273 (1977).
4. R A Guyer, Proc ULT Hakone Symposium, p 178 (1977).
5. J Weber, W Dietsche and H Kinder, Phys Lett 64A, 202 (1977).
6. W Dietsche and H Kinder, J Low Temp Phys 23, 27 (1976).
****

DISCUSSION

W. M. Saslow: If you permit an inelastic process going from solid to the interface that would solve the problem that experimentally the energy and frequency that comes out on the helium side is lower. It would also solve the problem of why momentum is not conserved parallel to the interface in the excitation process.

F. W. Sheard: Yes, you can destroy momentum conservation through inelastic processes, but it also messes up some of the other details of the theory. It's not clear at this stage how you would come out with a $\cos\theta$ angular distribution.

AGREEMENT BETWEEN EXPERIMENT AND THEORY IN THE KAPITZA CONDUCTANCE

BETWEEN CLEAN COPPER SAMPLES AND ^{4}HE FROM 0.3 to 1.3 K

A.W. Pattullo and J.C.A. van der Sluijs

School of Physical and Molecular Sciences,
University College of North Wales,
Bangor, Gwynedd, LL57 2UW, U.K.

The Kapitza conductance has been measured between annealed and electropolished clean copper samples and ^{4}He in the temperature range between 0.3 and 1 K. The results are in quantitative agreement with older results on similar samples between 1.2 and 2K[1] and with damped acoustic mismatch theory[2].

INTRODUCTION

It is of interest to study the Kapitza conductance of clean metal samples between 0.3 and 1 K, where literature results for copper[3,4] show a sharp rise of the transfer coefficient from 0.5 to 20% with rising temperature. The heat transfer mechanism is acoustic mismatch at low temperatures[5]. There is strong evidence[1] that most high temperature data in the literature are determined by impurity based excess heat transfer. For sufficiently clean samples coupling between bulk excitations in copper and helium via evanescent waves scattered on dislocations in the copper near to the interface should become important[2,6,7]. This damped acoustic mismatch model together with the dense helium layer at the interface[8] predicts a weak maximum in the heat transfer coefficient near 1 K, the position of which yields information on the nature of the helium layer.

EXPERIMENT

A steady heat current Q is sent through the metal to helium interface giving rise to a temperature difference ΔT across the interface. The Kapitza conductance is defined as $h_k = Q/\Delta T$. Because of the T^3 dependence of the Debye energy density we study

195

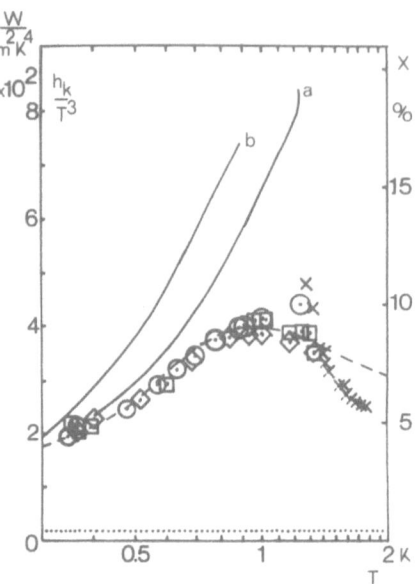

Fig. 1. Kapitza conductance h_k/T^3 and heat transfer coefficient
α as a function of the temperature T. Our data; x : data ref.1;
a : data ref.3; b : data ref.4; ----damped acoustic mismatch theory
for v = 0.27 and F = 1.3[2].

h_k/T^3 which is proportional to the heat transfer coefficient α
defined as the ratio of the transmitted over the incident heat flux
at the interface. The experimental cell is similar to that described
before[1] and is mounted on the evaporator of a ^{3}He refrigerator. A
large bore pumping line is provided into the sampe cell for UHV
baking without undue heat leak along the ^{4}He film. Full details
will be given elsewhere[9]. The samples consist of high purity
annealed copper which is electropolished to 150 μm[1].

     Some results are shown in fig.1 together with other experimental
and theoretical data[1-4]. Good agreement exists with the earlier
results by Rawlings and van der Sluijs[1] over the common temperature
range. Our results are significantly lower than those by Anderson
et al[3] and by Denner[4]. This is in line with the behaviour of
similarly prepared samples in the temperature range 1.2 to 2 K[1]
which all gave results below the literature data, apart from the very
clean data by Snyder[10] which reproduced in the common temperature
range. There is a distinctive maximum in our measured curve just
below 1K which coincides with the maximum in the theoretical curve
for a choice of the layer parameter F = 1.3 and the damping parameter v = 0.27. This choice gives also agreement within the experimental accuracy over the whole range.

CONCLUSIONS

The present results show a behaviour of the Kapitza conductance on clean metal samples which has not been seen before, suggesting that this is the underlying trend which becomes visible when the impurity based excess heat transfer is sufficiently reduced.

This conclusion is supported by the quantitative agreement between the present results and the theoretical calculations by Opsal and Pollack[2] based on the presence of a dense helium layer[8] and the scattering of evanescent waves in the copper on dislocations near to the interface[6,7]. The discrepancy between the results of ref.1 and the theory requires further investigation.

An interesting point is raised by the value of the F parameter. For solid layers of helium to be present in the dense helium layer F must exceed 1.3. It is therefore tempting to suggest that the presence of the maximum in our curve supports evidence obtained elsewhere[11,12] that no solid layer is formed at the interface but a high density disordered liquid-like layer, which seems to be present even when the bulk helium is solid.

We authors acknowledge with thanks support from the Science Research Council.

REFERENCES

1. K.C. Rawlings and J.C.A. van der Sluijs, J. Low Temp. Phys. 34, 215 (1979).
2. J.L. Opsal and G.L. Pollack, Phys. Rev. 9A, 2227 (1974).
3. A.C. Anderson, J.I. Connally and J.C. Wheatley, Phys. Rev. 135A, 910 (1964).
4. H. Denner, Proc. 6th Int. Conf. Cryog. Eng. Grenoble 1976 p.348.
5. I.M. Khalatnikov, Zh.E.T.F. 22, 687 (1951).
6. H. Haug and K. Weiss, Phys. Lett. 40A, 1 (1972).
7. R.E. Peterson and A.C. Anderson, Phys. Lett. 40A, 317 (1972).
8. L.J. Challis, K. Dransfeld and J. Wilks, Proc. Roy. Soc. A 260, 31 (1961).
9. A.W. Pattullo and J.C.A. van der Sluijs, to be published.
10. N.S. Snyder, J. Low Temp. Phys. 23, 257 (1979).
11. D.F. Brewer, A.J. Dahm, J. Hutchins, W.S. Prescott and D.N. Williams, J. Physique 39, C6-351 (1978).
12. S. Balibar, C. Laroche and D.O. Edwards, Phys. Rev. Lett. 42, 782 (1979).

****

DISCUSSION

H. Kinder: Have you ever used a better vacuum than 10^{-7} at the top
of the cryostat?

J. C. A. van der Sluijs: No. The equipment we have got is not able
to go lower. A lot of the residual gas will not condense on the
sample itself because in the experimental arrangement the sample remains relatively hot during the cooling procedure and so much of the
residual gas will condense on the wall of the tube.

SURFACE SUPERCONDUCTIVITY AND KAPITZA RESISTANCE

A. Ridner,   F.de la Cruz and E.N. Martínez

Centro Atómico Bariloche - Comisión Nacional de Energía
Atómica e Instituto Balseiro - Univ.Nac. de Cuyo
8400 - S.C. de Bariloche, Argentina

It has been shown[1] that surface superconductivity can be used
as a localized thermomether to study the temperature close to an
interface through which heat is being transported. Knowledge of the
temperature profile in the interface region provides evidence which
any theory of the heat transport mechanism must take into account.
For interfaces between pure lead and HeII we have previously found[1]
clear differences between extrapolated surface temperatures T_k and
T_m, measured within some 1000Å of the interface using a superconducting thermometer. (The reader is refered to ref.1 for details
on the experimental method).

Using the same technique, we have studied the interface between
lead and sapphire. We find in this case no measurable difference
between the extrapolated and surface temperatures, and,incidentally,
that the Kapitza resistance is the same for either direction of heat
flow. We find $R_k = 50\ T^{-3} K^4 cm^2/W$, for temperatures between 1.5K and
4K, half the value measured by Wolmeyer et al[2] for the same tempera
ture range. The difference could be due to a better lead-sapphire
contact in our samples.

We have investigated the temperature profile at the Pb-He
interface by means of alloying the lead, thereby changing the
coherence length and hence the size of the superconducting thermom
eter. We have measured three lead-thallium samples (nominal thallium
concentrations 2, 7.7 and 17.5 atomic percent), one lead-indium
(3 atomic percent indium), and one pure lead sample, all subjected
to the same surface treatment. The results for the 2 and 17.5% Pb-Tl
samples are shown in Fig.1, and they are typical of the rest. R_k
is defined in the usual way: the difference between the extrapolated
and helium bath temperatures $(T_k - T_{He})$, divided by the heat flux,
whereas in R_m the "surface temperature" T_m is used instead of T_k.

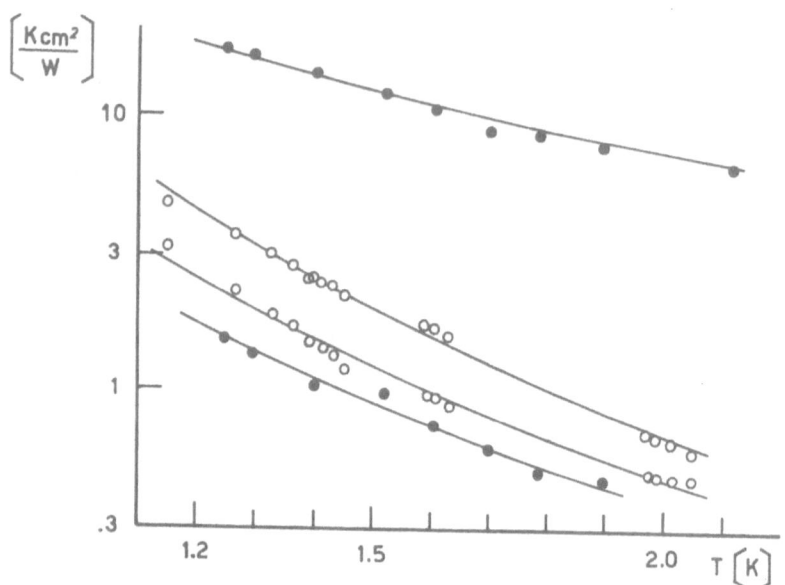

Fig.1. Thermal boundary resistances. Fullcircles: 17,5% Tl alloy;
opencircles: 2% Tl alloy. For each case, the upper points
are R_k, the lower R_m.

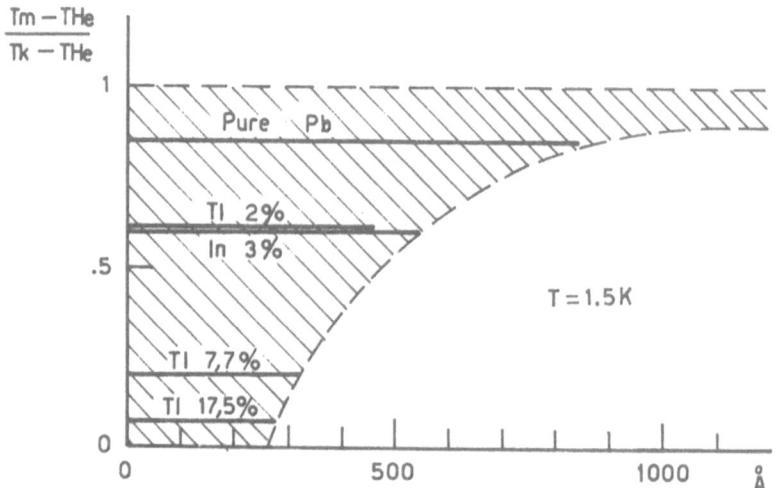

Fig.2. The ratio $(T_m-T_{He})/(T_k-T_{He})$ measured at 1.5K, as a function
of the coherence length in lead alloys with the given impurity
concentrations.

For each sample, T_m is a measure of the electron temperature in the metal within a distance from the interface of the order of the coherence length of the alloy. In Fig.2 we plot the ratio $(T_m-T_{He})/(T_k-T_{He})$ as a function of the distance from the interface. The result for each sample is plotted as a horizontal line from the interface to the sample's coherence length, showing the uncertainty in the position of the measured temperature. Any spatial profile of temperature should fall inside the shaded part of the diagram. It should be kept in mind that the phonon mean free path in lead at 1.5K is about 2 μm, much longer than any relevant lengths in Fig.2. Because of this, and their size, these profiles cannot be explained by Perrin's calculations[3] which predict very weak phonon- (and negligibly small electron-) temperature variations over distances of the order of the phonon mean free path. The electron mean free path, ℓ, would be the next choice for the characteristic lenght of the temperature profile, but it is disqualified by the fact that $(T_m-T_{He})/(T_k-T_{He})$ decreases with impurity concentration, whereas ξ/ℓ increases with decreasing ℓ. The pure lead sample, for example, has an electron mean free path of some 70 μm in the bulk, which is probably decreased at the surface, but the measured values of H_{c3} show that it is still much larger than the coherence length. Fig.2 would suggest a characteristic length not greater than some 500 Å, and different from either electron or phonon mean free paths.

It should be added that the extrapolated temperature T_k stays constant to one part in at least 2000 as the field is increased beyond H_{c2} to above H_{c3}. The existence of the superconducting sheath does not visibly affect either the thermal conductance of the metal or the Kapitza resistence of the interface. Tunneling experiments[4] show that the electron density of states at the metal surface is strongly affected by surface superconductivity, and the superconducting differential conductivity at zero bias, $\sigma_s(0)$, is reduced to half the normal-state value at fields just above H_{c2}. This indicates that in our case the eventual electron contribution to the Kapitza conductance is less than one thousandth of the phonon contribution.

A. Ridner acknowledges the support of a fellowship from Consejo National de Investigaciones Cientificas y Tecnicas, Republica Argentina.

References

1. A. Ridner, E. N. Martínez and F. de la Cruz, Phys. Rev. Lett. 35, 855 (1975).
2. M. W. Wolfmeyer, G. T. Fox and J. R. Dillinger, Phys. Lett., 31A, 401 (1970).
3. N. Perrin, J. L. Temp. Phys., 31, 257 (1978).
4. W. J. Tomasch, Phys. Lett., 9, 104 (1964); Y. Goldstein, Phys. Lett., 12, 169 (1964).

****

DISCUSSION

W. Eisenmenger: Do you have any idea what the phonon mean free path may be in these alloys?

F. de la Cruz: One order of magnitude longer than the coherence length - 1 μ.

W. Eisenmenger: One would expect such a temperature change in a layer of thickness of one phonon mean free path.

F. de la Cruz: That is true of you if you accept that you are going to use a sort of Boltzmann equation to describe the phenomon. I don't want to speculate too much on that. Many people are talking here about interactions near the surface. Now we know that our electrons are not coupled to the phonons, but it might be coupled to something else that is causing this high transfer between the helium and the phonons - some other modes, for example. I want to remind you that for the sapphire - which is a classical solid - we have no profile that we can see.

L. J. Challis: Your're saying that the temperature charges abruptly in a distance of 1000 Å, and that this is very short compared to the electron-phonon mean free path. Do you think there's any possibility that this is the sort of distance that a surface wave might be penetrating into the solid?

F. de la Cruz: You have to realize that there are tremendous gradients in a few hundreds of Å. This means that one atom is about 1 mK apart from another! But it's a problem with any superconducting bolometer, so this is not really too different. It's possible that it is related to the waves on the face of the crystals.

STATIC AND TRANSIENT ENHANCED POWER ANOMALY IN THE KAPITZA

CONDUCTANCE OF SILVER TO HELIUM FOUR

J.C. Bishop and J.C.A. van der Sluijs

School of Physical and Molecular Sciences,
University College of North Wales,
Bangor LL57 2UW, Gwynedd, U.K.

A λ transition is reported in the helium layer near to a
silver to helium interface. The enhanced anomalous power dependence observed before in copper is confirmed.

INTRODUCTION

The anomalous power dependence of the Kapitza conductance was
first reported by Cheeke et al for copper and lead[1]. Rawlings and
van der Sluijs first reported for a clean copper sample an enhanced effect with static and transient anomalies, confirming that
the effects were situated in the thin helium layer near to the
interface[2].

In the present paper we report on the study of the static and
transient enhanced anomalies for a silver to ^{4}He interface. The
basic measuring method is d.c. heat current on clean and annealed
samples. For details of techniques see earlier papers[2,3].

EXPERIMENTAL RESULTS

The anomalous power dependence of the Kapitza conductance
consists of a deviation from the linear relationship between the
heat current density Q through a helium to solid interface and
the temperature difference ΔT across the interface. In Fig. 1 we
show the experimental dependence of ΔT vs Q for two samples with
23 μm and 60 μm electropolishes respectively.

Fig. 1a demonstrates the separate existence of the onset of
bulk supercriticality at Q_{c1} and surface supercriticality at Q_{c2}

203

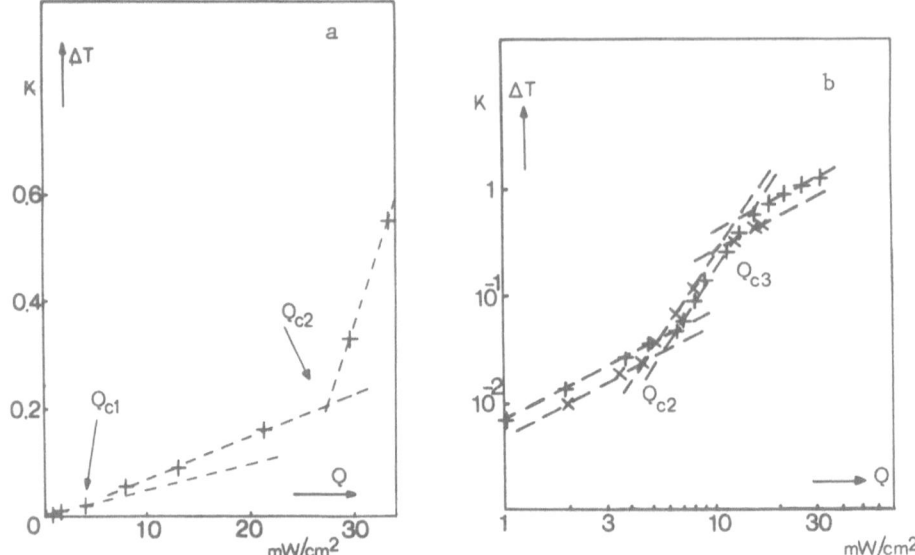

Fig.1   Static enhanced power anomaly. a. 23 μm sample, T = 1.34 K;
b. 60 μm sample, x T = 1.54 K; + T = 1.30 K. ---- sketch lines to
show anomaly; Q_{c1} onset of bulk supercriticality; Q_{c2} surface anomaly.

at 5 mW/cm^2 and 20 mW/cm^2 respectively. In fig 1b (log log scales)
the slopes of the curves show that $\Delta T \sim Q^1$ for $Q < Q_{c2}$ and $\Delta T \sim Q^3$
for $Q > Q_{c2}$. This indicates that for $Q > Q_{c2}$ supercritical superfluid exists distinct from the bulk fluid and associated with anomalously large temperature gradients.

A new anomaly exists when Q is increased to Q_{c3}, where the
silver temperature exceeds T_λ and ΔT becomes $\sim Q^1$. However the
bulk superfluid temperature has hardly been raised and therefore
the transition takes place in a thin layer near to the interface.

The temperature dependence of Q_{c2}, displayed in fig 2a, together
with that for Q_{c1} is similar to that for bulk supercriticality.

The transient behaviour found in copper is largely confirmed[2],
the transient time being less than that for copper: subscritical 4s,
supercritical silver 16s, copper 1000s. The large scale noise in
copper is not found, but for $Q > Q_{c3}$ fast small scale noise is
observed which may be due to bubble-forming in the HeI layer[4]. The
distinct step behaviour found in copper is confirmed (see fig 2b).
The evidence supports the view that the transients are due to vortex
line rearrangement in the supercritical helium layer[2].

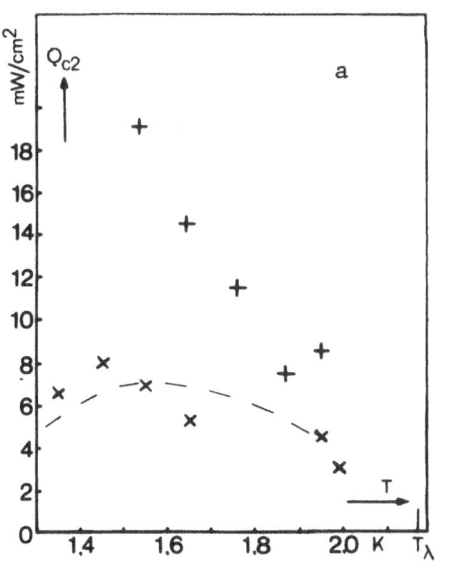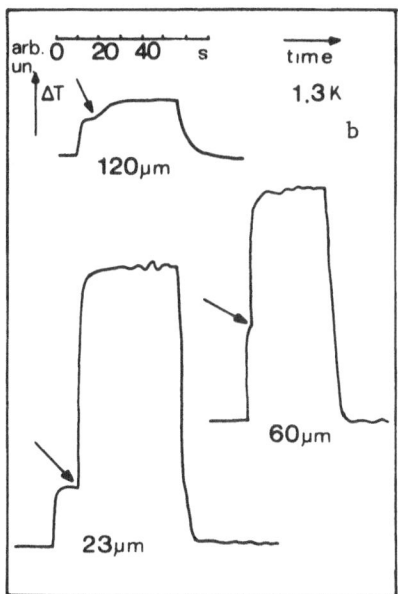

Fig. 2   a. Surface anomaly Q_{c2} vs temperature. + 23 μm sample; x 60 μm sample; ---- onset of bulk supercriticality Q_{c1}. b. Step behaviour in transients for $Q > Q_{c2}$.

CONCLUSIONS

The observation of the distinct anomalies Q_{c1}, Q_{c2}, and Q_{c3} indicates the presence of a supercritical layer of helium near the interface. Q_{c1} is bulk supercriticality because of sample indepen- dence and numerical agreement.   Localised heat transfer (hot spots) in a supercritical surface layer accounts for the large temperature gradients and the ΔT vs Q dependence.   The λ transition at Q_{c3} shows that the layer is at the silver temperature.   The shifts of Q_{c2} with sample treatment show that the phenomena are related to surface tex- ture.   On severely strained samples the effects disappear[5].   The transient behaviour, particularly the difference in transient times between copper and silver, supports these views[2].   The authors acknowledge support from the Science Research Council.

REFERENCES

1.   J.D.N. Cheeke, B. Hebral, J. Richard and R.R. Turkington, Phys. Rev. Lett. 32, 658 (1974).
2.   K.C. Rawlings and J.C.A. van der Sluijs, J. Low Temp. Phys. 33, 313 (1978); 34, 215 (1979).
3.   A.E. Alnaimi and J.C.A. van der Sluijs, Cryogenics 14, 599 (1974).
4.   S.S. Kutateladze and B.P. Arksentyuk, Cryogenics 19, 235 (1979).
5.   J.C. Bishop and J.C.A. van der Sluijs, to be published.
****

DISCUSSION

H. J. Maris: Is there any possibility that there could be a connection between the power-dependence you see and the things that
Kinder was mentioning this morning about saturation effects?

J. C. A. van der Sluijs: I would think so. It seems to me we might
be onto the same thing.

H. Kinder: I don't think that the saturation effects we have seen
at 25 GHz have anything to do with the supercriticality which you
have seen here, and I would like to mention that our powers are much
much smaller than these powers.

A NEW METHOD USING BRILLOUIN SPECTROSCOPY TO STUDY THE PROPAGATION

AND REFLECTION OF INDUCED PHONONS AT HYPERSONIC FREQUENCIES

R. Vacher*, J. Pelous*, W. Arnold and J.D.N. Cheeke

Max-Planck-Institute für Festkorperforschung
Heisenbergstrasse 1, 7000 Stuttgart 50, West Germany
Université de Sherbrooke, J1K 2R1, Québec, Canada

The method we describe here is based on the study of the intensity and the spectrum of light scattered by hypersonic waves in a solid. One end of the sample was exposed in a Ka-band waveguide to an rf field of frequency $\omega/2\pi \sim 35$ GHz generating phonons at the surface via piezoelectric excitation. By adjusting the direction of the incident laser light (wavevector $\vec{k}_o$) and the direction of the scattering (wavevector $\vec{k}_s$), the phonon beam gives rise to an anti-Stokes Brillouin line provided that the conservation of energy and momentum relations are fulfilled. The Stokes Brillouin line arises from the beam reflected at the opposite end of the crystal.

As shown in Fig. 1, the sample, an X-cut quartz with 4 mm diameter and 8 mm length, was enclosed in a cylindrical copper cell with two opposite glass windows. This cell was mounted in the sample chamber of a He cryostat. If the two faces of the crystal were perfectly parallel, the two lines should appear at exactly the same microwave frequency. Also multiple reflections would contribute to the intensity of these lines. In practice, the two faces of our sample were slightly non-parallel. This was sufficient to change the frequency shift of the two lines by a few MHz relative to each other and therefore the contribution of multiple reflections vanished.

The spectrum of the scattered light was analyzed by a triple-pass Fabry-Pérot interferometer with a free spectral range of 80 GHz and a finesse of 60. In general, the intensity I_S of the Stokes line must be smaller than that of the anti-Stokes, I_{AS}, due to (i) the reflection coefficient R of the phonons at the surface and (ii) due to the attenuation of the phonon beam in the crystal. At low temperatures, where the mean free path of the phonons is

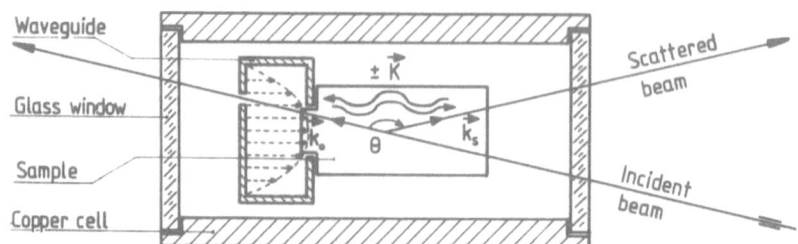

Fig. 1.   Sketch of the sample holder - Scattering geometry

much larger than the sample dimensions and with the sample in
vacuum, I_S should be equal to I_{AS}. In practise, I_S was always
found to be lower than I_{AS}, probably due to a diffraction of the
beam on the imperfect surface, and all our measurements were corrected by this residual loss. It should be noted that the intensity
of the light scattered by induced phonons was a hundred times
higher than that of thermal phonons.

As a test of the method, we measured the mean free path of 35
GHz longitudinal phonons in the X-direction of quartz between 15
and 40 K. In this case, the method is similar to that used by
Krüger[1] at room temperature and at lower frequencies. The cell was
evacuated and the difference between I_S and I_{AS} was then only due
to propagation losses. Two spectra recorded at 32 and 20 K, respectively are shown in Fig. 2a, whereas in Fig. 2b we show the inverse
mean free path deduced from the intensity ratio I_S/I_{AS}. The accuracy of the measurements is mainly limited by the length of the
scattering volume ($\sim$ 0.3 mm). The agreement with previous results
at different frequencies[2,3] is satisfactory, taking into account
the well-known frequency dependence of the attenuation.

However, at helium temperatures, the propagation damping becomes negligible and then the ratio I_S/I_{AS} is proportional to the
reflection coefficient of phonons at the surface. Therefore, the
method should allow us to measure the Kapitza resistance at the
helium-solid interface for monochromatic phonons of given polarization, at frequencies from 10 to 100 GHz. In a first test, we
have measured the reflection coefficient of 35 GHz longitudinal
phonons at the helium - quartz interface, and found R = 0.95 ± 0.05,
in good agreement with previous values.[4]

The frequency resolution of the method is independent of the
resolving power of the spectrometer, and is only due to the light
scattering process itself, via the energy and momemtum conservation relations. By scanning the microwave frequency, the spectrum
of the reflected phonons can analyzed with a resolution of less
than 5 MHz. Here, we found the same profile for both incident and

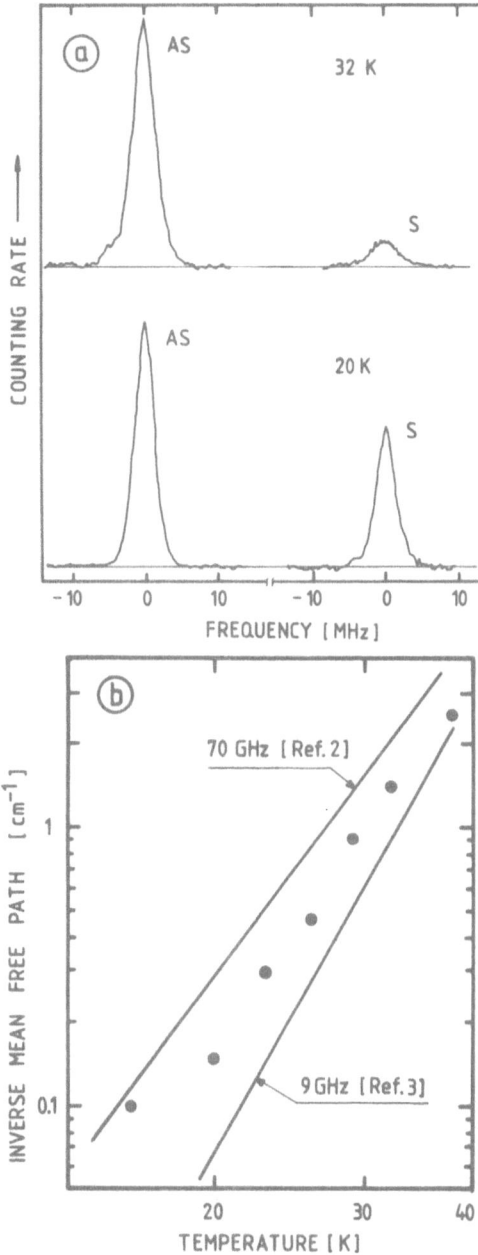

Fig. 2. (a) Spectra recorder in X-quartz showing propagation damping
(b) Measurements (•) of inverse mean free path of phonons
in X-quartz

reflected phonon beam under normal incidence. However, when applied to various experimental conditions, (transverse phonons, oblique incidence), this method should allow us to study in detail inelastic or quasi-elastic changes in the phonon frequency after reflection[5], and non-linear elastic energy transfer mechanisms.

It is a pleasure to thank Prof. K. Dransfeld and Dr. H. Sussner for valuable discussions.

REFERENCES

1. .J. Krüger and H.G. Unruh, Solid State Comm. 21, 583 (1977)
2. R. Nava, R. Arzt, I. Ciccarello and K. Dransfeld, Phys. Rev.
        134, A581 (1964)
3. J.B. Thaxter and P.E. Tannenwald, Appl. Phys. Lett. 5, 67 (1964)
4. E.S. Sabisky and C.H. Anderson, Solid State Comm. 17, 1095 (1975)
5. H. Kinder and W. Dietsche, Phys. Rev. Lett. 35, 578 (1974)

* Permanent address: University of Montpellier - France

Reflection Coefficient of Phonons at a Si-He[3] Interface: Effect

of Cryopumping the Si Surface[*]

C. Peters and J.P. Harrison

Physics Department, Queen's University
Kingston, Ontario, Canada

Abstract. Results are presented of a study of the effect of
prolonged cryopumping of a machine- and chemical-polished silicon
surface at room temperature on the reflection coefficient of
thermally generated phonons at the silicon-liquid He[3] interface.
. . . .

The relative importance of surface damage and surface contamination to Kapitza resistance has been in question in recent
years. Van der Sluijs et al[1] were able to show better agreement
with acoustic theory as the length of time that the solid surface
was maintained under vacuum was increased. This was presumed to
be due to desorption of surface contamination. Weber et al[2] showed
by using crystals cleaved in situ in He[4] at 1 K that surface contamination alone was not responsible for poor agreement with
acoustic theory but that contamination was an important contributing factor.

The experimental work presented here was an attempt to repeat
van der Sluijs' experiment but with a heat pulse technique. The
experiments were also a continuation of the work of Folinsbee and
Harrison[3] (FH) which had identified high transmission of energy
across the silicon-helium interface with diffuse scattering of
phonons at the boundary. Since the diffuse scattering was assumed
to be due to surface damage, an attempt to obtain a damage-free
surface was considered necessary before a dependence upon contamination could be investigated. The magnitude of the diffusely
reflected transverse pulse was used as a measure of surface damage.
Whereas the sample used by FH was mechanically polished, the two
silicon crystals used for the present work were first mechanically
polished and then chemically polished to remove the surface damage.

The experimental technique was as described by FH except that
the cryostat was modified so that whilst the sample could be cooled to
below 1 K by a dilution refrigerator, the sample could also be warmed
to room temperature while liquid helium was maintained in the dewar
surrounding the vacuum jacket.   The reflecting surface was connected
to the He^3 sample gas handling system via a ¼ inch line which included
a charcoal trap in the helium bath.   This system enabled cryopumping
of the reflecting surface whilst it was at room temperature and then
in situ cooling to below 1 K for the heat pulse measurements.

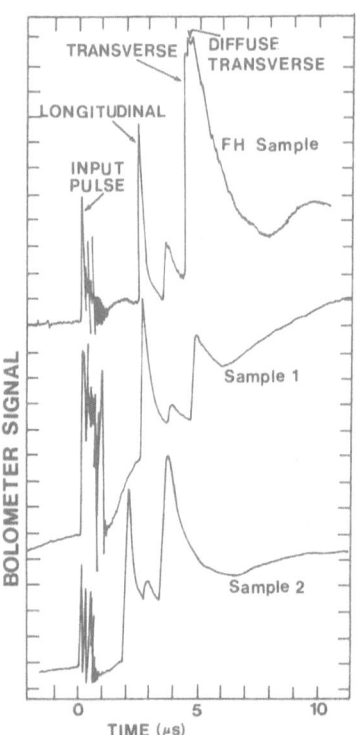

Figure 1.   Bolometer signals for heat pulses reflected from the
             <111> faces of three silicon samples.   In each case
             the pulse temperature was 6 K.

Sample 1 unfortunately cracked along one side prior to experimental work with it. It was not realized at the time that the crack
introduced significant damage into the bulk and surface of the crystal. The larger internal scattering when compared to the FH sample
(see figure 1) was attributed to the larger impurity content
(10,000 Ω cm vs 30,000 Ω cm). However the diffusely reflected
transverse pulse was significantly smaller. Sample 2 was from the
same single crystal as sample 1. The bolometer signal showed less
internal scattering than sample 1 and the diffusely reflected peak
was essentially missing. Samples 1 and 2 were therefore good can-
didates for a surface contamination study. As predicted the diffuse
reflection appeared to be associated with surface damage.

Figure 2 summarizes the results of the surface contamination
study. Sample 1 was cleaned with freon; sample 2 was used as
received and also after solvent cleaning. The reflection coefficients (ratio of pulse heights with helium at the surface and
vacuum at the surface) were about 0.9 and 0.7 for the longitudinal
and transverse phonons respectively. These compare with 0.95, 0.8
and ∿ 0.6 for the longitudinal, transverse and diffusely reflected
transverse phonons for the FH sample. There was little dependence
upon phonon frequency. The attempt to remove contamination

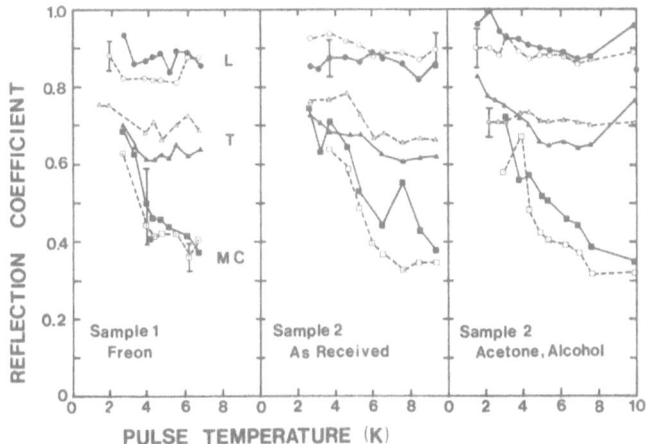

Figure 2. Reflection coefficients for longitudinal, transverse and
mode converted phonons. Open symbols, dashed lines were before
cryopumping; closed symbols, solid lines were after cryopumping.

(solvent residue, nitrogen or oxygen atoms, water molecules) was made by holding the sample at room temperature and cryopumping the surface with activated charcoal at 4 K for 3 days. As can be seen from figure 2 there was no significant effect of this cryopumping. If anything there was a small decrease in the transverse reflection coefficient. The reason for the null result is not clear to us. No attempt at measuring contamination before or after was made and so we do not know whether the cryopumping was effective. We conclude that this naive approach to a surface contamination study is inadequate and that an attempt at surface characterization is required.

*Research supported by N.S.E.R.C. and School of Graduate Studies and Research, Queen's University.

REFERENCES
1. J.C.A. van der Sluijs, K.C. Rawlings and J. Bishop, Phys. Rev. B 19, 4323 (1979).
2. J. Weber, W. Sandmann, W. Dietsche and H. Kinder, Phys. Rev. Lett. 40, 1469 (1978).
3. J.T. Folinsbee and J.P. Harrison, J. Low Temp. Phys. 32, 469 (1978).

PHONON CATASTROPHES AND THE KAPITZA RESISTANCE*

Peter Taborek and David Goodstein

Low Temperature Physics, 63-37
California Institute of Technology
Pasadena, CA   91125

In an anisotropic solid, the phase and group velocities of an elastic wave are not collinear. One consequence of this phenomenon is that the energy emitted from a point source may be defocused from some directions and strongly focused into others. Maris[1] has quantitatively analyzed this effect by introducing the phonon focusing factor $A = (d\Omega_k/d\Omega_V)$, the ratio of solid angles in k-space and group velocity space; it is implicitly assumed that the acoustic field has a $1/r$ spatial dependence.

Although the Maris focusing factor has been successfully used to interpret experiments on phonon propagation in crystals, it is not generally recognized that the focusing factor is based on a geometrical optics approximation to the asymptotic acoustic field. In most crystals, there are directions along which $d\Omega_k/d\Omega_V = \infty$; this unphysical result signifies the breakdown of the geometrical optics approximation, and the asymptotic field must be obtained from a more careful analysis of the wave equation.

Some of the necessary calculations have been outlined by Lighthill[2] and Buchwald.[3] A result which is derived in these papers is that the displacement field $u_j(z)$ along the z axis due to a point harmonic source with frequency ω_o and strength f_j at the origin is

$$u_j(z) = 2\pi i e^{-i\omega_o t} \int \frac{f_j}{(\partial G)/(\partial k_z)} e^{ik_z z} dk_x dk_y \quad (1)$$

where

$$G = \det \left| c_{ijm\ell}\, k_i k_\ell - \rho\omega_o^2\, \delta_{jm} \right|$$

and the integral is over the surface $G = 0$. Although the integral
cannot be carried out in closed form, the principle of stationary
phase can be used to obtain the geometrical optics approximation
to $u.(z)$. In the vicinity of the points of stationary phase on the
surface $G = 0$, k_z can be expanded to second order as

$$k_z = k_z^{\,o} + \alpha k_x^{\,2} + \beta k_y^{\,2} \qquad . \tag{2}$$

Substituting Eq. (2) into the integral Eq. (1) and replacing
the limits by $\pm\infty$ yields the approximation

$$u_j(z) \sim \frac{e^{ik_o z}}{z} \frac{1}{\sqrt{\alpha\beta}} \qquad . \tag{3}$$

The product $\alpha\beta$ is the Gaussian curvature at the stationary point.
Using this formalism, it is clear that the Maris focusing factor
is the inverse of the Gaussian curvature of the surface $G = 0$.
Eq. (3) is an excellent approximation to the far field except in
regions where $\alpha\beta$ is close to zero. If, for instance, $\alpha = 0$,
Eq. (3) blows up because Eq. (2) is no longer a good approximation;
so, a third order term must be included:

$$k_z = k_z^{\,o} + \beta k_y^{\,2} + \gamma k_x^{\,3} \qquad . \tag{4}$$

Using this approximation, the field is

$$u_j(z) \sim \frac{e^{ik_o z}}{z^{5/6}\, \beta^{1/2}\, \gamma^{1/3}} \qquad . \tag{5}$$

Note that the field is finite and depends on the third derivative term and that the spatial dependence is $z^{-5/6}$ rather than z^{-1}.

The vanishing of the Gaussian curvature along a closed curve in k-space on the $G = 0$ surface is a typical feature of most crystals. Such a curve separates regions of positive curvature from regions of negative curvature on the $G = 0$ surface. The group velocities associated with this curve sweep out a conical surface in the crystal on which the intensity is large but finite. These conical surfaces where the geometrical optics approximation for the field goes to infinity are known as caustics in classical wave theory, and can be investigated using mathematical catastrophe theory.[4]

Catastrophe theory shows that one should expect two types of "catastrophic" behavior along a caustic: 1) a smooth curve on which the expansion of Eq. (4) holds, corresponding to the coalescence of two geometrical optics rays, and known as a fold catastrophe; and, 2) isolated directions where a fourth order expansion is required, corresponding to three parallel rays, and called a cusp catastrophe. The spatial dependence along a cusp is $z^{-3/4}$.

The field due to a point source is thus dominated by a conical surface on which there may be cusps where the field is particularly intense. We have carried out detailed computer calculations to locate the caustic surface and the cusps in sapphire. These calculations have been used to interpret the diffuse signal in phonon reflection experiments. By diffuse signal we mean those parts of the signal which cannot be accounted for by assuming specular reflection from a flat reflecting surface. We find that the diffuse signal is very sensitive to the location of the caustic; in fact, the diffuse signal can be reduced by a factor of 10 by choosing a geometry in which the caustic does not intersect the reflection surface. Details of these experiments will be reported elsewhere.

*Supported by NSF grant DMR77-00036 and a grant from the JPL President's Fund.

References

1. H.J. Maris, J. Acoust. Soc. Am. 50:812 (1970).
2. M.J. Lighthill, Phil. Trans. Roy. Soc. A 252:397 (1960).
3. V.T. Buchwald, Proc. Roy. Soc. A 253:563 (1959).
4. M.V. Berry, Adv. Phys. 25:1 (1975).

****

DISCUSSION

H. Kinder: Horstmann and Wolter have done experiments where they
put the generator and detector on top of each other and they see
very sharp peaks from certain focussing directions.

P. Taborek: That makes very good sense from the theory, because in
that case the two caustics overlap exactly, and it's the correlation
of the caustics that will determine whether you see these sharp
features in one experiment or another. I think that's been the
source of some confusion with previous experiments. If you do the
experiment in one direction, you see this sharp non-specular peak,
if you do it in another, you don't. And if you do it at precisely
normal incidence, that's the ideal case where the diffuse scattering
is optimized.

Specular and Diffuse Phonon Reflection Processes at a Crystal Interface*

Peter Taborek and David Goodstein

Low Temperature Physics, 63-37
California Institute of Technology
Pasadena, CA   91125

We have performed high resolution phonon reflection experiments in sapphire using the heat pulse technique. Our results[1] show as many as eight sharp specular peaks; this structure in the reflection signal cannot be explained using isotropic elastic theory. This is somewhat surprising since the phase velocities of sapphire vary with direction by only 15%.

In order to interpret these results we have analyzed the problem of reflection of elastic waves in anisotropic media. As in isotropic media, the boundary conditions at a free surface stipulate that the parallel component of momentum is conserved, and that the normal stress must vanish

$$k_{||}^{in} = k_{||}^{ref} \tag{1}$$

$$\sigma_{ik} n_k = 0 \tag{2}$$

The physical content of Eq. (1) is illustrated in Fig. 1, which shows curves of constant ω for each phonon polarization in k-space at a crystal boundary. As shown in the figure, there are generally three outgoing waves that satisfy Eq. (1), so all modes may be coupled upon reflection. The amplitudes of these waves can be determined by demanding that the stress induced by the incoming wave is precisely relieved by the outgoing waves so that the total stress satisfies Eq. (2). The stress for each polarization is related to the strain $u_{j\ell}$ by

$$\sigma_{ik} = c_{ij\ell k}\, u_{j\ell} = \frac{1}{2} c_{ij\ell k}\, (e_j k_\ell + e_\ell k_j) \qquad (3)$$

where the polarization e_j must satisfy the eigenvalue equation

$$(c_{ij\ell m}\, k_j k_\ell - \rho\omega^2 \delta_{im})\, e_m = 0 \quad . \qquad (4)$$

For each $\underset{\sim}{k}$, Eq. (4) gives three orthogonal polarization vectors, which twist as $\underset{\sim}{k}$ changes. Except in planes of high symmetry where the twisting is somewhat restricted, all three outgoing waves are therefore necessary to balance the stresses.

    Although the above equations completely solve the problem of the reflection of a wave with given polarization and k-vector, the

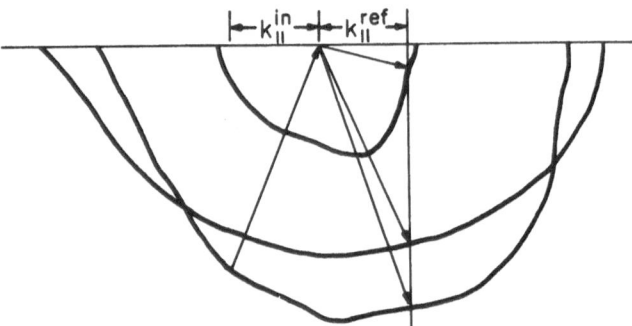

Fig. 1. Schematic polar plot of $\underset{\sim}{k}$ at constant ω in a particular crystallographic plane. The inner curve represents the quasi-longitudinal mode, the larger curves the two quasi-transverse modes. The geometrical construction shown here determines the k-vectors of the three reflected waves, which conserve $k_{||}$.

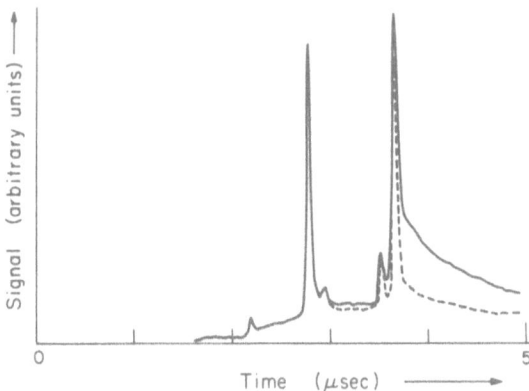

Fig. 2.   Reflection signal due to heater pulse at t = 0.   Dashed
curves show signal with helium present on reflection surface.
Note large diffuse tail which follows the sharp specular peaks,
and the large effect of helium on that bump.

detected signal is due to energy flux which is described by the
power flux vector

$$P_i = \frac{1}{2} \omega c_{ij\ell m} \, k_\ell e_j \, e_m \tag{5}$$

Since $\underset{\sim}{P}$ and $\underset{\sim}{k}$ are generally not collinear, the k-vectors which
result in energy transmission from generator to detector must be
found using iterative numerical techniques.   In general there are
nine solutions corresponding to the three possible ingoing and
outgoing polarizations.   The agreement between such calculations
and experimental results like those shown in Fig. 2 is very good;
see ref. 1 for more details.

     A considerable portion of the received signal is due to diffuse
($k_\parallel$ not conserved) processes which traverse longer paths and arrive
at later times than the specularly reflected phonons.   As shown
in Fig. 2, the diffuse signal seems to couple to the helium more
effectively than the specular component.   Our experiments show

that the magnitude of the diffuse signal is very sensitive to the crystallographic orientation of generator and detector. Detailed computer calculations, which will be described elsewhere, are used to analyze the diffuse signal; caustic surfaces on which the usual phonon focusing calculations are not valid are found to be particularly important.

*Supported by NSF grant DMR77-00036 and a grant from the JPL President's Fund.

Reference

1.   P. Taborek and D. Goodstein, J. Phys. C., in press.

HEAT TRANSFER BETWEEN POWDERED CMN AND SOLID ^{3}He

IN THE MILLIKELVIN RANGE

B. Hébral, G. Frossati, H. Godfrin, D. Thoulouze

Centre de Recherches sur les Très Basses Températures,
C.N.R.S., BP 166 X, 38042 Grenoble-Cedex, France

We present the results of heat transfer experiments between
CMN and solid ^{3}He from 4 to 30 mK. We have measured the recovery
time τ after small magnetization-demagnetization cycles (H∿5-25 G).
Our experimental cell contained 6.25 cm^3 of bcc (24 cm^3/mole) or
liquid (29 bars) ^{3}He and 22 g of powdered CMN (grain size < 40 μ,
pores 1-5 μ). The ^{3}He contained less than 50 ppm ^{4}He which was not
sufficient to cover the CMN surface (3.5 m^2) with one ^{4}He monolayer.
Through the relaxation of the CMN susceptibility we observed the
thermal recovery which appeared to be non exponential, probably due
to the distribution in the powder size. τ is taken from the initial
relaxation.

As seen in fig. 1, the recovery time is short and has the same
order of magnitude for solid and liquid ^{3}He. It varies as about
$T^{-1.4}$ for solid ^{3}He from 4 to 30 mK, and as $T^{-1.7}$ for liquid ^{3}He
from 5 to 15 mK. It is 1.5 to 2 times larger for bcc ^{3}He than for
liquid ^{3}He. The internal relaxation time in ^{3}He, τ_D, and in CMN, τ_D',
are given by $\ell^2 C/k$ where ℓ is the geometrical size, C the specific
heat measured in the same experiment[1] and k the thermal conductivity[2,3,4]. In our temperature range, τ_D and τ_D' are much shorter
(< 0.5 s) than the measured τ and so the classical thermal diffusivity in ^{3}He or CMN is not a limit to the heat transfer.

Relaxation times ranging around 100 s have been observed previously[5] below 15 mK between liquid ^{3}He and CMN and the corresponding thermal boundary resistance, R_K, varies as T. This was attributed to a magnetic coupling process between the electronic Cerium
spins and the nuclear ^{3}He spins, by-passing the parallel phonon mechanism. The latter gives $R_K \propto T^{-3}$ while, for a dipolar coupling
model, $R_K \propto \chi_{CMN}^{-1} \chi_{3He}^{-1}$ is predicted[6] (χ_i is the magnetic suscepti-

bility of medium i). Above the CMN ordering temperature (T $\approx$ 2 mK) χ_{CMN} follows a Curie law, the susceptibility of liquid ^{3}He is constant that of a Fermi liquid[2], and thus : $R_K \propto T$. For solid ^{3}He above the ordering temperature ($T_N \simeq$ 1 mK) χ_{3He} follows a Curie-Weiss law[7], so that the model predicts : $R_K \propto T^2$ above 6 mK[6,8].

The variation of our thermal resistance versus T is presented on fig. 2 as $\tau/\bar{C}$ using the classical two bath model : $\bar{C}=C_0C_3/(C_0+C_3)$ where C_0 and C_3 are the CMN and ^{3}He specific heat we have measured[1]. For solid ^{3}He it gives a constant resistance $R_K \approx 0.3$ s mK/erg For liquid ^{3}He, the thermal resistance decreases as $\approx T^{-1.5}$ from 5 to 10 mK and becomes constant above 10 mK.

Previous experiments[9] in the same temperature range with similar grains in contact with a liquid mixture 1 % ^{3}He in ^{4}He gave AR_K = 8×10^4 T^{-2} cm^2 K/W leading to $R_K \simeq$ 2.3 s mK/erg at 10 mK in this experiment : at least ten times larger than observed with the bcc or liquid ^{3}He. Evidently we need to consider another process in parallel with the classical phonon process to interpret the present measurements.

By contrast, extrapolating boundary results[10] obtained for large CMN crystals (AR_K = 55 T_K^{-3} cm^2 K/W) at the liquid ^{3}He-CMN interface or the theoretical values ($AR_K \sim$ 40 T_K^{-3} cm^2 K/W) at the solid ^{3}He-CMN interface to our powder dimensions would give R_K smaller than the temperature independent resistance we measured with bcc ^{3}He. This might be explained by considering a bottleneck effect[10,11] acting in series with the boundary phonon resistance. For our grain size (< 40 μ) and our Praseodymium concentration (500 ppm) the constant bottleneck resistance taken from higher temperature measurements[11], is in reasonable agreement with our results for solid ^{3}He. However the smaller values we observed with liquid ^{3}He need also another parallel mechanism to be understood.

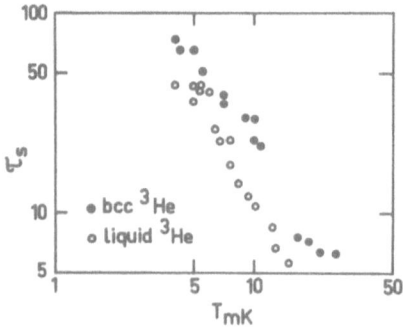

<u>Fig. 1</u> Recovery time at the interface CMN/liquid or solid ^{3}He.

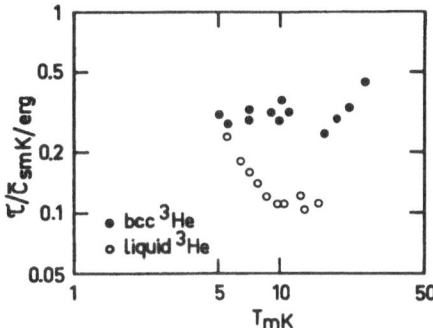

Fig. 2 Thermal resistance between CMN and liquid or solid ^{3}He.

The theoretical predictions of a magnetic dipolar coupling model for bcc ^{3}He[8] give : $AR_K = 10^{10}\, T_K^2$ cm^2 K/W. At 10 mK : $R_K = 3\times10^{-2}$ s mK/erg, 10 times smaller than the measured resistance. Our measurements imply that for solid ^{3}He a magnetic coupling would be much less efficient than calculated. This could be due to the rather large distance ($\sim$ 10 Å) between the Cerium ions and the ^{3}He spins at the interface.

For liquid ^{3}He, the dipolar magnetic model considering the coupling between the ^{3}He and CMN spins does not explain the $T^{-1.5}$ observed variation, but this dependence is expected for the coupling between liquid ^{3}He and ferromagnetically aligned spins. Such a ferromagnetic state has been proposed to explain the relaxation and susceptibility behaviour of the first layers of ^{3}He near the surface[6].

Results obtained with CMN grains ten times larger in diameter have been independently reported[12]. A constant resistance was also measured with bcc ^{3}He in the same temperature range but it was about one order of magnitude smaller. This may originate from the difficulty to estimate the surface area in this type of experiment and also from a different Pr concentration.

As a conclusion it appears that short thermal relaxation time constants are observed between CMN and solid ^{3}He in the milliKelvin range, and with the same order of magnitude as for liquid ^{3}He. A phonon process including a bottleneck inside the CMN might interpret the results with the solid but a new process, probably magnetic, has to be involved for the liquid. Nevertheless a classical dipolar coupling model does not describe the temperature dependence of the measured resistance.

REFERENCES

1. B. Hébral, G. Frossati, H. Godfrin, G. Schumacher, D. Thoulouze, J. de Phys. Lett. 40, L-41 (1979) and to be published.
2. J.C. Wheatley in "The Helium Liquids" edited by J.G.M. Armitage and I.E. Farquhar (Academic Press, 1975), p. 241.
3. G. Armstrong, A.S. Greenberg, J. de Phys. 39, C6-135 (1978).
4. J.E. Robichaux, A.C. Anderson, Phys. Rev. B2, 5035 (1970).
5. W.R. Abel, A.C. Anderson, W.C. Black, J.C. Wheatley, Phys. Rev. Lett. 16, 273 (1966).
6. M.T. Béal-Monod, D.L. Mills, J. Low Temp. Phys. 30, 289 (1978).
7. T.C. Prewitt, J.M. Goodkind, Phys. Rev. Lett. 39, 1283 (1977).
8. R.A. Guyer, J. Low Temp. Phys. 10, 157 (1973).
9. B. Hébral, G. Frossati, H. Godfrin, G. Schumacher, D. Thoulouze, Rev. de Phys. Appl. 13, 533 (1978).
10. J.P. Harrison, J.P. Pendrys, Phys. Rev. B8, 5940 (1973).
11. A.C. Anderson, J.E. Robichaux, Phys. Rev. B3, 1410 (1971).
12. Y. Morii, M.R. Giri, H. Kojima, Phys. Lett. 70A, 457 (1979).

A MACROSCOPIC APPROACH TO THE KAPITZA RESISTANCE

J.D.N. Cheeke

Département de physique
Université de Sherbrooke
Sherbrooke, Québec, J1K 2R1, Canada

and

H. Ettinger

C.R.T.B.T.
C.R.N.S.
38042 Grenoble, France

The macroscopic treatment of Khalatnikov[1] of the Kapitza resistance problem was modified by Challis, Dransfeld, and Wilks[2] to take into account impedance matching effects due to the absorbed layer of dense helium at the interface. This model was further modified by the present authors[3] to take into account the effect of dissipation in the layer, and qualitative agreement was found with experiment. In the present work we consider essentially the same model but we treat incidence from the solid, which enables us to obtain the transmission coefficient as a function of frequency, mode and angle, as well as to consider the effects of thin layers of contamination.

We suppose a perfect solid surface in contact with a bath of liquid helium, with a few Å of highly compressed helium at the interface. In this layer the phonon wave vector is written as $k = k' + ik''$, $k'' = \nu k'$ where ν is the absorption parameter. We also consider the possibility of thin layers (4 - 30Å) of a non-absorbing contaminant on the surface, and ice has been chosen as a representative candidate. Results for the transmission coefficient of the transverse mode $t_\perp$ at near normal incidence are shown in fig. (1). We see that above about 100 GHz both absorption in the layer and the presence of contaminant increase the transmission,

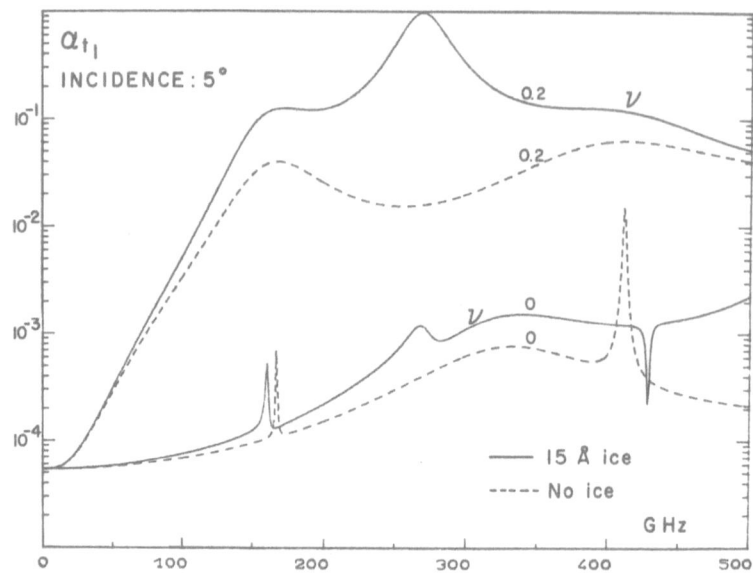

Fig. 1.   Plot of α for the tl mode as a function of frequency at 5°
          incidence for a Al_2O_3-15Å ice - 3.5 Å solid He at 400 atm-
          3.5 Å solid He at 40 atm-liquid He interface.

the former being particularly important. The two upper curves
correspond to typical experimental results for clean and dirty
surfaces.

     The effect of different thicknesses of small amounts of con-
tamination on $R_K T^3$ for copper surfaces are seen in fig. 2, for
$\nu = 0\,2$. It is seen that order of magnitude agreement with experiment could be obtained for d ∿ 8 or 9 Å . Finally we note that
the observation of anomalous transmission for quantum systems other
than helium could be accounted for in this model by the presence of
a ∿ 15 Å layer of impurity. Full details of all of the present
results have been given elsewhere[4] .

     It remains to justify the use of values of ν ∿ 0 1 in the
dense helium layer. Experimentally this corresponds to the results
and Anderson and Sabisky[5] for slightly thicker films in this frequency and temperature range. From a semi-classical viewpoint the
dense layer is highly inhomogeneous, leading to extremely rapid
changes in density and sound velocity. This would suggest phonon
mean free paths of the order of the layer thickness, which is more
than enough. From a quantum mechanical viewpoint we expect tunnelling of helium atoms[6] similar to the effects observed in disordered systems, together with a very short phonon mean free path
in this frequency range. A detailed QM calculation of the acoustic attenuation in the dense helium layer due to such effects would
make contact with our more macroscopic approach.

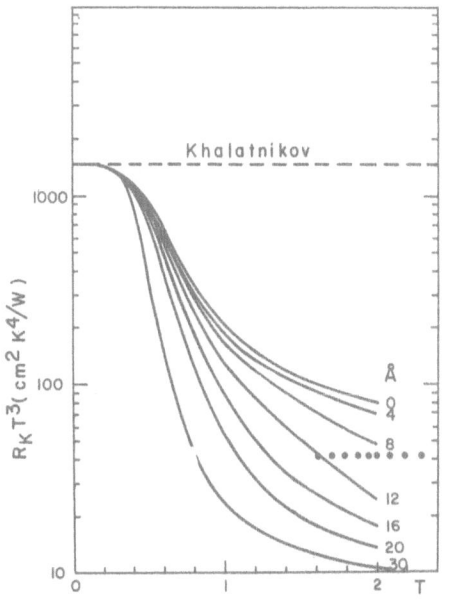

Fig. 2. Plot of $R_k T^3$ for copper as a function of temperature for different thickness of H_2O in angstroms. The curves were calculated for a copper -3.5Å ice -3.5Å solid He at 400 atm $-$ 3.5Å solid He at 40 atm $-$ liquid He interface. The experimental points are from Ref. 7.

REFERENCES

1. I.M. Khalátnikov, Zh, Eksp. Teor. Fiz $\underline{22}$, 687, 1952.
2. L.J. Challis, K. Dransfeld, and J. Wilks, Proc. Roy Soc. London $A\underline{260}$, 31, 1961.
3. D. Cheeke and H. Ettinger, Phys. Rev. Lett. $\underline{37}$, 1625, 1976.
4. D. Cheeke and H. Ettinger, Jour. Low Temp. Phys. $\underline{36}$, 121, 1979.
5. C.H. Anderson and E.S. Sabisky in Physical Acoustics, ed. W.P. Mason and R.N. Thurston (Academic New York 1971) Vol. 8.
6. T. Nakáyama, J. Phys. C $\underline{10}$, 3273, 1977.
7. N.S. Snyder, J. Low Temp. Phys. $\underline{22}$, 257, 1976.

****

DISCUSSION

H. Kinder: A central point of your model is that you need attenuation in the very first highly compressed layer. Is that true?

J. D. N. Cheeke: I think that would be true, although we have also
taken a rather less compressed layer but rather thicker and put the
attenuation in that uniformly, and that gives us essentially the
same result.

H. Kinder: But experimentally nobody finds a difference with just
one layer - one always needs more than one layer. If you assume

the attenuation only in the second layer, then the matching from the solid to that layer is not so good anymore.

J. D. N. Cheeke: We have studied both of those cases, and it doesn't change even qualitatively the semi-quantitative result. In other words you are taking two layers, one very compressed and the other very much less, or you take one thicker uniform layer gives essentially the same thing.

C. Elbaum: I am a little puzzled by the physical meaning of the attenuation in a monolayer unless one uses it in a strictly pheno-menological sense. After all, one is dealing with phonons of wave-length very much greater than this.

J. D. N. Cheeke: I think the basic justification for that is that Guyer and co-workers have shown that even for monolayer thicknesses of solid helium if you put in the relevant acoustic parameters then you can get away with a continuum type of approach.

INTERFACIAL THREE-PHONON PROCESSES AND

ANOMALOUS INTERFACIAL ENERGY TRANSPORT

W. M. Saslow

Physics Department
Texas A&M University
College Station, TX 77843

This work was done in collaboration with my former graduate student, M. E. Lumpkin.[1,2] It was motivated by the phonon pulse measurements of Wyatt et al,[3] who observed that high-frequency phonons can break the so-called "critical cone" on passing across the interface between NaF and liquid ^{4}He. A simple mechanism for cone-breaking is presented by the interfacial three-phonon process: a phonon incident from, e.g., NaF splits into two phonons emergent in liquid ^{4}He.

The most striking aspect of the Kapitza problem is its "fuzziness"; that is, there are many qualitative and semi-quantitative facts, but few really quantitative ones. This means that any theory must involve a mechanism that is very general and very flexible, so that it can be adapted to "hard" data when that becomes available. Initially, a theory must be convincing primarily because it can qualitatively explain a wide variety of experiments; at a later stage it must also provide semi-quantitative agreement with experiment; finally, it must provide detailed quantitative agreement with experiment. Interfacial three-phonon processes satisfy the first two criteria, and provide hope for satisfying the third.

We limit our discussion to the channel in which a phonon from the classical material decays into two phonons in the quantum material. Numerous other channels, such as one in which one of the final state phonons is in the quantum material, are also possible.[1,2] Now, note that phonon velocities in classical materials significantly exceed phonon velocities in quantum materials. If such a process were to occur within the bulk (with conservation of energy and all three components of quasi-momentum), it would

have a great deal of phase space. Since it occurs at an interface, with conservation only of the two momentum components along
the interface, it has even more phase space (because there is one
less restriction).

In addition to phase space considerations, we also need a
matrix element. Because the interaction dominantly takes place
at the interface, and because the first layer of ^{4}He is believed
to be solidified,[4] we use the same type of phenomenological cubic
anharmonic interaction as used in the bulk, except that it is
(in the continuum limit) multiplied by a delta function to localize it at the interface. Thus the interaction energy density is
$H' = C_{\alpha\beta\gamma\delta\epsilon\mu} (\partial_\alpha U_\beta)(\partial_\gamma U_\delta)(\partial_\epsilon U_\mu)\delta(z)$, where U_β is the displacement,
C is an energy per unit area, and $\delta(z)$ localizes the interaction
to the z=o plane. In terms of Lennard-Jones parameters, C is of
the order of ϵ/σ^2. In the long-wavelength limit, if the three
phonons have roughly the same momentum k, H' varies as $k^{3/2}$.

We now employ Fermi's Golden Rule to obtain the transmission
rate. If the first Born approximation may be used, then the momentum dependence of the matrix element arises from H'^2, or k^3.
Since the available phase space varies as k^3, and since k must
scale as the frequency ω of the incoming phonon, the effective
transmission coefficient w_3 for this process varies as k^6, or ω^6.
At low ω the process doesn't matter, whereas at high ω it becomes
so large that the first Born approximation becomes invalid. In
that case a renormalized theory must be used, which is expected
to lead to a saturated value for w_3. In the vicinity of onset,
the process has an ω^6 variation as its signature.[1,2]

We note that Refs. 5-7 treat the first layer of ^{4}He as a
liquid, which it is not, obtaining a different form for H', which
has too small an amplitude to be significant for the Kapitza
problem.

In Refs. 1 and 2 we extensively discuss how interfacial
three-phonon processes are consistent with experiment. This mechanism: 1) does not strongly depend upon whether the "quantum"
material is liquid or solid[8]; 2) permits inelastic (frequency-
shifting) phonon scattering[9,10]; 3) permits mode-conversion[11,12];
4) is consistent with 12 Å thick layers of liquid ^{4}He behaving
like bulk liquid ^{4}He[9,11] (the mechanism "turns on" at such high
frequencies that the characteristic wavelengths are quite small);
5) has a wide variety of symmetry-allowed matrix elements with
which to fit the details of the angular emission and angular absorption data[3] (the experiments of Weber et al[13] can be interpreted
in terms of the relative amplitudes of certain matrix elements);
6) predicts that phonons emitted outside the "critical cone" will
be lower in energy than phonons within the "critical cone"[14]; 7)
requires a coupling C≈1-10 eV/A to provide a semi-quantitative fit

to the data of Ref. 1, with a Lennard-Jones estimate giving $C \approx .2-2$ eV/A; 8) is consistent with the fact that the mechanism "turns on" at a higher frequency (near 100 GHz) for H_2,[15] than for ^{4}He (20 GHz)[16]. Most important, the mechanism works only at "classical-quantum" interfaces, because the large difference in phonon velocities gives an enormous amount of phase space.

I now suggest a new tunable high-frequency phonon source. Consider a picosecond (10^{-12} sec) pulsed laser. Using fiber optics, break it into (e.g.) 100 beams. Put each beam through its own tunable phase-shifter (e.g. KDP) adjusted so that the beams are delayed by a fixed amount of time T with respect to one another. Recombine and focus the beams on a thin film evaporated onto the material to be studied. If the relaxation time τ of the thin film, and the transit time t_f for sound to cross the film, are smaller than T, then the film should undergo a sharp thermal expansion when hit by each sub-pulse. The succession of 100 sub-pulses at intervals of T should produce a "phonon hammer" whose cycle frequency $f=T^{-1}$ is defined to within 1%, and can be as high as 500 GHz. In practice, τ and t_f will introduce broadening also. Another scheme for producing a tunable phonon source might utilize a heat pulse, Bragg-scattering it off an atomically well-defined crystal face, and picking out the desired frequency by observing from the appropriate direction.

REFERENCES

1. W.M. Saslow and M.E. Lumpkin, Sol. St. Comm. 29, 395 (1979).
2. M.E. Lumpkin and W.M. Saslow, Phys. Rev. B (to be published, 1979).
3. A.F.G. Wyatt, G.J. Page, and R.A. Sherlock, Phys. Rev. Lett. 36, 1184 (1976).
4. S.E. Polanco and M. Bretz, Phys. Rev. B17, 151 (1979).
5. I.M. Khalatnikov, Zh. Eks. i Teor. Fiz. 22, 687 (1952).
6. F.W. Sheard, R.M. Bowley, and G.A. Toombs, Phys. Rev. A8, 3835, (1973).
7. T.J. Sluckin, F.W. Sheard, R.M. Bowley, and G.A. Toombs, J. Phys. C8, 3521 (1975).
8. J.T. Folinsbee and A.C. Anderson, Phys. Rev. Lett. 31, 1580 (1973).
9. H. Kinder and W. Dietsche, Phys. Rev. Lett. 33, 578 (1974).
10. W. Dietsche and H. Kinder, J. Low Temp. Phys. 23, 27 (1976).
11. C.J. Guo and H.J. Maris, Phys. Rev. Lett. 29, 855 (1972).
12. J.S. Buechner and H.J. Maris, Phys. Rev. Lett. 34, 316 (1975).
13. J. Weber, W. Sandmann, W. Dietsche, and H. Kinder, Phys. Rev. Lett. 40, 1469 (1978).
14. A.F.G. Wyatt and G.N. Crisp, J. de Physique 39, Colloque C6, 244 (1978).
15. J.S. Buechner and H.J. Maris, Phys. Rev. B14, 269 (1976).
16. E. Sabisky and C.H. Anderson, Sol. St. Comm. 17, 1095 (1975).

DISCUSSION

P. G. Klemens: Did you use your scaling arguments to verify that you would get an ω^5 dependence in ordinary 3-phonon interactions?

W. M. Saslow: I really didn't do that.

P. G. Klemens: It would be a nice check.

AN EXACTLY SOLVABLE MODEL FOR THERMAL BOUNDARY RESISTANCE IN A NONHARMONIC SYSTEM[*]

William M. Visscher and M. Rich

Los Alamos Scientific Laboratory, Univ. of California

Los Alamos, New Mexico 87545 USA

Theories of the thermal boundary resistance (Kapitza resistance if one side of the boundary is occupied by liquid helium) have been based on the assumption that the lattice heat current is given by

$$J = \sum_{\alpha} \hbar\omega_{\alpha} v_{\alpha} n(\alpha,x) \qquad (1)$$

where v_{α} is the group velocity and $n(\alpha,x)$ is the density of phonons α at point x, or that the density matrix is diagonal in the local phonon representation. Eq. (1) can be justified for a homogeneous system far from boundaries with small temperature gradients. Then it is assumed that the part of $n(\alpha,x)$ which describes phonons propagating toward the boundary, which is at the $x = 0$ plane, is a Bose-Einstein equilibrium distribution. This is probably a good approximation for $|x| \gg \ell$ (phonon mean free path), but $n(\alpha,x)$ for $|x| \ll \ell$ is what is needed for the next step in the calculation, to match phonons across the boundary to find a transmission coefficient t_{α} enabling one to write, for $|x| \ll \ell$:

$$n(\alpha,x) = \begin{cases} t_{\alpha}\, n_{BE}(\alpha,T_L) & v_{\alpha} > 0 \\ t_{\alpha}\, n_{BE}(\alpha,T_R) & v_{\alpha} < 0 \end{cases} \qquad (2)$$

Substitution of (2) into (1) yields the familiar result of the acoustic mismatch theory.

We have used the self-consistent reservoir (SCR) model to check Eqs. (1) and (2). The SCR model[1] simulates a real solid in which phonons interact with each other and with other systems (electrons, spins) by a harmonic solid in which each atom interacts with its own thermal reservoir. Self-consistency is attained by adjusting reservoir temperatures so that no heat flows to or from any reservoir except those at the surfaces of the solid.

The Heisenberg equations of motion for the system are a set of coupled harmonic equations with Langevin terms involving a frictional dissipation and a random force acting independently on each atom. Quantum mechanics imposes severe constraints on the autocorrelation of the stochastic force, which depends strongly on the temperature.

Expressions for the time-averaged expectation values of normal products of pairs of position and momentum operators satisfy certain inhomogeneous linear equations which can be solved under the self-consistency constraint to yield heat currents and kinetic temperatures.

The steady-state values of these expectation values, $X_{ij} = \langle x_i x_j \rangle$, $Y_{ij} = \langle p_i p_j \rangle$, and $Z_{ij} = \langle x_i p_j \rangle$, can be thus obtained for simple models of one dimensional (1d) and of 3d lattices. From Y_{ii} one can obtain the kinetic temperatures; an example is shown in Fig. 1. From Z_{ii} may be obtained the heat current, and it also provides a test of Eq. (1), because if the density matrix is diagonal in the phonon representation, then it is easy to show that Z_{ij} depends only on i-j. An example of a 6 × 6 portion of Z_{ij} evaluated in the SCR model is given in Table 1. The case shown here is from a 3d lattice at high temperature; at low temperature the correlation of phonon phases is not so evident. Eq. (1) is wrong in some cases; the criteria for its correctness are not understood.

From the heat currents and the SC temperatures a thermal boundary resistance R_K may be deduced. One would expect, if the acoustic mismatch theory applied here, that R_K should depend only on the acoustic impedances and sound velocities in the two media. This is approximately true in the 1d lattice, but not for the 3d case, in which $R_K \sim \lambda^{-1/2}$, where λ is the coupling to the SCR's, a measure of the nonharmonicity of the system.

The method used to solve the set of linear equations mentioned above in the 3d case restricts λ to have the same value in both media. This hinders the direct application to physical systems, which would involve fixing the central and noncentral force constants and λ on the two sides of the interface by fitting sound velocities and thermal conductivity (which in both the low and high-temperature limits turns out to be proportional to λ^{-1}).

Without this restriction realistic values of λ can be used, which for metals at low temperatures will be proportional to T because then the reservoir interaction simulates the electron–phonon interaction. For insulators it will depend exponentially on T at high temperature, simulating anharmonicities, and will be independent of T at low temperatures, simulating boundary and impurity-scattering. The problems of removing the constraints on λ and of estimating it theoretically from first principles are being studied.

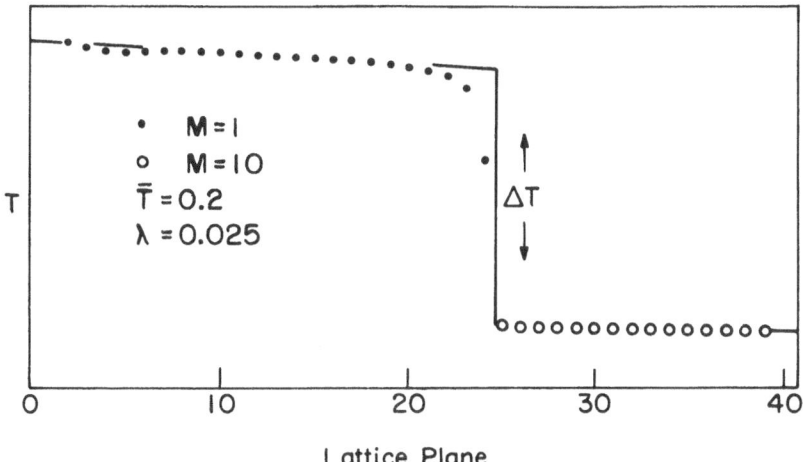

Fig. 1. Self-consistent temperatures for a simple cubic system of 40 planes of atoms; 24 with M = 1, 16 with M = 10. Average temperature is 0.2 $\hbar\omega_D$. λ = 0.05 engenders a mean free path whose effects are manifested by the deviation from linearity in the M = 1 portion. Temperatures of reservoirs 1 and 40 are fixed (off-scale here).

Table 1. A segment of the Z_{ij} matrix times 10^6 in natural units. This simple cubic system has 20 planes of atoms; the first 12 have M = 1, the remainder M = 10. λ = 0.1 so this segment is not far from the boundary in terms of mean free paths. The average temperature is 10 $\hbar\omega_D$.

i j	5	6	7	8	9	10
5	0	13.36	4.85	4.41	4.45	4.35
6		0	13.36	5.54	5.34	5.42
7			0	13.36	6.29	6.44
8		Skew-		0	13.36	7.17
9		symmetric			0	13.36
10						0

* Work performed under the auspices of the United States Energy
  Research and Development Administration.
[1] William M. Visscher and M. Rich, Phys. Rev. A (August 1975).

DISCUSSION

H. J. Maris: By a phonon in this context you refer to the normal modes of the whole system, and so that includes both sides of the interface?

W. M. Visscher: By a phonon I mean $e^{i\vec{k}\cdot\vec{x}}$.

H. J. Maris: But extending both sides of the interface?

W. M. Visscher: Yes.

A. C. Anderson: If you calculate the Khalatnikov value for the same system, how do the magnitudes compare?

W. M. Visscher: Since this does have such a strong dependence on λ, which is something which is orthogonal to the Khalatnikov theory, you can't really compare in the three-dimensional case. In the one-dimensional case you can compare and they're about a factor of 2 different.

ABSORPTION OF SURFACE PHONONS BY ADSORBED HELIUM SYSTEM ON AN INHOMOGENEOUS SURFACE

T. Nakayama* and F.W. Sheard**

* Engineering Science Department, Hokkaido University,
  Sapporo, Japan
**Physics Department, Nottingham University, Nottingham,
  UK

In recent years, considerable interest has been paid to the dynamics of adsorbed atoms on a solid surface, particularly light atoms where quantum effects are expected to be important. The quantum states of a single helium atom adsorbed on a static ideal surface have been studied in detail[1]. More recently the interaction of an acoustic surface wave (ASW) with a gas of noninteracting mobile adsorbed helium atoms has been analysed[2].

However, except at very low surface coverages, interactions between helium atoms play an important role in determining the characteristics of the adsorbed layer. Moreover a real crystal surface prepared by polishing in some way, is unavoidably inhomogeneous due to surface imperfections or impurities. This leads to randomness in the sites and in the binding energies of the adsorbed atoms. Such a randomness allows the possibility that localized tunnelling states exist in the helium layer, as are found in bulk amorphous materials[3]. Although there is no direct structural evidence for this, measurements of the specific heat of adsorbed helium on copper substrates have shown a strong influence of surface inhomogeneities[4,5] and, for submonolayer films, Schottky-like contributions have been found. The anomalous Kapitza thermal conductance has also been interpreted[6] in terms of a coupling between bulk phonons and the tunnelling states of helium atoms in the compressed layers adjacent to the solid surface. Recent experiments using quasimonochromatic phonons[7] have shown that this coupling is apparently weaker for a crystal cleaved under helium than for polished samples, which suggests a reduced density of tunnelling defects for a clean, atomically flat surface.

In this paper we propose the use of high-frequency ASW to study

the dynamic properties of adsorbed helium atoms in films of one or
two monolayer thickness. The physical content of the model is that
certain helium atoms may tunnel between localized potential minima
and the energy splitting E between the two lowest levels varies randomly from site to site. A similar model[3] is used to describe the
low temperature properties of amorphous bulk materials. In the presence of the ASW the displacement of the solid surface perturbs the
tunnelling states via the change in the interaction potential V between the helium atoms and the solid. We consider a coupling proportional to the normal displacement of the surface. The interaction
matrix element for a transition between the tunnelling states with
emission or absorption of a surface (Rayleigh) wave quantum is then

$$M = g\left(\frac{\hbar}{2\rho c_R S}\right)^{\frac{1}{2}} \frac{\gamma}{K^{\frac{1}{2}}} \left(\frac{1 - \eta^2}{1 + \eta^2}\right)$$

where ρ is the solid density, S the surface area and c_R the Rayleigh
wave velocity. The quantities $\gamma^2 = 1 - (c_R/c_t)^2$, $\eta^2 = 1 - (c_R/c_\ell)^2$,
$K = (\gamma - \eta)(\gamma - \eta + 2\gamma\eta^2)/2\gamma\eta^2$ depend on the longitudinal and transverse sound velocities c_t, c_ℓ in the solid. The constant g is the
matrix element of the potential slope $\partial V/\partial z$ between the localized
atomic states. This displacement coupling is certainly important but
for a rough or impure surface in which some helium atoms may be partially embedded in the surface a coupling to the deformation will also
occur.

We have calculated the damping and dispersion of a high-frequency ASW due to the resonant absorption and emission processes. It
is assumed that $\omega\tau_d \gg 1$ where ω is the acoustic frequency and τ_d the
relaxation time of the two-level defects. The damping, expressed as
the inverse relaxation time of the surface phonon, is given by

$$\tau^{-1} = \frac{\pi g^2 \gamma^2}{2\rho c_R K} \left(\frac{1 - \eta^2}{1 + \eta^2}\right)^2 n(\hbar\omega) \tanh\left(\frac{\hbar\omega}{2k_B T}\right),$$

where $n(\hbar\omega)$ is the number of tunnelling defects per unit surface area
per unit energy range for the splitting $E = \hbar\omega$. If a constant density of states n_o up to $E_{max} \gg k_B T$ is assumed (as in glasses[3]) then
for $\hbar\omega \ll k_B T$ the velocity change due to the resonant interaction is
given approximately by

$$\frac{\Delta c}{c} = \frac{n_o g^2 \gamma^2}{2\rho\omega K} \left(\frac{1 - \eta^2}{1 + \eta^2}\right)^2 \left(-1 + \ln\frac{2k_B T}{E_{max}}\right).$$

The ℓnT temperature variation is similar to that found for bulk waves
in amorphous solids[3] but the inverse frequency dependence, which
arises from the displacement form of the coupling is quite different.

We have estimated the mean density of states $n_0 \sim 10^{16}$ cm^{-2} eV^{-1} from considerations of the likely width of the energy distribution and the available number of states per unit area. For g we have taken the derivative of the attractive van der Waals potential[2] at the energy of the ground state of a single helium atom, $\partial V/\partial Z \sim 50$ K Å^{-1}. This gives an attenuation coefficient $\ell^{-1} = (c\tau)^{-1} \sim 0.3 \times 10^4$ cm^{-1} for 2 GHz at 0.1 K, which is comparable with that for bulk phonons in glasses. Thus experiments on the attenuation and velocity dispersion of ASW appear to be quite feasible and would give valuable information on the dynamic properties of helium monolayers adsorbed on an inhomogeneous substrate.

REFERENCES
1. F J Milford and A D Novaco, Phys Rev A4, 1136 (1971).
2. G Benedek and G P Brivio, J Phys C 9, 2709 (1976).
3. S Hunklinger and W Arnold, Physical Acoustics, eds R N Thurston and W P Mason (Academic, New York), Vol 12, p 155 (1976).
4. D W Princehouse, J Low Temp Phys 7, 287 (1972).
5. J G Daunt and P Mahadev, Physica 69, 562 (1973).
6. T Nakayama, J Phys C 10, 3273 (1977).
7. J Weber, W Sandmann, W Dietsche and H Kinder, Phys Rev Lett 40, 1469 (1978).

****

DISCUSSION

H. J. Maris: I don't believe that your way of estimating that coupling constant is appropriate. You can show analytically that as the splitting between two tunneling levels becomes very small, which is what we're talking about here in considering the attenuation of the sound wave, the matrix element has to tend to zero. In fact it tends to zero as ω^2. So taking a constant matrix element is very unreasonable. In fact if you look at that matrix element where you have $\partial V/\partial z$ you have nearly exact cancellation from the contribution from the left hand and the right hand well in that case. I think that this gives you an attenuation which varies with a different frequency dependence - a higher frequency dependence - and I think it should make the overall effect smaller. Now, on the other hand if we go to a different sort of coupling process which is more like the glass coupling, which is really coupling to the strain field not to the displacement, then I think it is legitimate to take a constant matrix element, and it would certainly seem to me that from what we know about the Kapitza resistance results it might very well be that a surface wave travelling on a surface with a monolayer of helium should have a rather reasonable attenuation value. So I intuitively believe that the answer you have of 10^{-4} or something in that ball park sounds very reasonable, but I do not believe the mathematics.

F. W. Sheard: Yes, I accept that my guess at the coupling constant
is quite arbitrary because we haven't taken into account the details
of the double-well potential.

H. J. Maris: It's very very sensitive at low frequency.

L. J. Challis: In the earlier talk that you gave you were coupling
phonons to a non-localized system which might well have been, as you
suggested, a spin-wave in these tunneling states. Would you like to
comment on the relative attenuation of surface waves by the localized
tunneling states and by the spin-waves?

F. W. Sheard: I want it both ways. I want spin waves this morning
and localized states this afternoon. The reason for discussing non-
localized states is to provide the wide-angular distribution. This
is simply suggesting an experimental means for studying what actually
is going on there, and in fact I haven't attempted here to combine
my tunneling states into non-localized waves simply because we wanted
to carry through the calculation in close analogy with amorphous bulk
materials to see what the order of magnitude was. I agree with you
that I should do it with non-localized waves, but we simply didn't
do that partly because we did two completely independent calculations.

LOW TEMPERATURE THERMAL CONDUCTIVITY OF KTaO$_3$ AND KTN SINGLE CRYSTALS

A.M. de Goër, B. Salce and L.A. Boatner[*]

S.B.T., Centre d'Etudes Nucléaires de Grenoble
B.P. 85 X, 38041 Grenoble Cedex, France
[*]O.R.N.L., Sol. State Div., Oak Ridge,Tennessee 37830,USA

The mixed crystals KTa$_{1-x}$Nb$_x$O$_3$ (KTN) are ferroelectrics with a transition temperature T$_c$ which depends on the Nb content x. The phase diagram for small x, established by dielectric and elastic measurements[1,2], shows that crystals with x < 0.8% remain paraelectric at all temperatures. We have measured the thermal conductivity K of several crystals with x varying from 0 to 3%, from 1.3 to 200 K and, in some cases, down to 80mK. The standard stationary flow method was used and the error in K is thought to be less than 7%. The single crystals were grown by means of a flux technique, using a high temperature (1430°C) reaction of Ta$_2$O$_5$, Nb$_2$O$_5$ and an excess of K$_2$CO$_3$. The material, which was contained in a platinum crucible, was held at the maximum temperature for approximately 12 hours, and then slowly cooled (1° to 3°C/hour) between 1430 and 950°C. Self-nucleation usually resulted in the production of 1 to 4 large grains. Cooling from 950°C to room temperature was accomplished in 8 to 10 hours and hot H$_2$O was then used to remove the flux in order to free the crystals. The samples were cut, using a wire saw, as parallelepipeds with cross-sections about 2 x 2 mm^2 and length 8 to 16 mm along 100 axis. Typical compositional inhomogeneities can be estimated to about 0.3%[2].

Three KTaO$_3$ crystals (x = 0) containing different impurities have been measured and the results are shown in fig. 1(a). The "pure" sample contains some Ti as residual impurity. The introduction of Cu or Ag does not affect noticeably the thermal conductivity (not at all above 100 K) and the existence of a minimum near 6 K appears to be an intrinsic feature. The Cu-doped sample has been measured down to 80mK (insert of fig. 1) and it is seen that K/T^3 reaches the Casimir limit at the lowest temperatures. Above 100 K, our results agree with those given by Steigmeier[3], who found that K(T)

243

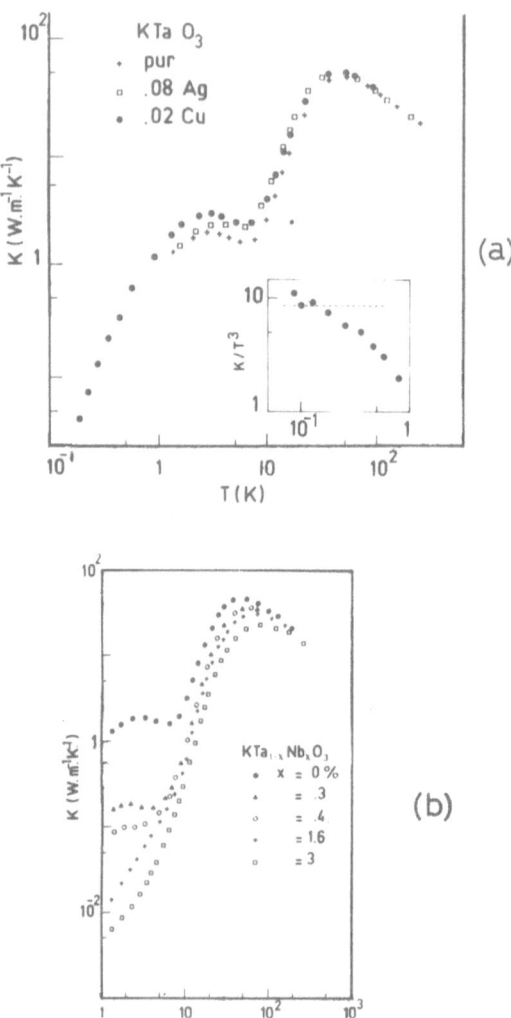

Fig. 1 - Thermal conductivity as a function of temperature.

below 100 K was very much sample dependent. Several KTN crystals
have been measured and some of the results are shown in fig. 1(b).
The thermal conductivity decreases with x increasing, in the whole
temperature range and especially below 10 K, even for the non ferroelectric crystals with x < 0.8%. There is no specific anomaly in
K at T_c in the ferroelectric crystals (T_c = 23 K for x = 1.6% and
37 K for x = 3%). For x = 3%, two brother samples give the same
results within 10% ; the value of K at 1 K is about 10^3 times smaller than the Casimir limit. One sample with x = 1.6% has been measured recently down to 80mK and K is still less than the Casimir
limit by a factor $\simeq$ 50 at the lowest temperature. A feature observed in all KTN samples is that the temperature dependence of K is
stronger than T^3 in the range 8-17 K.

We have tried to analyze quantitatively the KTaO₃ results with
the hypothesis of a strong scattering of the acoustic phonons by the
soft transverse optic mode[4] but so far no fit could be obtained.
We note that the specific heat data[5] shows a large discrepancy with
the calculated Debye contribution so that the question of the existence of a phase transition below 10 K remains opened. The very
strong phonon scattering observed in the KTN crystals is apparently
not related to the appearance of ferroelectricity : it is present
in crystals with small x and the values of K near 10 K are about
10^2 times smaller than those measured on ferroelectric BaTiO₃
crystals[6]. It is plausible that the phonon scattering is related
to the interaction of the acoustic phonons with the soft mode, which
is sensitive to the Nb content, but a suitable theoretical model
is lacking ; on the other hand, extrinsic phenomena due to residual
impurities or clustering (in the doped samples) cannot be completely
excluded in the present state of this work.

REFERENCES
1 L.A. Boatner, U.T. Höchli et H. Weibel, Helv. Phys. Acta 50 : 620
  (1977).
2 U.T. Höchli, H.E. Weibel and L.A. Boatner, Phys. Rev. Letters
  39 n° 18 : 1158 (1977).
3 E.F. Steigmeier, Phys. Rev., 168 n° 2 : 523 (1968).
4 H.H. Barrett and M.G. Holland, Phys. Rev. B 2, n° 8 : 3441 (1970).
5 W.N. Lawless, Phys. Rev. B 16 n° 1 : 443 (1977).
6 A.J.H. Mante and J. Volger, Physica 52 : 577 (1971).
                              ****

DISCUSSION

B. H. Armstrong: Do you know how the magnitude of the thermal
conductivity you measure compares to the Casimir boundary scattering
value in the low temperature regime?

A. M. de Goer: It is quite small. This means that the pure crystal
recovers the Casimir value at below 1.2°K and the doped crystals are
about 1000 times smaller at the lowest temperature than this.

PHONON SCATTERING BY NITROGEN AGGREGATES IN

INTERMEDIATE TYPE NATURAL DIAMONDS

J. W. Vandersande

Physics Department, University of the Witwatersrand

Johannesburg, South Africa

The low temperature thermal conductivity of crystals can and has been used to determine the approximate size and concentration of inclusions, precipitates and defects produced by irradiation damage.[1,2,3,4,5] This property makes thermal conductivity a very powerful tool eventhough it involves an average over the phonon spectrum. Large aggregates (the general term used here to indicate inclusions, precipitates etc.) scatter phonons with either a Rayleigh cross section (a relaxation rate proportional to the fourth power of the frequency), when $qa < 1$, or with a constant cross section (a frequency independent relaxation rate), when $qa > 1$ (called geometrical scattering), where q is the phonon wave vector $(2\pi/\lambda)$ and a is an average linear dimension of the aggregate. The average size of the aggregates is determined from the temperature at which the scattering changes from geometrical to Rayleigh. This changeover occurs when $qa \simeq 1$. Walton[3] and Guenther and Weinstock[4] considered a to be the radius of the aggregate whereas Schwartz and Walker[2] considered it to be the diameter. Unfortunately not one of these groups determined the size of their aggregates by a direct method, such as electron microscopy, to confirm their conductivity results. Only indirect methods were used.[1,2] There thus seems to be some confusion as to how to determine the sizes of aggregates from experimental results. Of course, the changeover region is not well defined and should depend to some degree on the differences in density and elastic constants between the aggregates and the host lattice. Since these differences are very likely not known or difficult to calculate in most cases, they are usually omitted when determining the sizes of the aggregates. The question of whether a is the radius or the diameter is thus of importance. The work reported here is an attempt at answering that question by

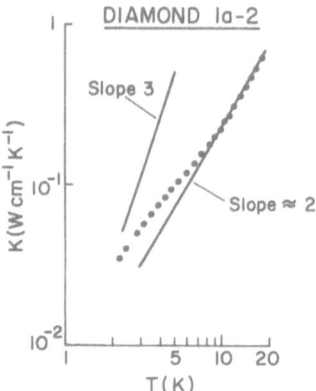

Fig. 1. A plot of thermal conductivity vs T for diamond Ia-2

determining the mean size of aggregates by both thermal conductivity measurements and by direct observation in the electron microscope.

The experimental procedure has been described elsewhere.[6] The
two diamonds used (called Ia-2 and Ia-3) were classified according
to their ultraviolet and infrared absorption spectra and were
found to be intermediate between types I and II. Diamond Ia-2 was
examined in the electron microscope by Evans at Reading University.
He found a large number of dislocation loops, which were from about
50 nm to 300 nm long and around 10 to 100 nm wide, in the cube
planes with a $\frac{1}{2} < 110 >$ type Burgers vector. Decorating and inside
these dislocations he found aggregates, believed to consist of
nitrogen,[7] of about 50 Å in diameter and with a concentration of
about 10^{14} per cm^3.

Fig. 1 shows the experimental results for diamond Ia-2. The
results for diamond Ia-3 were almost identical to those for Ia-2
and have not been drawn in. The Debye model for thermal conductivity with the usual combined relaxation-time assumption was used to
fit the results. At these low temperatures (diamond has its
conductivity maximum at about 80K) only boundary, isotope, impurity
point-defect, dislocation and aggregate scattering need be considered. For aggregates that are approximately spherical, Schwartz
and Walker[2] used the following very simple aggregate scattering

relaxation rates:

$$\tau^{-1}_{aggregate} = C\omega^4 \text{ for } qa < 1 \text{ (Rayleigh scattering)}$$

$$\tau^{-1}_{aggregate} = Nv\pi a^2/4 \text{ for } qa \gtrsim 1 \text{ (geometrical scattering)}$$

where a is the effective diameter, N is the concentration of the aggregates, q is the phonon wave vector $(2\pi/\lambda)$, v is an appropriately averaged sound velocity (the Debye velocity here) and $C = N\pi a^6/4v^3$. The relaxation rates for the other scattering processes as well as the complete theoretical model have been discussed elsewhere.[6] The computer fit, which has not been drawn in for reasons of clarity, was very good and only missed going through three points.

Choosing the middle of the changeover region to be between $qa=1$ and $qa=1.5$, the mean diameter determined from the fit was found to be between 44 and 65 Å with a corresponding aggregate concentration of between 1.7×10^{14} per cm^3 and 0.75×10^{14} per cm^3. These values agree very nicely with those determined from the electron micrographs. Had a been chosen as the radius of the aggregates then there would have been no such agreement. From the experimental results above about 7K in fig. 1 it can be seen that true geometrical scattering is not observed since the results do not lie on a line with slope 3, which would correspond to a conductivity proportional to T^3 and thus to a constant scattering cross section. These results lie on a line that corresponds to a conductivity proportional to approximately T^2 which is characteristic of scattering by dislocations. This would seem to agree nicely with the large number of dislocations observed spectroscopically in the diamond as well as the large dislocation term $(\tau^{-1}=7.6 \times 10^{-7}\omega)$ needed in the fit. The conductivity above 7K seems to be a mixture of aggregate and dislocation scattering.

To conclude, it thus appears that in this particular case of nitrogen aggregates in diamond the elastic properties of the aggregates need not be known to determine their average size by means of thermal conductivity measurements. Before this conclusion can be generalized to cover any kind of aggregate and host more experimental work would have to be done.

I am very grateful to Professor T. Evans for the micrographs.

References
1. J. M. Worlock, Phys. Rev. 147, 636 (1966).
2. J. W. Schwartz and C. T. Walker, Phys, Rev. 155, 969 (1967).
3. D. Walton, Phys. Rev. 157, 720 (1967).
4. R. A. Guenther and H. Weinstock, J. of Appl. Phys. 42,3790 (1971)
5. J. W. Vandersande, Phys. Rev. 15, 2355 (1977).
6. J. W. Vandersande, Phys. Rev. 13, 4560 (1976).
7. T. Evans, Diamond Research 1978, 17 (1978).

DISCUSSION

J. K. Wigmore: Is this data then saying that when nitrogen impurities go into diamond that they form clusters of this size, or is the
50 Å the peak of some sort of random distribution that you would get
by just putting them randomly in the lattice?

J. W. Vandersande: That's an interesting point. A lot of work has
actually been done on nitrogen defects in diamond and from work I
have done and some other people have done it looks like this 50 Å
comes up over and over again. We really don't know. They exist in
very small groups, but this is the first time we've seen big groups.

P. G. Klemens: I understand from microscopic investigations of
neutron irradiated sapphire at Los Alamos there are larger aggregates
and they too happen to be 50 Å.

CORRELATED THERMAL CONDUCTIVITY AND IONIC THERMOCURRENT

MEASUREMENTS TO STUDY THE SUZUKI-LIKE PHASE NUCLEATION IN KCl:Pb

M. Locatelli*, R. Cappelletti**, E. Zecchi*

* L.C.P., S.B.T., C.E.N.G. 85X 38041 Grenoble Cedex
**Istituto di Fisica dell'Universita Parma, Italy

In a previous work (1), we have used low temperature thermal conductivity to detect **Suzuki-like-phase** (S.L.P.) occlusions in Pb doped KCl which grow, for increasing annealing time at 220°C, at the expense of simpler defects such as impurity-vacancy (I.V.) dipoles. We concluded in (1) that these occlusions are not spherical in shape but have a typical dimension of a few tens of **nanometers**.

In the present work, we have extended these studies to get information on the concentration, shape and dimensions of the S.L.P. occlusions. Using standard methods for thermal conductivity (1) and ITC (2) experiments we have studied several samples made at Parma and heat treated simultaneously for both types of measurements.

Results and Analysis

The thermal conductivity results given in fig. 1 as K/T^3 versus T curves show two dips : one at high temperatures and attribued to I.V. dipoles and one at low temperatures **due to** S.L.P. occlusions.

Results are analysed using the Debye model for phonon conduction as in (1). The total mean free path $l(\omega)$ is given by $l_{(\omega)}^{-1} = \Sigma l_i^{-1}$ where l_i^{-1} describe different scattering mechanismes ; i.e. scattering at boundaries, dislocations, other phonons, points defects, I.V. dipoles and occlusions. For the first four contributions, we use the expressions derived in (1) stationary that given in (3):

$$l_{I.V.}^{-1} = \frac{c}{v} B \exp(-6.76\, T/T_0)\, \omega^4 T^2/(\omega^2 - \omega_0^2)^2 \quad \text{where } c \text{ is the}$$

I.V. dipole concentration determined by ITC measurements, v the
velocity of sound, T_D the Debye temperature and ω_o a resonance
frequency attribued to the I.V. dipole, $\omega_o = T_o = 29$ K, and B is a
constant.

Concerning the last contribution, previous work (1), (2), suggested that the S.L.P. occlusions are possibly cylindrical. Other
non-spherical occlusions have been considered in (4) and using
similar arguments we propose here a phonon scattering model for

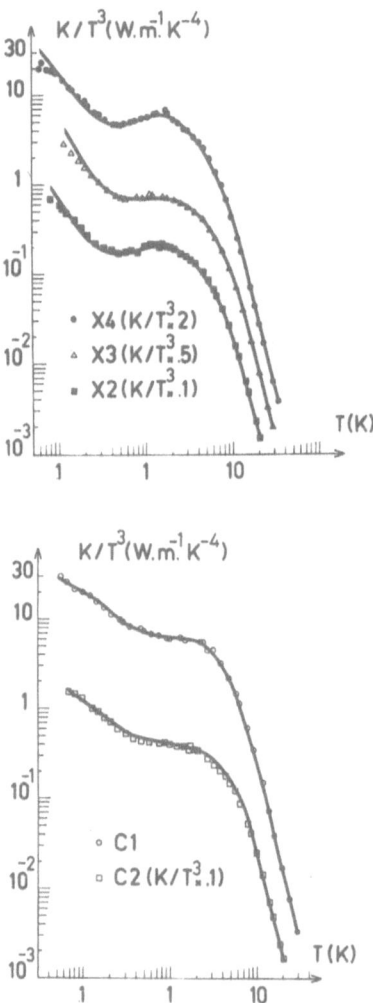

Fig. 1 : K/T^3 versus T, experimental points and calculated curves

cylindrical occlusions (of height H and diameter d) to account for
the new data. At low frequencies, we expect Rayleigh-type scattering and the scattering cross-section σ_H is taken to be proportional to the square of the scattering volume $V = \pi H d^2/4$ and proportional to $(\omega/v)^4$. If the occlusion is cylindrical, it has been
shown (see for instance W. BARBER and al. (5)) that this must be
modified to $(\omega/v)^n$ where $n \leq 3$ so that

$$\sigma_H = g \ V^2 \ (\omega/v)^3 \ \text{for} \ \frac{H\omega}{v} < b$$

The cross-over condition is given by $b = 1.5$ and g is a dimensional constant to take into account the ω^3 dependence of σ_H.

For higher frequencies, we assume that σ_H decreases and describe this by a gaussian multiplier

$$\sigma'_H = \sigma_H \exp \left(- \left(\frac{\omega - (bv/H)}{c}\right)^2 \right) \ \text{for} \ \frac{\omega H}{v} > b$$

At still higher frequencies, we define a scattering cross-
section σ_d associated with the diameter d and a cross-over condition $a = 1.5$.

For $\omega d/v \geqslant a$, we **expect** a geometrical cross-section $\sigma_d = d \ H$

For $\omega d/v < a$, we **assume** $\sigma_d = d \ H \ (d\omega/av)^4$

The combined cross-section $\sigma_c = \sigma_d + \sigma_H$ gives the usual expression for the mean free path by $l^{-1} = N_c \ \sigma_c$ where N_c is the
number of cylindrical objects per cm^3.

The curves of fig. 1 are calculated with this expression in-
clued in $\sum_i l_i^{-1}$.

The overall agreement with experimental results is quite
good. Some residual disagreement at the lowest temperatures is taken as a indication that there, the frequency dependence should be
perhaps $(\omega/v)^n$ with $n < 3$ in agreement with (5). The concentration
and shape parameters deduced are summarized in table 1.

Table 1

Sample	Pb ppm(ITC)	annea- ling h	IV dipoles ppm (ITC)	Nc $cm^{-3} \times 10^{-13}$	H nm	d nm	c $s^{-1} \times 10^{-4}$
C1	90	24	15	1	20	1.5	2
C2	90	200	5	1	20	1.7	8
X4	32	103	4	6	35	1.5	1
X3	32	300	1	5	20	1.8	1
X2	32	700	<1	4	45	2	1

In conclusion, it seems that cylindrical (7) S.L.P. occlusions grow in KCl:Pb during annealing at the expense of I.V. dipoles but their concentration remains constant.

References

(1) M. LOCATELLI, Journal de Physique, C7, 12-37, 322 (1976)

(2) R. CAPPELLETTI and A. GAINOTTI, Journal de Physique, C7, 12-37, 316 (1976)

(3) M. LOCATELLI, R. CAPPELLETTI, E. ZECCHI, Phonon resonances induced by Pb in KCl Studied by means of thermal conductivity and ionic thermocurrents technique, this conference

(4) M. ABOU GHANTOUS, K. GUCKELSBERGER, M. LOCATELLI, To be published in J. Phys. Chem. Solids

(5) P.W. BARBER and DAN JIN WANG, Appl. Opt 17-5, 797 (1978)

(6) Assuming a cylindrical shape of the occlusions, we get good agreement with experimental data. The torus-like shape used in (4) for LiFγ gives also an overall agreement but perhaps less so than the cylinders, which are consistent with I.T.C. data.

PHONON RESONANCES INDUCED BY Pb IN KCl STUDIED BY MEANS OF THERMAL CONDUCTIVITY AND IONIC THERMOCURRENT TECHNIQUES

M. LOCATELLI*, R. CAPPELLETTI**, E. ZECCHI**

*L.C.P.,S.B.T.,C.E.N.G., 85X 38041 GRENOBLE CEDEX

** Istituto di Fisica dell'Università PARMA (Italy)

In order to study the evolution of Suzuki-like phase occlusions in KCℓPb (1), (2), by thermal conductivity measurements at low temperatures, we have to consider the phonon interactions with divalent ions Pb^{2+}. These interactions were studied many times (3) and theoretical models were proposed. One model is based on one-phonon scattering due to the mass defect (4) with an extension proposed by M. WAGNER (5) to include two-phonon interactions. A second is based on one-phonon scattering due to force constant changes in the presence of a vacancy (6). In (3) J.W. SCHWARTZ et al. concluded that phonon scattering is mainly due the cationic vacancy; the divalent ion acts when its size differs from that of the host ion. Nevertheless they do not exclude the possibility of two-phonon processes as treated by M. WAGNER (5).

Attempts to fit our experimental thermal conductivity results on the basis of these models were not satisfactory, and the purpose of this work is the search for an appropriate phenomenological model. Using standard methods for thermal conductivity (1) and ITC (2) experiments, we have studied a few samples with Pb concentration ranging from 1.2 to 144 ppm. These samples were properly quenched in order to avoid (or minimize) the presence of precipitates which change strongly the thermal conductivity (1).

Thermal conductivity results are analysed using the Debye model for phonon conduction as in (1). The total relaxation time $\tau(\omega)$ is given by

$$\tau^{-1}_{(\omega)} = \sum_i \tau_i^{-1} \text{ where } \tau_i^{-1} \text{ describe different scattering mechanisms;}$$

i.e. scattering at boundaries, dislocations, point defects $(A\omega^4)$, other phonons, Pb^{2+}. For the first four contributions we use the equation derived in (1). For the Pb^{2+} contribution the following model is used. The starting point is the general inverse relaxation time for elastic processes (7)

$$\tau_E^{-1} = D\omega^4/(\omega^2-\omega_o^2)^2 \quad ,$$

where D is proportional to the concentration of defects and ω_o a resonance frequency.

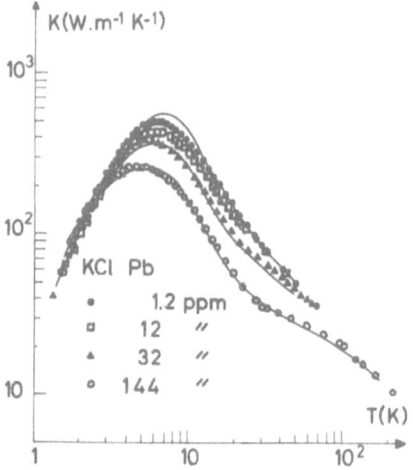

Fig. 1: K(T) exp. pts and cal. curves.

As the dip on thermal conductivity, fig. 1, is observed in a temperature range where phonon-phonon interactions are important and these interactions are temperature dependent, we introduce a temperature dependence, as suggested in WAGNER's model (5).

The following expression gives the best fits to our results

$$\tau_{Pb^{2+}}^{-1} = \exp(-6.76\ T/T_D)\ T^2\ \tau_E^{-1}$$

T_D is the Debye Temperature (230 K for KCl).

The values of D and A (point defect coefficient) are plotted in figure 2 versus the concentration of simple defects such as impurity-vacancy (I.V.) dipoles determined by ionic thermocurrents (I.T.C.) and versus the overall Pb concentrations determined by optical absorption. Linear relations are obtained between D,A and the concentration; good agreement is obtained over the whole range of Pb concentrations.

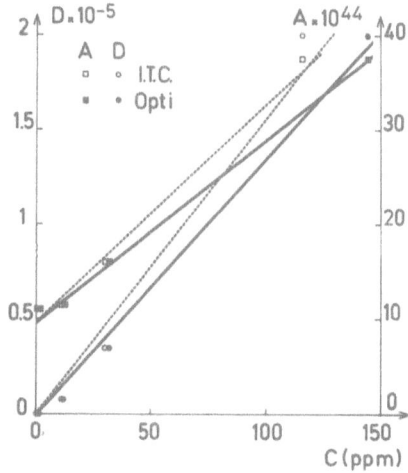

Fig. 2:   D and A versus C_{Pb} (ITC and optical)

In summary, we have proposed a phenomenological expression for phonon scattering by Pb^{+2} in KCl, this expression being related to phonon-phonon processes. The resonance frequency $\omega_o = T_o = 29$ K confirms the role of the cationic-vacancy in resonant phonon-scattering (3). The linear relation between A (point defect coefficient) and the total Pb concentration seems to indicate that the divalent ion acts as a point defect.

References

(1) M. Locatelli, R. Cappelletti, E. Zecchi, Correlated Thermal

Conductivity and Ionic Thermocurrents Measurements to Study
the Suzuki-like Phase Nucleation in KCl, paper presented at
this conference.

(2) R. Cappelletti, A. Gainotti, Journal de Physique, C7, 12-37,
316 (1976).

(3) J.W. Schwartz and C.T. Walker, Phys. Rev. 155, 3, 959 (1967).

(4) C.W. McCombie and J. Slater, Proc. Phys. Soc. (London) 84, 499
(1964).

(5) H. Wagner, Phys. Rev. 131, 1433 (1963).

(6) J.A. Krumhansl, in Proceedings at the International Conference
on Lattice Dynamics, Copenhagen, 1963, edited by R.F. Wallis
(Pergamon Press, Inc., New York, 1965), p. 523.

(7) R.O. Pohl, in Localized Excitations in Solids, edited by R.F.
Wallis, (Plenum Press, New York, 1968), p. 434.

SCATTERING OF PHONONS BY VACANCIES

C.A. Ratsifaritana and P.G. Klemens

Dept. of Physics and Institute of Materials Science
University of Connecticut
Storrs, CT   06268

Scattering of phonons by substitutional atom has been treated by several authors[1], both by perturbation theory, and by self consistent solutions of the wave field (Green's function technique). We aim to find the scattering cross section of a vacancy, because this defect is important in radiation damage and for non-stoichiometric compounds. There are some conceptual difficulties which prevent the straightforward application of the usual theory. We shall be interested in the long-wavelength limit, because the long waves are important for the thermal conduction both at low and at high temperatures.

We shall use perturbation theory and the Green's function technique[2,3] both in the continuum limit and for a simple discrete model. We shall find that the result of perturbation theory in an appropriately chosen continuum model seems quite adequate, and that the distortion of the lattice about the vacancy can be disregarded.

A vacancy is a place where an atom is missing, and where the linkages between that atom and all its neighbors are removed. Thus the perturbation Hamiltonian has two contributions: the kinetic energy of the missing atom and the potential energy of the missing linkages. Since each linkage is shared by two atoms, the missing potential energy is that of two atoms. Using the virial theorem, we have

$$H' = -\frac{3}{2}\frac{M}{G}\sum_{q,q'}\omega^2 \alpha^*(q')\alpha(q)\,(\underline{\varepsilon}\cdot\underline{\varepsilon}')$$

(1)

and in the usual way the relaxation rate[4] becomes

$$\frac{1}{\tau} = 9 \frac{a^3}{G} \frac{\omega^4}{4\pi \, v^3}$$

(2)

Here M is the atomic mass, a^3 the atomic volume, G the number of atoms per defect, ω and q the phonon frequency and wave vector, ε the polarization direction, $a(q) \, a^*(q)$ the annihilation and creation operators describing the time dependent displacement and v the phonon velocity.[4]

Distortion, together with anharmonicity, changes the force constants of all the linkages. This problem has been treated by Carruthers[5] in the continuum limit, who showed that the perturbation resides right at the origin. In a discrete model, this means that the perturbation resides at the nearest linkages emanating from the defect. In the case of the vacancy, these are the missing linkages. Since missing linkages cannot be modified by anharmonicity, one is justified in disregarding distortion effects when calculating phonon scattering by a vacancy.

We have also treated distortion on a discrete model, a simple cubic lattice. We assumed a distortion field of the same form as in the continuum case (displacement proportional to r^{-2}), calculated the change in force constants due to anharmonicity and found indeed that the distortion effects cancel almost completely if the nearest linkages are removed, while the inclusion of nearest linkages recovers closely Carruthers continuum result.

So far we have only used perturbation theory, which is justified at low frequency, provided there are no low-lying resonances. To explore that possibility, we have used Green's function method on the same model of a vacancy in a simple cubic lattice.[3] We have verified that for this model the resonances all lie above $\omega_D/2$. This means that vacancies do not lead to giant scattering cross section at low frequencies. The appearance of very strong low frequency scatterings found for the thermal conductivity of LiF with F-centers[6] and of neutron irradiated MgO[7] thus have to be explained in terms of other defects than simple vacancies. However, it must be emphasized that our discrete model only admits forces between nearest neighbors, and that the consideration of long range electrostatic forces might lead to a different picture.

In summary, the perturbation theory, using the removal of the kinetic energy of one atom and the potential energy of two atoms, should provide a good description of phonon scattering by vacancies.

We have applied this simple result to three cases: Ca^{++} plus a vacancy in KCl,[8] carbon vacancies in ZrC[9] and tin vacancies in SnTe.[10] The first was measured some time ago by Slack.[8] We have reanalyzed his results in terms of the Callaway theory,[11] which was then not available. From the difference between the pure specimen and the one of lowest Ca content (0.6×10^{-4} per K atom), we obtain

$$\frac{1}{\tau} = A\omega^4 \tag{3}$$

where $A = 5.8 \times 10^{-43}$ sec^3. Our simple theory would give 5.4×10^{-43} sec^3 for the vacancy. The Ca^{++} substitutional atom is a relatively weaker scatterer, but has to be added.

The other cases refer to results at high temperatures but with several high concentrations of defects so that the important frequency regime for point defect scattering is still below $\omega_D/2$. With

$$\frac{1}{\tau} = A\omega^4 = \frac{3a^3\omega^4}{G\pi v^3} S^2 \tag{4}$$

we find in the case of the carbon vacancy in ZrC that the observed scattering number S^2 is .98, while the present theory gives 1.24. For the tin vacancy in SnTe the experiments yield $S^2 = .89$ while theoretically $S^2 = .75$.

REFERENCES

1. Berman, R. and J.C.F. Brock, Proc. R. Soc., A289:46 (1965).
2. Lifshitz, I.M., Nuovo Cim., 3SVI:716 (1956).
3. Elliott, R.J. and D.W. Taylor, Proc. Phys. Soc., 83:189 (1964).
4. Klemens, P.G., in "Thermal Conductivity", Vol. 1, Ch. 1, R.P. Tye, ed., Academic Press, London (1969).
5. Carruthers, P., Rev. Mod. Phys., 33:92 (1961).
6. Pohl. R.O., Phys. Rev. Letters, 8:481 (1962).
7. Kupperman, D.S. and H. Weinstock, J. Low Temp., 10:193(1973).
8. Slack, G.A., Phys. Rev., 105:832 (1957).
9. Taylor, R.E., and E.K. Storms, in "Thermal Conductivity 14," p. 161, P.G. Klemens and T.K. Chu, eds., Plenum Press, New York (1975).
10. Damon, D.H., J. Appl. Phys., 37:3181 (1966).
11. Callaway, J., Phys. Rev., 113:1046 (1959).

****

DISCUSSION

A. M. de Goer: What is the ingredient that you have to put into the theoretical calculation of S^2?

P. G. Klemens: In simple systems you will have the same value of S^2 for all systems because the kinetic energy and the potential energy are related by the virial theorem. But in compounds it varies a little bit because $\Delta M/M$ depends upon which type of atom is missing.

EVIDENCE OF QUASI-LOCAL PHONON MODES

AROUND DISLOCATIONS IN CRYSTALS

Y. Hiki and F. Tsuruoka

Tokyo Institute of Technology, Oh-okayama, Meguro-ku, Tokyo 152, Japan

It has been shown by some authors that lattice vibrational modes called the quasi-local vibrations should exist in crystals containing point defects or dislocations.[1] Quasi-local mode is the vibrational state spatially localized near the defects and the vibrational frequencies are inside the band of ideal lattice. Such vibrational mode is important since it must have direct and significant effects on the properties of the crystals at low temperatures. It is noted that, in the case of dislocations, the quasi-local vibrations are distinct from the low-frequency wavy vibration, or the fluttering, of dislocation lines. Some experimental evidences of the quasi-local modes in crystals containing dislocations will be shown in the following.

Specific heat. It was found that low-temperature specific heat of Cu-Al alloy crystals containing high density of dislocation showed an anomalous behavior.[2] Specific heat of metals is composed of the electronic and the lattice components: $C = \gamma T + \alpha T^3$. As can be seen in Fig. 1 (a), the C/T-vs-T^2 plot shows a deviation from linearity in the case of swaged specimens. We expect that there exists in the crystal an extra phonon mode associated with dislocations, and we also assume the phonon density of state $D(\omega) \propto \omega^m$; $\omega_1 \leq \omega \leq \omega_2$. Then one obtains the expression for the lattice specific heat, which is the sum of the contributions from the usual phonons with the ω^2 distribution and the extra phonons with the above distribution. The expression was fitted to the experimental data by adjusting the parameters m, ω_1, and ω_2. It was decided that the most likely distribution of the extra phonons was the Einstein type one, being shown in Fig. 1 (a). It was also found that the specific heat of deformed LiF crystals was anomalous.[3] The C-vs-T^3 plot shows a nonlinearity as illustrated in Fig. 1 (b). In this case, results are well explained when extra phonons with the Debye distribution are assumed.

Thermal conduction. Since the quasi-local mode phonons localize

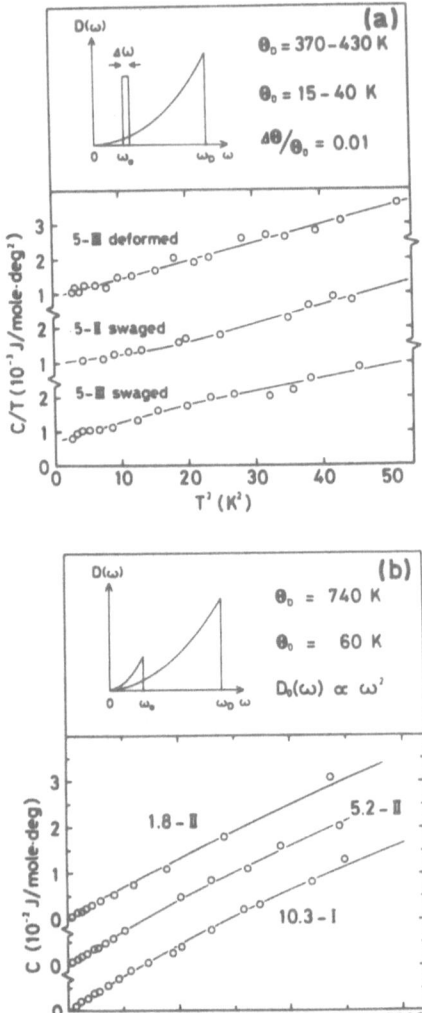

Fig. 1. Temperature dependence of specific heat C. (a) C/T–vs–T² plots for deformed and swaged Cu – 5 at.% Al crystals. (b) C–vs–T³ plots for LiF crystals deformed by 1.8, 5.2, and 10.3 %. The curves in the figures are theoretical ones.

around dislocations, flow of thermal phonons in crystal may be scattered by dislocations accompanied by the quasi-local phonons, and so
thermal resistivity may arise. After analyzing the thermal diffusivity of deformed LiF crystals, we found a new kind of thermal phonon
relaxation besides the relaxation arising from the fluttering of dislocation lines.[3] We are now developing a theory concerning the thermal resistivity due to the scattering through three-phonon processes
between thermal phonons and quasi-local phonons. The idea is similar
to the case of dislocation damping described in the following.

    Ultrasonic attenuation. We have measured the attenuation of
longitudinal sound at frequencies of 5 - 45 MHz in hcp ^{4}He crystals
at temperatures between 1.3 and 2.3 K. The main origin of the attenuation in our crystals was considered to be the overdamped resonance of dislocations, and the experimental results were analyzed
on this basis.[4] One of the most important quantities concerning the
dynamical properties of dislocations is the damping constant B.
A dislocation moving with a speed v in a crystal suffers a viscous

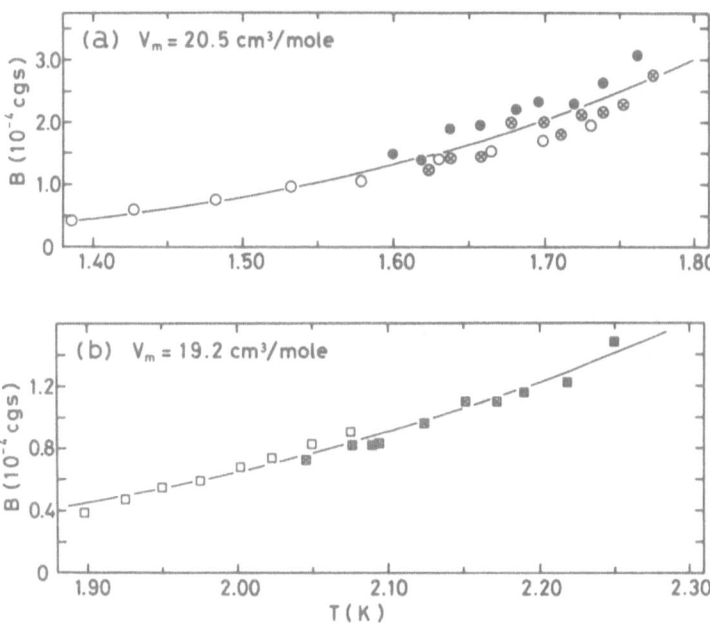

Fig. 2.   Temperature dependence of damping constant for dislocation
          motion B in hcp ^{4}He crystals with different molar volumes.
          The curves represent the results of the parameter fit.

force F = Bv per unit length. The damping constant can be determined
from the attenuation measurement, and the results are as shown in
Fig. 2. The magnitude and also the temperature dependence of B in
this temperature range cannot be explained by the existing theories.[4]
We considered as follows. When a dislocation moves, the quasi-local
phonons around dislocation absorb and emit thermal phonons. Probabilities of the two processes are different, and a certain amount of
energy is transferred from the quasi-local phonons to the crystal.
Then the energy of moving dislocation is dissipated. The damping
constant B is proportional to the number of quasi-local phonons N
$\propto$ exp($-\varepsilon$/T). The data in Fig. 2 show the exponential temperature
dependence, and the value of ε is 12 - 14 K. The absolute values
of B can also be calculated and are compared with the experiments.[4]

     Other experiments. Results on the temperature dependence of the
electrical resistivity due to dislocations in metals[5] and infrared
spectra of deformed ionic and semiconductor crystals[6] indicate the
existence of quasi-local mode phonons near dislocations.

REFERENCES

1.   I. M. Lifshitz and A. M. Kosevich, Rep. Prog. Phys. 29:217 (1966);
     Ya. A. Iosilevskii, JETP Lett. 7:22 (1968); G. G. Taluts, A. Ya.
     Fishman, and M. A. Ivanov, Sov. Phys. - Solid State 13:3016 (1972).
2.   Y. Kogure and Y. Hiki, J. Phys. Soc. Jpn. 39:698 (1975).
3.   Y. Kogure, H. Kaburaki, and Y. Hiki, in this issue.
4.   F. Tsuruoka and Y. Hiki, Phys. Rev. B, Oct. 1 (1979).
5.   G. I. Kulesko, Sov. Phys. - JETP 45:1138 (1977);
     T. Endo and T. Kino, J. Phys. Soc. Jpn. 46:1515 (1979).
6.   B. M. Tashpulatov, V. I. Vettegren', and I. I. Novak, Sov. Phys.
     - Solid State 20:111 (1978).

LOW-TEMPERATURE THERMAL PROPERTIES OF

LITHIUM FLUORIDE CONTAINING DISLOCATIONS

Y. Kogure*, H. Kaburaki, and Y. Hiki

Tokyo Institute of Technology, Oh-okayama, Meguro-ku,
Tokyo 152, Japan
* Present Address: Department of Physics,
University of Illinois, Urbana, Illinois 61801, U.S.A.

Thermal conductivity κ and thermal diffusivity $D = \kappa/\rho C$, where ρ is the density and C is the specific heat of the material, of Harshaw LiF crystals have been measured in the temperature range of 1.5 - 15 K by the temperature-wave method developed by the present authors.[1] Specimens with the (100) faces and $0.2 \times 0.3 \times 6$ cm in size were annealed at 750° C for 2.5 hr in argon atmosphere and then side faces were roughened with fine emery. Crystal dislocations were introduced in the specimen by compression along its axis by a loading machine. At that time, sides of the specimen were supported by brass blocks to prevent buckling, and the ends were covered with indium sheets. The amount of the deformation was 0.42 - 10.3 % in reduction of length. Then we shaped a carbon-film heater (H) and two thermometers (T_1, T_2) around the sides of the specimen by painting thin layer of Aquadag, and H, T_1, and T_2 were distant from an end of the specimen by about 0.3, 0.8, and 1.8 cm. The other end of the specimen was lapped with soft thin copper foil and covered with Apiezon N grease and mounted to a copper block attached to the bottom of a small chamber in a cryostat. The chamber filled with liquid helium acted as a heat sink. Temperature of the specimen was controlled by pumping the liquid helium and by adjusting the current in a heater wound around the heat sink. An AC current with the frequency of around f = 1 kHz was supplied to H, and a temperature wave propagating along the specimen to the heat sink was produced. The thermal diffusivity of specimen is determined from the measured phase difference $\Delta\phi$ of the wave between T_1 and T_2 as $D = (2\pi f)(\Delta x/\Delta\phi)^2$, where Δx is the distance between T_1 and T_2. In the present experimental condition, static temperature gradient is built up along the specimen. The thermal conductivity of specimen is determined from the temperature difference ΔT between T_1 and T_2 as $\kappa = Q_o(\Delta x/\Delta T)$, where Q_o is the power supplied to H.

Examples of our experimental data on the temperature dependence
of D and κ are shown in Fig. 1.   The gas kinetics expression for the
phonon thermal conduction κ = $\frac{1}{3}Cv^2\tau$, where v is the phonon velocity
and τ is the relaxation time, is written as D = $\frac{1}{3}(v^2\tau/\rho)$.   Namely,
the temperature dependence of D directly corresponds to the change of
the relaxation time with temperature.   The temperature dependence of
κ is, however, strongly affected by the change of the specific heat.
Accordingly, it is considered that the relaxation times for various
scattering processes can be determined more reliably and accurately
by using the diffusivity data.   We choose the relaxation rate for the
phonon scattering as[2] τ^{-1} = (v/L) + $0.32x^4T^4$ + $35xT^4$ + τ_D^{-1}, where L
is the dimension of crystal, and x = $\hbar\omega/k_BT$.   The individual rates
represent scattering by boundaries, isotopes, N-processes, and dislocations.   The U-process term can be omitted in the present tempera-

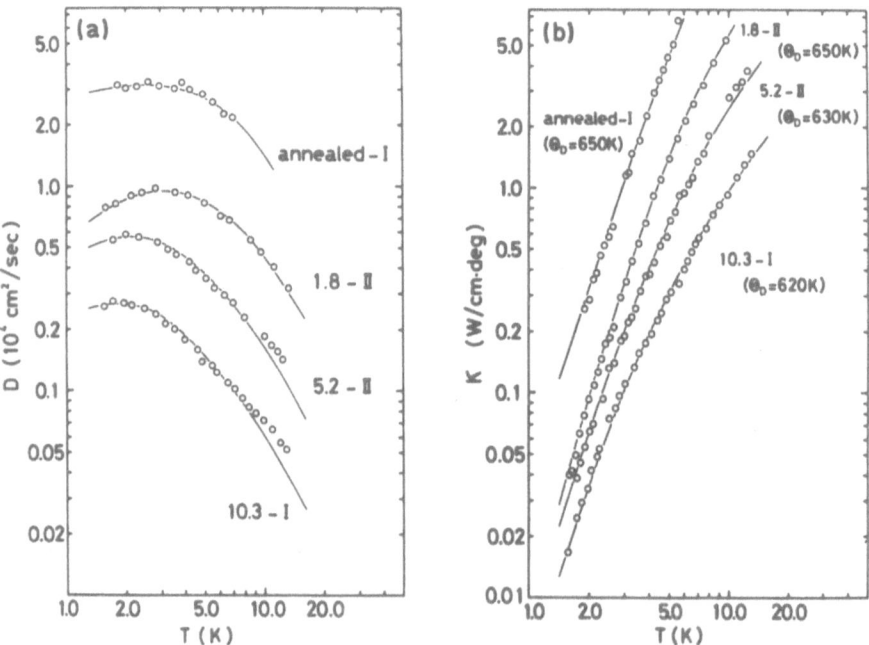

Fig. 1.   Temperature dependence of (a) thermal diffusivity D and
            (b) thermal conductivity κ of annealed and compressed LiF
            crystals.   The data are specified by per cent reduction in
            length and specimen number by Arabic and Roman numerals.
            The curves represent the results of the parameter fit.

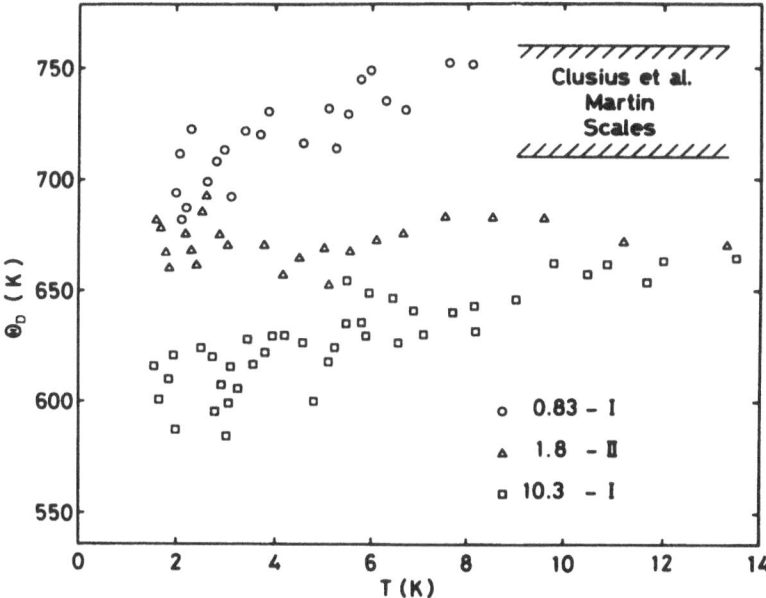

Fig. 2.   Debye temperature Θ_D of deformed LiF crystals.   The level
of the values reported by other authors is also shown.

ture range.   When the data of D for deformed crystals are compared
with those for annealed crystal, one can see a decrease of D at low
temperatures superposed on the overall decrease of the values.   We
adopted the expression $\tau_D^{-1} = F\omega^{-n} + A\omega^m$ as the frequency dependence
of the relaxation rate for the dislocation scattering, and the diffusivity data were fitted to the Callaway's formulas.[2]   It was found
that the case of n = 1 and m = 2 gave the best fit to the data.   The
curves in Fig. 1 (a) are the fitted ones.   The determined τ_D^{-1} was
used to calculate the values of κ.   The curves in Fig. 1 (b) show
the results.   At that time, it was necessary to adjust the values
of Debye temperature Θ_D to produce good fits.   Specific heat of the
crystals can be calculated from the experimental values of D and κ.
Debye temperature is determined as a function of temperature, and the
results are as shown in Fig. 2.   Decrease of Θ_D can be seen in the
deformed crystals compared with the existing data for undeformed
crystals.[3]   This can be explained as due to quasi-local mode phonons
around dislocations.[4]   The $A\omega^2$ term in τ_D^{-1} may be originated from
the scattering of thermal phonons by the quasi-local phonons, while
the $F\omega^{-1}$ term represents the scattering by vibrating dislocations.[2]

REFERENCES

1. Y. Kogure and Y. Hiki, Jpn. J. Appl. Phys. 12:814 (1973).
2. R. Berman, "Thermal Conduction in Solids," Clarendon Press,
    Oxford (1976); and references cited there.
3. W. W. Scales, Phys. Rev. 112:49 (1958); and cited references.
4. Y. Hiki and F. Tsuruoka, in this issue.

AN ULTRASONIC STUDY OF THE IONIC CRYSTAL β-AgI

J. H. Page[*] and J.-Y. Prieur

Laboratoire d'Ultrasons[†], Université Pierre
et Marie Curie, Tour 13, 4 place Jussieu,
75230 PARIS CEDEX 05, France.

Recently there has been much interest in the superionic (α) and wurtzite (β) phases of silver iodide. Below T_c = 420 K, the β phase is stable and although the ionic conductivity is nearly four orders of magnitude less than it is in the α phase, it is none-the-less higher than in many ionic compounds. The conductivity is due primarily to the hopping of silver ion vacancies created by the formation of Frenkel defects.[1]

We have made ultrasonic attenuation and velocity measurements[2] in β-AgI as a function of temperature at selected frequencies between 7.5 and 540 MHz. Longitudinal ultrasonic waves propagating along the c-axis were used. The variation of the attenuation is shown in Fig. 1 and was measured to an accuracy of 1 dB. The corresponding change in the elastic constant c_{33} is shown in Fig. 2.

The ultrasonic waves are coupled to the movement of the silver ions via a mechanism first considered by Hutson and White[3] for ultrasonic interactions in piezoelectric semiconductors. Since β-AgI is piezoelectric, an ultrasonic wave propagating in the crystal creates an internal electric field which in turn influences the hopping of the silver ions, resulting in a modification of the attenuation and velocity. As shown by Page and Prieur[2], this leads to the following expressions for the ultrasonic attenuation and elastic constant:

*Present address: Department of Physics, Queen's University, Kingston, Ont., Canada, K7L 3N6

†Associated with the Centre National de la Recherche Scientifique.

$$\alpha = \frac{\kappa^2}{v_s} \frac{\omega^2 \tau_c}{1 + \omega^2 \tau_c^2} \tag{1}$$

$$C = C_A \left\{ 1 + \kappa^2 \frac{\omega^2 \tau_c^2}{1 + \omega^2 \tau_c^2} \right\} \tag{2}$$

Here $\tau_c = \varepsilon \varepsilon_o / \sigma_v$ is the conductivity relaxation time where ε is the dielectric constant and σ_v is given by the usual Arrhenius law for ionic conduction. κ is the piezoelectric coupling constant, v_s is the velocity of sound and C_A includes the contribution to the elastic constant due to phonon-phonon interactions.

The solid curves in Fig. 1 are obtained from Eq. (1) with taken to be an adjustable scaling constant and all other parameters obtained from the literature.[1] The agreement with experiment is good at low frequencies where the position of the maximum in the attenuation is predicted exactly by the model and does not depend on any adjustable parameters. We find that $\kappa^2 = 0.088$, about twice the value reported by Fjeldly and Hanson.[4]

The curves for c_{33} in Fig. 2 are given by Eq. (2) where c_A is taken to decrease linearly with temperature, $c_A = c_o - \gamma T$, as

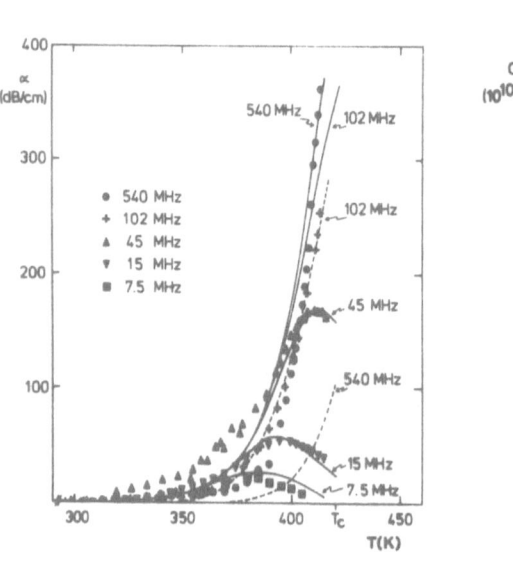

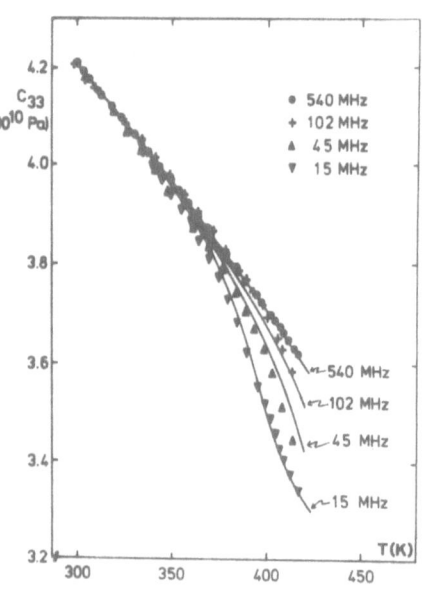

Fig. 1. Longitudinal ultrasonic attenuation parallel to the c-axis in β-AgI.

Fig. 2. Temperature dependence of c_{33} in β-AgI.

expected above the Debye temperature.[5] With $c = 5.20 \times 10^{10}$ Pa and $\gamma = 0.45 \times 10^8$ Pa K^{-1}, good agreement is found, the deviation from a linear temperature dependence being determined by the parameters κ and τ_c whose values were already known from the attenuation measurements.

Finally it should be noted that for the experimental data in Fig. 1 there is a range of temperature over which the high frequency attenuation is less than it is at low frequencies. This indicates that there is a screening effect which is not accounted for by Eq. (1). Our results may be explained if the ionic conductivity is taken to have a frequency dependence of the Drude type, $\sigma(\omega) = \sigma(0)/(1 - i\omega\tau)$, with a temperature-dependent relaxation time $\tau = \tau_o \exp(E_A/k_BT)$. By analogy with the work of Rice and Roth[6] for superionic conductors we suggest[2] that τ may be related to the lifetime of the vacancies. With $\tau_o = (1.0 \pm 0.3) \times 10^{-12}$s and $E_A = 0.3 \pm 0.03$ eV, the dashed curves shown in Fig. 1 were obtained. (At 45 MHz and below, the dashed curves are not shown since they are virtually identical to the solid curves). The fit is good for the 102 MHz data, although the dashed theoretical curve at 540 MHz is well below the experimental points, indicating that at high frequencies there is an additional attenuation mechanism which is probably associated with the phase transition at 420 K. Further experiments at 715 MHz and 1.0 GHz have been attempted and although the data is poor, the results are consistent with this interpretation.

We may conclude on the basis of this work that ultrasonic measurements may be used to investigate the ionic conductivity of piezoelectric compounds such as β-AgI in a frequency range which is not normally accessible by direct methods.

REFERENCES
1. G. Cochrane and N.H. Fletcher, Ionic Conductivity and Low Frequency Dispersion in Hexagonal Silver Iodide, J. Phys. Chem. Solids 32: 2557 (1971).
2. J.H. Page and J-Y. Prieur, Ultrasonic Investigation of the Piezo-electric Ionic Crystal β-AgI, Phys. Rev. Lett. 42:1684 (1979)
3. A.R. Hutson and D.L. White, Elastic Wave Propagation in Piezo-electric Semiconductors, J. Appl. Phys. 33:40 (1962).
4. T.A. Fjeldly and R.C. Hanson, Elastic and Piezoelectric Constants of Silver Iodide, Phys. Rev. B 10:3569 (1974).
5. H.B. Huntington, The Elastic Constants of Crystals, in: "Solid State Physics", F. Seitz and D. Turnbull, eds., Academic, New York (1958), Vol. 7, p. 213.
6. M.J. Rice and W.L. Roth, Ionic Transport in Super Ionic Conductors: a Theoretical Model, J. Solid State Chem. 4:294 (1972).

****

DISCUSSION

T. Ishiguro: Have you ever tried to measure the attenuation of the sound mode which is not piezoelectrically active?

J. H. Page: We've only made measurements for longitudinal waves along the c-axis. It would be interesting to make further measurements. The crystal is very soft - it's sort of the consistency of butter that's come out of a not very cold refrigerator. So it's rather hard to make measurements at higher frequencies with transverse waves, but it still should be possible to make some measurements at lower frequencies.

T. Ishiguro: What kind of bonding material did you use?

J. H. Page: The bonding material was glycerol for most of the measurements, but we also used a rather liquid araldite without any setting compound.

PHONONS AND PHASE TRANSITIONS IN QUASI-ONE-DIMENSIONAL CONDUCTORS

Takehiko Ishiguro

Electrotechnical Laboratory
Tsukuba Research Center, Sakuramura
Ibaragi 300-31, Japan

INTRODUCTION

Crystals with the atomic or molecular arrangements such that the electrons are constrained to move preferentially in one direction are called quasi-one-dimensional (1-D) conductors. The physics and chemistry of these materials have attracted interest and a number of conference proceedings or texts are describing the unusual features of these materials.[1]

In this paper the properties or roles of phonons in the 1-D conductors are described. The 1-D electronic systems are intrisically unstable against the periodic lattice distortion (PLD) as has been pointed out by Peierls. On the other hand we know that purely one-dimensional systems cannot undergo phase transitions at finite temperature. The actual 1-D system is surrounded by three-dimensional (3-D) host lattice. This reconciles the contradiction and derives some unusual consequences. There are various kinds of 1-D conductors, but we take up two typical crystals; the mixed valence square planar compounds such as $K_2Pt(CN)_4Br_{0.3} \cdot 3.2H_2O$ (KCP) and the charge transfer complex tetra-thiafulvalenium-tetracyanoquinodimethanide (TTF-TCNQ). These two 1-D materials undergo metal-insulator transitions at ~ 60 K and ~ 230 K for TTF-TCNQ and KCP, respectively, but the high anisotropy in conductivity holds in the whole temperature range.

PROPERTIES OF HOST LATTICE

Before describing the unusual consequences which characterize the 1-D conductors, we discuss the properties of the host lattice or elastic behaviors. The structures of the 1-D conductors are

complicated in first sight. For example, the chemical composition
of TTF-TCNQ molecule is represented by $C_6S_4H_4$(TTF)$-C_{12}N_4H_4$(TCNQ)
and unit cell of the crystal includes two molecules and the number
of the freedom of atomic motion is more than 200. However, most
of them are ascribed to the intramolecular modes which have the
excitation energy of 0.1$\sim$1 eV and do not contribute to low energy
lattice properties, with which we are concerned. The structure of
TTF-TCNO is regarded as a bundle of columns which consist of
segregated stacks of TTF and TCNQ planar molecules. Concerning
with KCP, the planar Pt(CN)$_4$ complexes are stacked in such a way
that the Pt atoms form linear chains with lattice spacing of
2.89 Å, while the spacing between adjacent chains is 9.87 Å.

These structures recall us those of polymers in which the
intrachain forces are considerably larger than the interchain
interactions, and, as a consequence, the high anisotropies in the
compressibility, the elastic constants and the thermal expansion
coefficients have been observed. However, contrary to this
expectation, the lattice properties of the 1-D conductors are
different from those of polymers as will be described below at
some length.

The specific heat of TTF-TCNQ at low temperatures (1.5-4.2 K)
is expressed by the T^3 law (T is the temperature) indicating that
3-D Debye phonons contribute dominantly to the thermal properties.[2]
On the other hand the specific heat of KCP has T-linear term in
addition to the T^3 term above 0.15 K.[3] The origin of the T-linear
term cannot be attributed to the intrachain vibration, which
requires high excitation energy and may appear above the Debye
temperature. The contributions of metallic electrons or 1-D
antiferromagnetic spin waves are ruled out also by considering the
dc conductivity and the magnetic susceptibility. A surviving
possibility is a modification of the linear specific heat observed
in glasses; Br or H_2O may induce disorders as can be seen from
the chemical composition, but it is not clear yet if the tunneling
process in the disordered structure is valid in KCP.

By means of the vibrating reed technique, Barmatz et al.
Measured the Young's moduli of TTF-TCNQ and KCP.[4] They obtained
the finite values of the moduli indicating that the interstack
bonding forces cannot be ignored. The measurement of the elastic
moduli in TTF-TCNQ has been reexamined by two groups[5,6] and it
turned out that the modulus perpendicular to the stack axis is
larger than that parallel[6], in contradiction to the ordinary
polymer crystals. This indicates that the interstack mechanical
binding forces are comparable with, or stronger than, the
intrastack ones. This feature has been supported also by the
measurements of the compressibility[7] and the thermal expansion
coefficients[8]. The compressibility along the stack axis is almost
2 times of that perpendicular whereas the thermal expansion

coefficient along the stacks is almost 5 times of that perpendicular to the stacks. These results tempt us to consider that the dominant binding forces are not the Van der Waals forces but the Coulomb attraction between the opposedly charged TTF and TCNQ molecules.

On the other hand the compressibilities[9] and the longitudinal elastic constants[10] of KCP are not so unusual as in TTF-TCNQ: The compressibility along the chain axis is smaller than that perpendicular, whereas the elastic constants are comparable with each other. However, the thermal expansion along the chain axis is larger than that perpendicular,[11] indicating some unusual anharmonicity of the material. In fact the unusual elastic softening has been found in the shear modulus in KCP[10] irrespective to structural phase transition. This softening was explained recently in terms of the variation of the screening effect between the Pt-chains associated with the rotational freedom of the contained water molecules.[12]

PHONON ANOMALY DUE TO 1-D ELECTRONIC SYSTEM

The excitement of the study of 1-D conductors in view of phonon physics is due to the giant Kohn anomaly. When we consider the 3-D electron system with the spherical Fermi surface, the electrons can absorb the phonons satisfying the momentum and the energy conservation relations. By taking account of the Pauli's exclusion principle we can show that those phonons propagating along one direction with the wavenumber less than $2k_F$ (the diameter of the Fermi sphere) are absorbed. These absorptions have been demonstrated in degenerate semiconductors and semimetals.[13] In 1-D conductors whose Fermi surface is flat, the interacting phonons propagating along the 1-D axis are limited to those with $\sim 2k_F$ wavenumber, and an anomaly is expected to appear in the phonon absorption at the limited wavenumber region around $2k_F$.

The anomaly associated with the phonons with the $2k_F$ wavenumber can be observed also in the screening effect in the electron systems. When the 1-D electron system is disturbed by some potential variation associated with the lattice waves with wavenumber q along the 1-D axis, the linear response function χ_q at 0 K is given as

$$\chi_q \sim \ln |(q+2k_F)/(q-2k_F)| .$$

Associated with this logarithmic divergence with respect to $q \sim 2k_F$, the renormalized phonon frequency is softened at $q \sim 2k_F$.

In KCP, the giant Kohn anomaly in the logitudinal acoustic phonon mode has been observed distinctively by means of the X-ray and the neutron scattering.[14,15] In the 3-D momentum space, the

anomaly was observed on a plane normal to the 1-D axis. This fact
means that those 3-D phonons which have a wavevector component of
$2k_F$ along the 1-D axis are softened, and indicates that those
phonons interact with the 1-D electrons on the individual columns.
(Note that the momentum conservation along the perpendicular
directions need not be satisfied for the interaction with the 1-D
electrons.)

In TTF-TCNQ, the similar anomalies which have induced lots of
arguments have been observed.[16-19] Figure 1 shows the temperature
dependence of the X-ray scattering intensity as a function of
wavenumber η. According to the ordinary kinematical theory, the
intensity I of the scattered X-ray is given by

$$I \propto (\vec{e} \cdot \vec{s})^2 / \omega^2$$

where $\vec{e}$ is the polarization vector of the distortion, $\vec{s}$ is the
scattering vector and is the frequency of the phonons. Then if
the phonon is softened sharply at $\sim 2k_F$, I contains a peak. Figure 1
shows that two peaks are formed at $0.295b^*$ and $0.41b^*$ where b^* is
the reciprocal lattice spacing along the stacking axis. By
scanning the recipocal zones the following facts are found:
(1) The peaks at $0.295b^*$ are found irrespective to the wavenumbers
along the a and c axes above 53 K where the conductor shows
metallic behavior. This indicates that the distortion is 1-D

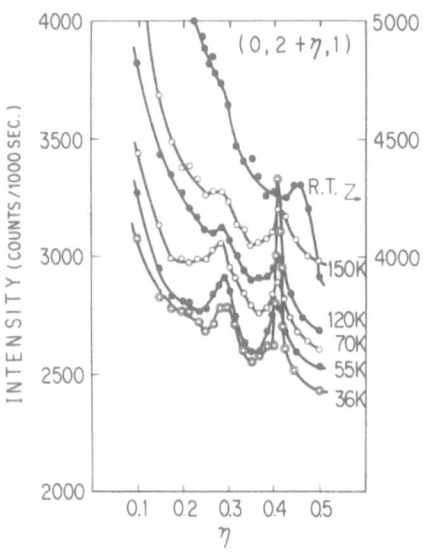

Fig. 1.   Observed intensity of the scattered X-ray in the (021)
zone of TTF-TCNQ at several temperatures (Ref. 18)

without the ordering in the transverse directions. Further the
polarization of the distortion has turned out to be parallel to
the c axis dominantly.
(2) The peaks at $0.41b^*$ have the 1-D behavior above 53 K also,
but the polarization is longitudinal.

By combining the result of the neutron scattering study[19],
which can extract the dynamical behavior, the following facts are
derived: The peak at $0.295b^*$ is ascribed to the softening of the
transverse phonon mode at $\sim 2k_F$, whereas the phonon softening has
not been found for the peak at $0.41b^*$ and this wavenumber
corresponds to $4k_F$ ($=(1-0.41)b^*$) located in the reduced zone.

According to the aforementioned description, the anomaly at
$2k_F$ was ascribed to the Kohn anomaly. This conclusion, however,
was criticized by taking up the fact that, because of the narrow
band width of the order of 0.1 eV, the electronic systems may not
have the well defined Fermi surface.[20] In particular the well
developed peak near $4k_F$ which is observable even at room temperature
is difficult to be related to the Fermi surface. Then, Kondo and
Yamaji interpreted the $4k_F$ scattering by assuming that the
intrastack Coulomb interaction induces the Wigner crystallizations
of the charges. By extending the Wigner crystal scheme Yamaji
pointed out that the planar molecules are polarized by localizing
the electrons at their frontiers, which tempt to order in
antiferroelectric way within the stacks. The polarized molecules
can couple to the polarization wave mode just like the spin wave
and the transverse acoustic phonons with $2k_F$ which couple with this
polarization mode are softened.

The initial interpretation of the phonon anomaly in TTF-TCNQ
ascribed to the Kohn effect has been reexamined and the contribution
of the polarization wave has been proposed as an alternative
mechanism. In any case, it is true that the dynamic PLD whose
wavenumber is determined by the electronic system does exist and
the PLD is associated with the charge density waves (CDW), whose
contributions to the transport phenomena in 1-D conductors are
collecting lots of arguments.

PHASE TRANSITIONS OF 1-D CONDUCTORS

In a metallic one-dimensional conductor, we can show that
uniform charge density is unstable against the formation of PLD
resulting in the CDW. Although the 1-D conductors such as KCP
and TTF-TCNQ are of 1-D electronic configuration, the lattice as
the stage of the electronic system is 3-D as described before.
Therefore the 1-D conductors undergo the phase transitions,
forming the static PLD; the superlattice.

In KCP, the metal-insulator transition is ascribed to the
formation of the static PLD which are consolidated near 120 K by
developing the correlations among the neighbouring chains; the 3-D
superstructures whose periodicity along the chain axis is determined
by $2k_F$ are formed. The lattice dynamics associated with the
structural phase transition have been studied in detail by means
of the neutron scattering.[22]

In TTF-TCNQ the occurence and origin of the metal-insulator
transition at 53 K and the additional phase transitions at 49 K
and 38 K have attracted attention. From the structural view
point the 3-D ordering begins to grow among the neighbouring CDW
as decreasing the temperature. Concerning with the X-ray scattering
shown in Figure 1, the scattering peaks grow rapidly below 53 K
but we should specify the transverse wavenumber also to describe
the peak zone; this is the consequence of the 3-D ordering in the
superstructure. The successive phase transitions are related to
the variation in the transverse periodicity. Plausible explanations
of the succesive transitions are based on the consideration that
TTF and TCNQ stacks which form the static PLD at different
temperatures correlate each other through the Coulomb interaction.[23]

The phase transition of 1-D conductors have been discussed in
the light of the Peierls transition, i.e. the structural transition
induced by the stabilization of the 1-D electronic system, or the
freezing of phonons at the Kohn anomaly. For TTF-TCNQ this scheme
was criticized recently by pointing out that the Peierls transition
temperature which is calculated on the microscopic basis by using
the realistic parameters becomes too low to explain the
experiments.[21] The problem is open.

CONCLUDING REMARKS

So far I described briefly the 1-D conductors by placing the
emphasis on the lattice dynamical features, and by taking account
of the recent results. As repeated, the 1-D conductor is the 1-D
electronic system on the stage of 3-D lattice. On the other hand
it has been shown recently that the linear chain mercury compound
$Hg_{3-\delta}AsF_6$ contains the mechanically one-dimensional system of Hg.[24]
Unusual mechanical properties are expected in these system: For
example, macroscopic fluctuations about the displacement of
constituent atoms are observable.[24,25]

Finally I should mention on the role of the ultrasonic means
for the study of 1-D conductors from two points. It is well known
that the critical behavior of the elastic constants and the
attenuation can give useful thermodynamic information to clarify
the nature of the phase transition. In TTF-TCNQ, the elastic
hardening appears in the longitudinal modes below 53 K.[5,6] On the
other hand the electrons in the 1-D system cannot absorb ultrasonic

waves whose wavenumber is much less then $2k_F$ as far as we take account of the one-particle process.[26] When we take account of the collective modes such as the CDW, theoretical study predicts that appreciable attenuation occurs through second order process.[27] Thus the ultrasonic study will be useful to clarify the properties of the CDW.

The author should like to acknowledge many informative conversations with Drs. S. Kagoshima, J. Kondo, K. Yamaji, H. Sumi and H. Anzai.

REFERENCES

1. For example, "Highly Conducting One-Dimensional Solids" J. F. Devreese, R. P. Evrard and V. E. van Doren, ed., Plenum Press, New York (1979), "Molecular Metals" W. E. Hatfield, ed., Plenum Press, New York (1979).
2. T. Wei, S. Etemad, A. F. Garito and A. J. Heeger, Low temperature specific heat of TTF-TCNQ, Phys. Letters 45A: 269 (1973).
3. W. A. Reed, F. S. L. Hsu, R. J. Schultz, J. E. Graebner and H. J. Guggenheim, Low Temperature specific heat of a one-dimensional system: $K_2Pt(CN)_4Br_{0.3} \cdot 3(H_2O)$, Phys. Rev. Letters 34: 473 (1975).
4. M. Barmatz, L. R. Testardi, A. F. Garito and A. J. Heeger, elastic properties of one-dimensional compounds, Solid State Commun. 15: 1299 (1974).
5. T. Ishiguro, S. Kagoshima and H. Anzai, Elastic property of TTF-TCNQ, J. Phys. Soc. Japan 42: 365 (1977).
6. T. Tiedje, R. R. Haering, M. H. Jericho, W. A. Roger and A. Simpson: Temperature dependence of sound velocities in TTF-TCNQ, Solid State Commun. 23: 713 (1977).
7. D. Debray, R. Millet, D. Jerome, S. Barisic, J. M. Fabre and L. Giral, Neutron diffraction study of the compressibility of TTF-TCNQ under hydrostatic pressure, J. de Phys. (Paris) Lett. 38: 227 (1977).
8. A. J. Schultz, G. D. Stucky, R. H. Bleesing and P. Coppens, The temperature dependence of the crystal and molecular structure of TTF-TCNQ, J. Am. Chem. Soc. 98: 3194 (1976).
9. L. V. Interrante and F. P. Bundy, The electrical conductivity behavior of $K_2Pt(CN)_4Br_{0.3} \cdot 2.3H_2O$ at high pressures, Solid State Commun. 11: 1641 (1972)
10. H. Doi, H. Nagasawa, T. Ishiguro and S. Kagoshima, Lattice softening in $K_2Pt(CN)_4Br_{0.3} \cdot 3H_2O$, Solid State Commun, 24: 729 (1977)
11. W. Sturm, S. Drosdziok and H. Happ, Thermal expansion of the one-dimensional conductor $K_2Pt(CN)_4Br_{0.3} \cdot xH_2O$ and $K_2Pt(CN)_4 \cdot xH_2O$, Solid State Commun. 16: 485 (1975).
12. S. Kurihara, H. Fukuyama and S. Nakajima, A model of elastic anomalies in $K_2Pt(CN)_4Br_{0.3} \cdot 3H_2O$, to be published in J. Phys. Soc. Japan
13. D. Huet and J. P. Maneval, Image of the Fermi surface and

screening effects in phonon attenuation, Phys. Rev. Letters 33: 1154 (1974), T. Ishiguro, K. Kajimura, S. Kagoshima, H. Tokumoto and J. Kondo, Quantum oscillation in absorption of ballistic heat pulse in bismuth, J. Phys. Soc. Japan 39: 1547 (1975).

14. R. Comes, N. Lambert, H. Launois and H. R. Zeller: Evidence for a Peierls distortion or a Kohn anomaly in one-dimensional conductors of the type $K_2Pt(CN)_4Br_{0.30} \cdot xH_2O$, Phys. Rev. B8: 571 (1973).

15. B. Renker, H. Rietschel, L. Pintschovius, W. Gläser, P. Brüesch, D. Kuse and M. J. Rice: Observation of giant Kohn anomaly in the one-dimensional conductor $K_2Pt(CN)_4Br_{0.3} \cdot 3H_2O$, Phys. Rev. Letters 30: 1144 (1973).

16. F. Denoyer, R. Comes, A. F. Garito and A. J. Heeger: X-ray-diffuse-scattering evidence for a phase transition in TTF-TCNQ, Phys. Rev. Letters 35: 445 (1975).

17. S. K. Khanna, J. P. Pouget, R. Comes, A. F. Garito and A. J. Heeger: X-ray studies of $2k_F$ and $4k_F$ anomalies in TTF-TCNQ, Phys. Rev. B16: 1468 (1977).

18. S. Kagoshima, T. Ishiguro and H. Anzai: X-ray scattering study of phonon anomalies and superstructures in TTF-TCNQ, J. Phys. Soc. Japan 41: 2061 (1976).

19. G. Shirane, S. M. Shapiro, R. Comes, A. F. Garito and A. J. Heeger: Phonon dispersion and Kohn anomaly in TTF-TCNQ, Phys. Rev. B14: 2325 (1976).

20. J. Kondo and K. Yamaji, Density correration of classical 1-D electron gas with reference to the $4k_F$ anomaly in TTF-TCNQ, J. Phys. Soc. Japan 43: 424 (1977).

21. K. Yamaji, New lattice softening mechanism in the quasi-one-dimensional TCNQ conductors, to be published in J. Phys. Soc. Japan.

22. K. Carneiro, G. Shirane, S. A. Werner and S. Kaiser: Lattice dynamics of $K_2Pt(CN)_4Br_{0.3} \cdot 3.2D_2O$ (KCP) studied by inelastic neutron scattering, Phys. Rev. B13: 4258 (1976).

23. S. Kagoshima, Roles of cation stacks in ordering of charge density waves in TseF-TCNQ: Comparison with TTF-TCNQ, in "Molecular Metals" E. Hatfield ed., Plenum Press, New York (1979).

24. J. M. Hastings, J. P. Pouget, G. Shirane, A. J. Heeger, N. D. Miro and A. G. MacDiarmid, One-dimensional phonons and "Phase-ordering" phase transition in $Hg_{3-\delta}AsF_6$, Phys. Rev. Letters 39: 1184 (1977)

25. J. Kondo, Macroscopic fluctuation of one-dimensional chain, Bussei 12: 17 (1971) in Japanese.

26. T. Tiedje and R. R. Haering, Theory of ultrasonic attenuation in one and two dimensional metals, Can. J. Phys. 54: 1454 (1976).

27. Y. Suzumura, The conductivity and the longitudinal sound attenuation above the Peierls transition point, Solid State Commun. 21: 537 (1977).

AN ACOUSTIC SOFT MODE FOR KCP

J.D.N. Cheeke and D. Houde

Département de physique,
Université de Sherbrooke
Sherbrooke, Québec, J1K 2R1

and

N.K. Hota
Collège Militaire Royal de St-Jean,
St-Jean, Québec

There has been great interest over the last decade in pseudo one and two dimensional crystals. While remaining relatively isotropic acoustically, these systems frequently exhibit phase transitions, sometimes accompanied by soft modes, due to the highly anisotropic electron phonon interaction. Our present concern, KCP, is a classic case where a giant Kohn anomaly, and associated charge density waves occur at room temperature in the chains of Pt ions, which are uncorrelated due to fluctuations [1]. At lower temperatures, progressive 3D ordering sets in, which is arrested at $\sim$ 80 K with coherence lengths $\xi_{//} \sim 100 \, d_{//}$ and $\xi_{\perp} \sim 3 d_{\perp}$ where $d_{//}$, $d_{\perp}$ are the Pt-Pt distances parallel and perpendicular to the chains respectively, as seen by neutron [2] and X-Ray scattering [3]. Ultrasonic measurements [4] also showed some unusual behaviour in this range, and in the present work we extend these and correlate the results to the partial 3D Peierls ordering.

The crystals were grown by one of us (N.K. Hota) in the form of small blocks, and a 10 MHz $LiNbO_3$ wave transducer was epoxied to a grown face. The orientation was chosen to excite C_{66}, the transverse mode with propagation and polarisation vectors perpendicular to the Pt chains. A sensitive CW swept frequency method was used for the measurements and precautions were taken during cooldown to avoid water loss. The results are shown in Fig. (1); we obtain good agreement with the pulse echo measurements of Doi et al. [4] and ourselves where comparison is possible. Our CW measurements

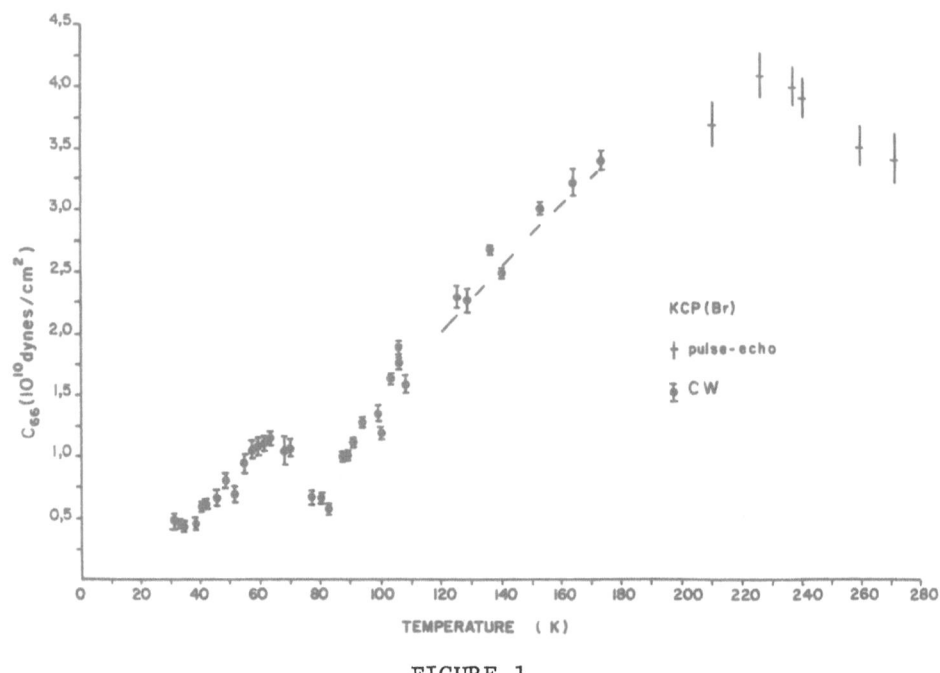

FIGURE 1

extend to both lower and higher temperatures. We consider here
only the results above ∿ 70 K, the 3D Peierls ordering regime which
has been of most interest up to now.

The maximum observed at 240 K correlates well with one observed here for the electrical conductivity [5], while the minimum at
80 K corresponds to the partial completion of 3D Peierls ordering
as observed by neutron and X-ray scattering. The elastic constant
C_{66} is directly related to inter chain coupling, and in the light
of the known partial chain alignment due to the Peierls effect and
the absence of anomalies in the other elastic constants, this suggests that C_{66} is the acoustic soft mode associated with the 3D
Peierls ordering. No temperature dependent effects are seen in the
relevant phonon spectrum by neutrons [6], which indicates that the
scattering is essentially a zone center phenomenon. Such effects
have been observed in the pseudo one dimensional A15 superconductors and we look in this direction for appropriate theoretical models.

A detailed calculation of the temperature dependence of the

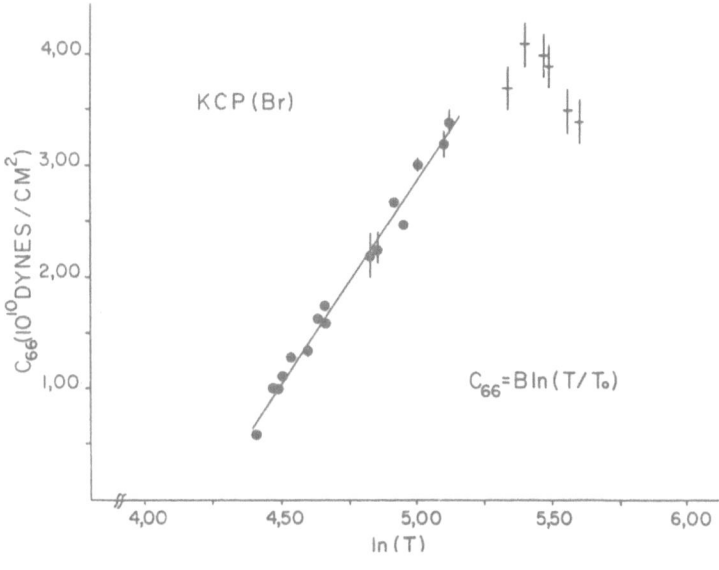

$$C_{66} = B \ln(T/T_o)$$

FIGURE 2

elastic constants for pseudo—one dimensional electron chains with
inter chain interactions has been given by Gorkov and Dorokhov [7].
When the Fermi level is high in the Brillouin zone, as for KCP,
they predict a Peierls type instability, with a temperature dependence above the transition going as ln T. This relation is verified
for our results as shown in Fig. (2), where a "transition" temperature T ∿ 69 K fits to this relation.

The present results give encouraging confirmation to the picture of progressive 3D Peierls ordering in KCP. Why is the transition never fully achieved? The experimental measurements are known
to be notoriously sensitive to the water content, and it may be
that precise control of this for a perfect crystal would give a
sharp transition. Or maybe this is a self blocking mechanism,
where the high temperature one dimensionality which leads to 3D
ordering at lower temperatures is progressively weakened due to the
consequent loss of unidimensionality! This remains an open question.

REFERENCES

1. G.A. Toombs, Phys. Reports 40, 3, 181, (1978).
2. J.W. Lynn, M. Iizumi, G. Shirane, S.A. Weiner,
   and R.B. Saillant, Phys. Rev. B12, 1154, 1975.
3. R. Comès, M. Lambert and M.R. Zeller, Phys. Stat. Solidi (b)
   (58, 587 (1973).
4. H. Doi, H. Nagasawa, T. Ishiguro and S. Kagashima, Solid State
   Communications 24, 729, (1977).
5. H.R. Zeller and A. Beck, J. Phys. Chem. Solids, 35, 77, (1974).
6. K. Carneiro, G. Shirane, S.A. Werner and S. Kaiser, Phys. Rev.
   B13, 4258, 1976.
7. L.P. Gorkov and O.N. Dorokhov, Jour. Low. Temp. Phys. 22, 1, 1976.

ANHARMONIC EFFECTS IN ULTRASONIC PROPAGATION NEAR THE λ-TRANSITION OF ^{4}He

H. Kwun, A. Hikata and C. Elbaum

Department of Physics and Metals Research Laboratory,
Brown University, Providence, Rhode Island 02912, U.S.A.

We have carried out the first measurements of second harmonics of a longitudinal ultrasonic wave generated in pressurized liquid ^{4}He near the λ-transition. The measurements yield the behavior of the anharmonic coupling coefficient, C, which represents the non-linearity of the system, near the critical region. The importance of these anharmonic terms near the critical region for the under-standing of critical phenomena has been stressed recently by Gitterman[1].

The pressure and temperature ranges covered in this study are 10 atm. < P < 25 atm. and $5 \times 10^{-4} < |t| < 5 \times 10^{-2}$ respectively, where $t \equiv T/\bar{T}_\lambda - 1$. The experimental results are initially processed in terms of a nonlinear wave equation (in one dimension, x), for fluids in general, which can be expressed as[2]

$$\ddot{u}\rho_o(1 + \partial u/\partial x)^{(B/A + 2)} = A(\partial^2 u/\partial x^2) \qquad (1)$$

where u is the displacement, ρ_o is the density of the undisturbed medium, A and B are temperature dependent coefficients of a power series expansion of pressure in terms of density changes, defined as follows:

$$P - P_o = P_e = A(\Delta\rho/\rho_o) + \frac{1}{2} B(\Delta\rho/\rho_o)^2 + \dots \qquad (2)$$

where P_e is the pressure change due to the acoustic wave.

When an initially plane sinusoidal wave of amplitude $u_{1,0}$ and frequency ω_1 is generated at position x=0 of the nonlinear medium to which eq. (1) applies[3], the second and higher harmonics are produced as the fundamental wave propagates. The amplitude of

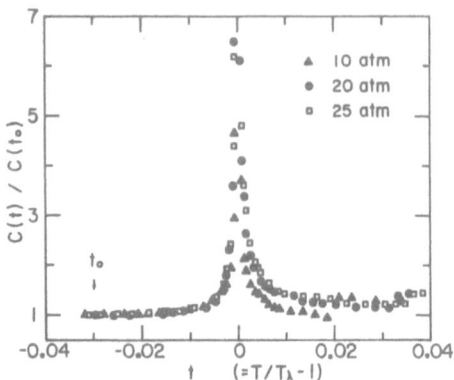

Fig. 1.   Temperature dependence of coupling coefficients for three,
           different pressures 10, 20 and 25 atm. The data were
           normalized to the values at t_o = -3 x 10^{-2}.

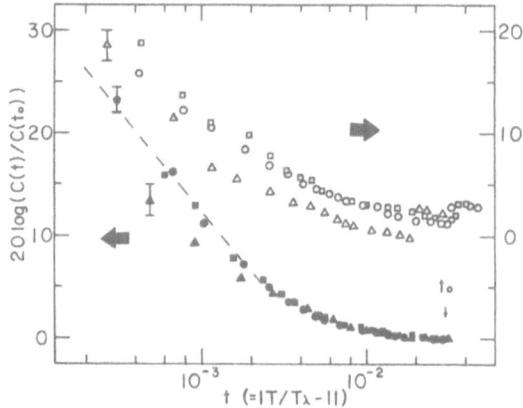

Fig. 2.   The same data as Fig. 1 plotted in log-log scale. Δ-10
           atm.; O-20 atm.; □-25 atm. Solid symbols are for T<T_λ
           and open symbols are for T>T_λ. For clarity, the scale for
           the data T>T_λ is shifted upward. A dashed line of slope
           -1 is included for reference.

the second harmonic, u_2, at propagation distance x in a dissipative medium may be approximated by[4,5]

$$u_2 = Cu_{1,0}^2 k^2 [\exp(-2\alpha_1 x) - \exp(-\alpha_2 x)]/(\alpha_2 - 2\alpha_1) \tag{3}$$

where $C = \frac{1}{4}(1 + \frac{1}{2}B/A)$ is the coupling coefficient, k is the wave vector, α_1 and α_2 are effective attenuation coefficients of ultrasonic waves at frequencies ω_1 and ω_2 ($= 2\omega_1$), respectively, and B/A is given by[2]:

$$B/A = 2\rho_o v_o (\partial v/\partial P)_{T,\rho_o} + 2(\beta Tv_o/C_p)(\partial v/\partial T)_{P,\rho_o} \tag{4}$$

Here β is the thermal expansion coefficient, v is the sound velocity, C_p is the specific heat at constant pressure. Thus if the values of u_2, $u_1 = u_{1,0} e^{-\alpha_1 x}$, α_1 and α_2 are determined experimentally, one can deduce from eq. (3) the coupling coefficient C.

The experimental arrangements are as follows. Two Lithium-Niobate transducers whose fundamental resonant frequencies are 10 MHz and 20 MHz, respectively, are used. They are separated by a distance of approximately one cm horizontally and are mounted on the sample cell, which is made of a brass block approximately $2 \times 2 \times 3$ cm^3. The pressure of the sample is maintained within $\Delta P/P < 2.5 \times 10^{-3}$ throughout the measurements.

Relative values of C obtained experimentally are shown as a function of temperature in Fig. 1 and Fig. 2. The C(t,P) were normalized against C(t$_o$,P) where t$_o$ was chosen arbitrarily to be t$_o$ = (T$_o$ - T$_\lambda$)/T$_\lambda$ = -3 x 10^{-2}. As seen, C(t,P) increases rapidly as temperature approaches the transition. In the temperature range $|t| < 2 \times 10^{-3}$, C(t) (Fig. 2) varies approximately as $|t|^{-1}$. Both above and below T$_\lambda$, C(t,p)/C(t$_o$,P) depends slightly on pressure. The measured change in C over the temperature range $3 \times 10^{-3} \lesssim t \lesssim 3 \times 10^{-4}$ was compared with the value calculated using the thermodynamic expression, eq. (4), in which the singularities in the thermodynamic parameters (their temperature dependence near the λ-transition) are included. It is found that the experimentally determined change in C is about a factor of ten larger than the value calculated from the above expression. A phenomenological interpretation of the results will be published separately.

In conclusion, the coupling coefficient, C, which represents the nonlinearity of the medium, has been found to diverge as $\sim |t|^{-1}$ near T$_\lambda$. Therefore, in any attempt at a proper understanding of such features as ultrasonic propagation near the λ-transition of ^{4}He, the nonlinear terms should be taken into account.

This research was supported in part by the National Science

Foundation under Grant DMR77-12249 and through the Materials
Research Laboratory of Brown University.

References

1. M. Gitterman, Rev. Mod. Phys. $\underline{50}$, 85 (1978).
2. R. T. Beyer and S. V. Letcher, Physical Ultrasonics, (Academic
    Press, 1969), chapter 7; O. V. Rudenko and S. I. Soluyan,
    Theoretical Foundation of Nonlinear Acoustics, (Consultants
    Bureau, New York and London, 1977).
3. The hydrodynamic equations (1) and (2) are considered to be
    valid in the temperature range near the transition where the
    inequality $q\xi \ll 1$ is fulfilled. Here q is the wave number of
    the probe and ξ is the correlation length of the fluctuations.
    For 10 MHz ultrasonic waves used as the fundamental wave in
    the present study, the above condition is satisfied up to
    $|t| \simeq 10^{-5}$.
4. A. L. Thuras, R. T. Jenkins and H. T. O'Neil, J. Acoust. Soc.
    Amer. $\underline{6}$, 173 (1935).
5. A. Hikata, B. B. Chick and C. Elbaum, J. Appl. Phys. $\underline{36}$, 229
    (1965).

INVESTIGATION OF PHASE TRANSITIONS BY MEANS OF

WAVEVECTOR REVERSED ULTRASOUND

K. Fossheim and R.M. Holt

Department of Physics, The Norwegian Institute
of Technology, 7034 Trondheim, Norway

In view of the great interest attached to soft mode driven instabilities and other phase transitions, and due to the well known difficulties often caused by insufficient acoustic quality of the crystals investigated, we have developed a new ultrasonic technique for such demanding measurements. The technique, which is based on application of the nonlinear electroacoustic echo effect, has been applied in the investigation of the 185 K phase transition in $KMnF_3$.

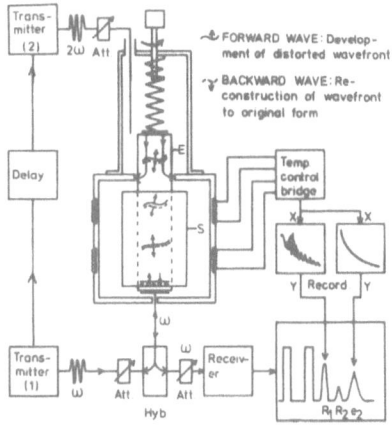

Fig.1. Echo technique for investigation of attenuation in solids. The electroacoustic echo is generated in BGO ($Bi_{12}GeO_{20}$) (E), and attenuated in the specimen (S). Also shown: cavity, rf equipment, temperature controls, and recording of reflections (R) and echo e_2 .

        Very briefly stated the method can be explained as follows,
with reference to Fig.1.: An ordinary acoustic pulse of frequency
ω is transmitted through the specimen   (S) in the forward
direction, whereafter it enters into an echo-active crystal (E)
attached to the far end of the specimen. Here the acoustic pulse
is exposed to a homogeneous electric field pulse of frequency 2ω ,
generated in a wide band tunable helical cavity recently developed
by the authors[1]. Through a nonlinear mixing process a backward
wave (echo) is generated with a wavevector -$\vec{k}$, exactly the
opposite of that ($\vec{k}$) of the forward wave at all points on the
wavefront. Whenever distortions of the forward wavefront have
occurred these are restored during backward propagation.
Experimental evidence for these claims is given in Figs. 2a and 2b.
Fig. 2a shows the signal level of the second reflected pulse vs.
temperature,   T-T_c , measured relative to the transition
temperature. The curve was recorded during a fast (1.5 min/K)
temperature sweep through the transition. Very strong inter-
ference effects are noted. Fig. 2b gives corresponding data recorded simultaneously with those of Fig. 2a , using the echo
method. The improvement is quite striking.

        Utilizing the technique outlined above we have undertaken an
experimental investigation of the 185 K phase transition in $KMnF_3$.
The principal aim of this work is to study more closely the
critical temperature dependence, including crossover, reported
some years ago[3], and to investigate the puzzling frequency dependence (~$\omega^{1.3}$) found by several workers[3,4].

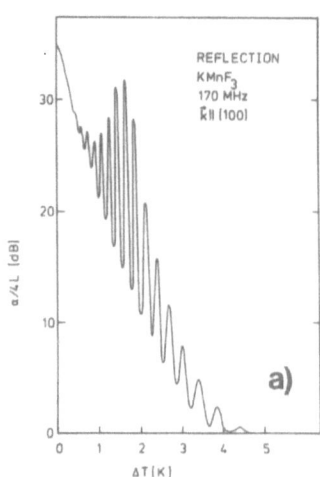
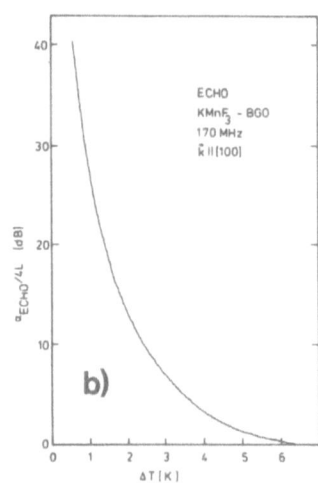

Fig.2.   a)Recording of the amplitude of reflected acoustic
            pulse in $KMnF_3$.
         b)Simultaneous recording of echo, attenuated in $KMnF_3$.

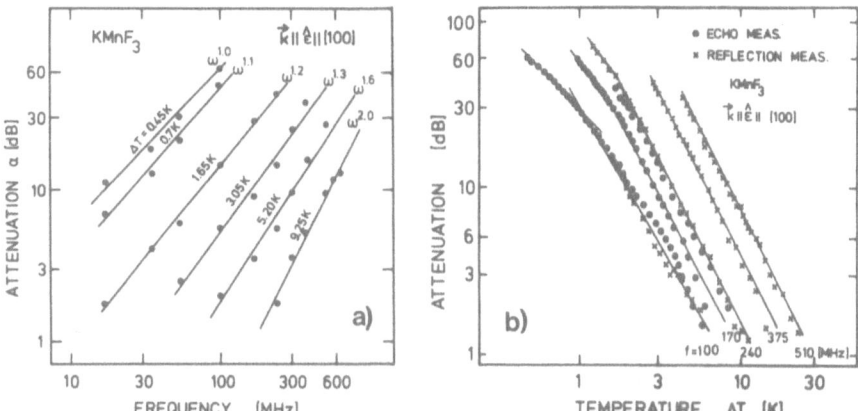

Fig.3. a) Frequency dependence of attenuation near T_c in KMnF$_3$
        measured at various ΔT . b) Temperature dependence of
        attenuation at various frequencies, measured by the
        echo method (●) and by pulse reflection (x).

As seen from Fig.3a the critical exponent for the temperature
dependence of the attenuation of longitudinal waves is ρ= 1.87±0.07
for ΔT>1.4-1.8K . From Fig.3b we find a frequency dependence which
approaches a ω^2-law well above T_c . In this region therefore
$\alpha \sim \omega^2 \cdot \varepsilon^{-1.87}$ close to that predicted by Schwabl[5] , and in very
good agreement with Muratas results[6], obtained by renormalization
group methods. Close to T_c we find $\alpha \sim \omega^{1.1} \varepsilon^{-1.24\pm0.10}$. The
frequency dependence in this regime is not explained by existing
theories, while the temperature exponent is of the magnitude given
by Schwabl. Domb et. al[7] found a single exponent ρ = 1.3 for
$\vec{q} \| \hat{\varepsilon} \|$ [123]. However, theoretical results for such directions are
not available. In conclusion the possibility of a dimensional
crossover remains open, and will be further investigated.

REFERENCES

  1. R.B. Thompson and C.F. Quate, Appl.Phys.Letters 16
      295 (1970)
  2. K. Fossheim and R.M. Holt, J.Phys.E: Scientific
      Instr. 11 892 (1978)
  3. K. Fossheim, D. Martinsen and A. Linz, Anharmonic
      Lattices, Structural Transitions and Melting, ed.
      T. Riste, Noordhoff 1974, p.141
  4. M. Furukawa, Y. Fujimori and K. Hirakawa, J.Phys.Soc.
      Jap. 29 ,1258 (1970)
  5. F. Schwabl, Phys.Rev. B 7 , 2038 (1973)
  6. K.K. Murata, Phys.Rev. B 13 , 4015 (1976)
  7. E.R. Domb, H.K. Schurmann and T. Mihalisin, Phys.Rev.
      Letters 36 , 1191 (1976)
                              ****

DISCUSSION

W. Arnold: Does the non-linear coupling itself change at the phase transition?

K. Fossheim: This is exactly one of the reasons we used two different crystals. We had the sample which is not used for echo purposes and the echo is produced in a separate crystal without any phase transition.

C. Elbaum: I would like to comment on Arnold's remark. In fact, in some systems at least the unusual behavior of the non-linear coupling coefficient is documented and tomorrow there will be at least one example presented - the situation near the λ-point in helium.

H. Kinder: Is it obvious that the echo is linear in the sound wave amplitude?

K. Fossheim: We have checked that and it is.

W. Arnold: It's not linear at very high powers - it falls over. Then you have to take into account higher orders - they are out of phase.

K. Fossheim: May I make one remark here? You might think that since this is an echo phenomenon and it's generated as a non-linear process that this would give you fairly small signals. But if you notice the dB scale here we measured attenuation in the specimen of over 60 dB, so it's a very powerful effect in other words.

STIMULATED PHONON EMISSION

Wolfgang Grill

Technische Hochschule Darmstadt

Festkörperphysik, West Germany

INTRODUCTION

Two methods to achieve a stimulated emission of phonons with frequencies in the THz-regime are described here. In both cases the generation of phonons is achieved by coupling to electronic states in doped dielectric crystals. The impurity ions are excited by pulsed laser light and an almost complete conversion of the absorbed energy is obtained, allowing the excitation of monochromatic phonons with a power of 10 to 100 W.

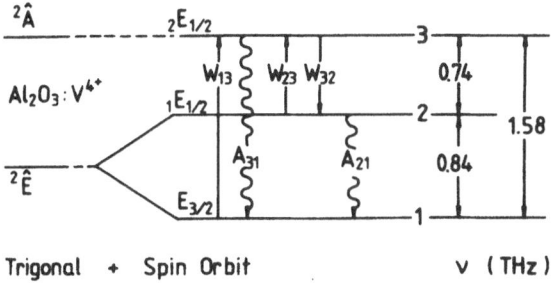

Fig. 1.   Energy level diagram of $Al_2O_3:V^{4+}$.

PHONON EMISSION BY STIMULATION IN A THREE-LEVEL ELECTRONIC SYSTEM

The three lowest-lying electronic states [1,2,3] in $Al_2O_3:V^{4+}$ can be used for the stimulated emission of phonons[4]. The energies and group theoretical assignments of the electronic states of V^{4+} ions in Al_2O_3 can be taken from Figure 1.

The group theoretical selection rules for electric-dipole tran-sitions among these states are known [5] and are identical with the selection rules for the emission of longitudinal phonons propagating along the C-axis of the Al_2O_3 crystal [4] (Table 1).

Table 1: Selection rules for photon absorption and phonon emission

	$E_{1/2}$	$E_{3/2}$
$E_{1/2}$	↓ + ‖	↓
$E_{3/2}$	↓	‖

If the electronic state at 1,58 THz (2E 1/2) is to exceed the thermal population, the absorbed laser power P must exceed the losses out of the state which results in

$$P = W_{13}N_1h\nu_{13} > A_{31}N_3h\nu_{13} ,$$

where A_{ij} denotes spontaneous and W_{ij} stimulated processes and N_i is the number of ions in the ith energy state. A_{32} can be neglected re-lative to A_{31} because of the branching ratio for the transitions $(3 \rightarrow 1)/(3 \rightarrow 2) = 10$ if one follows the analysis by Blume et al.[6]. If losses out of the 1E 1/2 state are neglected the threshold for inversion is reached when $N_3 = N_{2T}$ where $N_{2T} \cong 9 \times 10^{-6} N_o$ is the thermal population at 3,7°K, the temperature of the crystal. The number of V^{4+} ions in the excitation volume $N \cong 1 \times 10^{16}$ is deter-mined by the geometry of the excitation and the concentration of V^{4+} ions in the crystal.

This leads to a calculated threshold of P = 31 W which is in fair agreement with the observed onset of stimulated phonon emission.

The rate equations for the number of ions N_i under steady-state conditions are:

$$dN_3/dt = 0 = W_{13}N_1 + W_{23}N_2 - (A_{31} + W_{32})N_3 \, ,$$

$$dN_2/dt = 0 = -(W_{23} + A_{21})N_2 + (W_{32} + A_{32})N_3 \, ,$$

$$N_1 + N_2 + N_3 = N_0 \, .$$

Neglected are A_{32} relative to A_{31} and W_{31}, W_{21}, W_{12} because the inversion is much harder to achieve or the stimulating field is not present. For inversion ($N_3 - N_2 > 0$) it is required that $W_{21} - A_{21} > W_{32} + A_{32}$. This reduces to $A_{21} > A_{32}$ since $W_{ij} = W_{ij}$ in this case of equal grades of degeneracy.

The condition $A_{21} > A_{32}$ is met due to the relative density of terminal states, provided the transition matrix elements are roughly equal. Therefore one can expect that inversion can be sustained under steady state conditions and the dominance of stimulated phonon emission is only dependent on sufficient pumping power.

In the experiments a rectangular shaped crystal of $7 \times 7 \times 17$ mm^3 is used, partially doped with 0,3% (by weight) vanadium of which about 5% is converted by X-ray radiation into V^{4+}. At one surface the crystal is attached to a liquid helium bath. All other surfaces are kept in vacuum.

The radiation of an optically pumped pulsed far infrared laser at 1,57 THz (52,6 cm^{-1}) is focussed into the crystal just inside the doped region at the inner surface to the clear crystal.

A Sn superconducting bolometer of 1mm^2 cross section is used to detect the generated phonons at the clear end of the crystal.

The laser pulse has a half width of $\approx$ 100 nsec but at low laser power no distinct phonon pulses are observed, due to the strong resonant scattering in the doped region of the crystal, which leads to a broad steplike pulse of transverse phonons. A weak signal mainly due to scattering longitudinal phonons is also observed. Due to the known branching ratio the phonons should have a frequency of 1,58 THz (52,8 cm^{-1}).

At higher laser powers a strong longitudinal phonon peak L is observed which consists due to selection rules only of 0.74 THz (24,7 cm^{-1}) phonons. This signal is contributed to the stimulated emission of phonons from the 2E 1/2 level to the 1E 1/2 level. A different peak appears for the transverse phonons (T) which is weaker and broader and can not so easily be identified since the se-

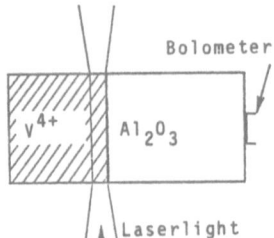

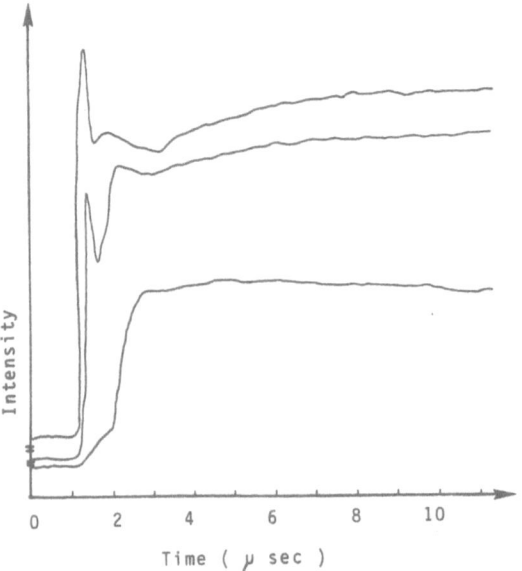

Fig. 2.   Phonon intensity, as measured at the bolometer, as a
          function of elapsed time after the onset of the laser pulse
          and as a function of laser intensity. The phonon signal is
          shown in terms of an arbitrary linear scale. The laser in-
          tensity increases relative to the bottom curve as 1:1.9:2.
          The three curves have been arbitrarily displaced from each
          other.

lection rules put in this case no restriction on the transition.
But this peak is most likely due to the stimulated emission of
0,74 THz (24,7 cm^{-1}) phonons which are, caused by their slower group
velocity, stronger scattered than the longitudinal phonons and show
therefore a broader peak, due to small number scattering.

     The unavoidable mass defect scattering of the vanadium ions
present in the generation volume will allow the build up of super-

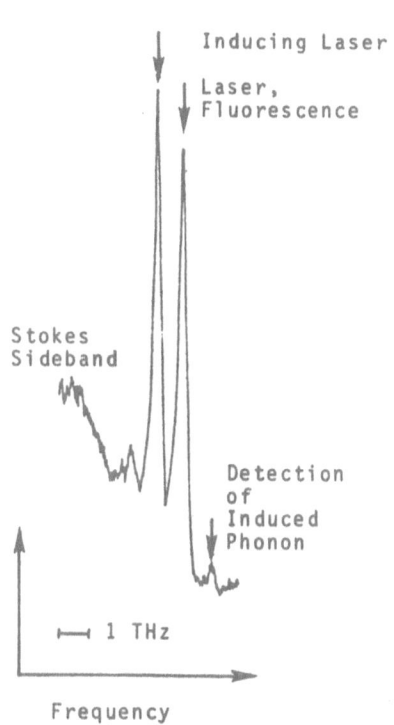

Fig. 3. Spectrum and energy
level system for the stimulated emission and detection
of phonons in the vibronic
sideband of $SrF_2:Eu^{2+}$

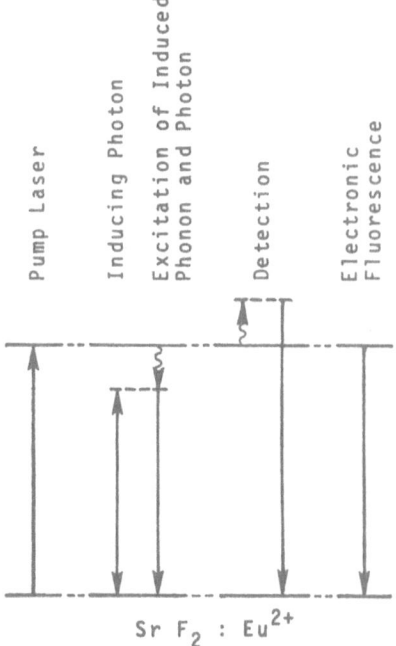

radiance only in a cavity of the size of the mean free path.
Since the laserradiation can not be focussed to illuminate a volume
small enough for that purpose ($\approx 0,1 \times 0,01 \times 0,01$ mm^3) the effect
of superradiance which would show a strong directionality for the
stimulated phonon emission could not be observed. But such experiments should become possible by cutting the crystal properly.

PHONON EMISSION BY STIMULATION OF TWO-QUANTUM PROCESSES

Vibronic sidebands to luminescent transitions at probe ions
placed in dielectric crystals do exist due to the fact that not
only the direct transitions are possible, but as well two-quantum
processes involving the absorption or emission of a phonon and a
photon in a combined transition [7]. Transitions involving the emission and absorption of phonons are known as the Stokes and anti-
Stokes sidebands and it has been shown that the observation of
emitted light in the anti-Stokes sideband can serve as a phonon
spectrometer [8]. These experiments indicate that at phonon power
levels of 10^{-1} W the emission of light in the Stokes sideband,
which is due to stimulation by a phonon, becomes already observable,
whereas photon stimulated processes are not observed. This is
caused by the ratio of the density of terminal states of phonons
and photons of $\approx 10^5$.

If the power of the stimulating photon field is put to a level
of 10^4 W one should expect that photon stimulated transitions in
the vibronic sideband become observable. The level system in Figure 3 shows the principle of the emission of phonons by stimulation
of a two quantum process.

The electronic level is either directly or indirectly pumped
and the crystal is put in a strong photon field, which will not
directly interact with the electronic level system of the crystal,
since the quantum energy is too low. But the presence of the light
will stimulate a transition of the electronic state in the Stokes-
sideband, generating a phonon with an energy equal to the difference of the electronic level and the energy of the stimulating
photon.

The excited phonons will further stimulate the same transition
or couple in the anti-Stokes sideband, allowing the spectroscopical
detection of the phonons generated by the original process.

The phonons generated in this way will be monochromatic and
their frequency can be tuned over the complete range of phonon frequencies by the use of dye lasers for the stimulating field. But
since additional momentum can be transferred to the crystal at the
site of the impurity and since the light and phonon field will not
be in phase at different sites, one can only expect the excitation
of incoherent phonons as long as the process is dominated by the

stimulation of the light and not by the stimulation due to the already excited phonons.

   Nevertheless this method allows the generation of tunable monochromatic phonons with up to 50 W power with the dye lasers presently available for these experiments.

   The measurements shown here have been performed in $SrF_2:Eu^{2+}$
doped with 0,1 mol % Eu and a nitrogen laser pumped dye laser system
is used to directly pump the electronic level and to supply the
radiation for the stimulating photon field. The crystal was kept at
liquid helium temperature and high speed photon counting with 1 nsec
resolution has been used to detect the signals.

   The experiments show that monochromatic phonons can be generated at high power levels by the use of stimulated phonon emission.
Beside of the use of these phonons for measurements in the field of
lattice dynamics, one can obtain direct information on the nature
and dynamical properties of the interaction of phonons with electronic states at the site of impurities in crystals.

ACKNOWLEDGEMENT

   The experiments in 3-level systems have been performed together
with W.E. Bron at the Indiana University and have been published in
detail [4]; experiments on 2-quantum processes have been performed at
the Technische Hochschule Darmstadt and are still in progress and
unpublished so far.

REFERENCES

1. R.M. Macfarlane, J.Y. Wong, and M.D. Sturge, Phys. Rev. 166,
       250 (1968).
2. J.Y. Wong, M.J. Berggren, and A.L. Schawlow, J. Chem. Phys. 49,
       835 (1968).
3. R.R. Joyce and P.L. Richards, Phys. Rev. 179, 375 (1969).
4. W.E. Bron and W. Grill, "Stimulated Phonon Emission", Phys. Rev.
       Lett. 40, 1495 (1978).
5. D.S. McClure, J. Chem. Phys. 36, 2757 (1962).
6. M. Blume, R. Orbach, A. Kiel, and S. Geschwind,
       Phys. Rev. 139, A314 (1965).
7. K.K. Rebane, "Impurity Spectra of Solids" (Plenum, New York,
       1970);
    M.H.L. Pryce, "Phonons in Perfect Lattices and in Lattices with
       Point Imperfections", edited by R.W.H. Stevenson (Oliver and
       Boyd, Edinburgh, 1966);
    G. Chioratti, "Theory of Imperfect Crystalline Solids" (Trieste
       Lectures)(International Atomic Energy Agency, Vienna, 1971).
8. W.E. Bron, W. Grill, "Phonon Spectroscopy I. Spectral Distri-

bution of a Phonon Pulse", Phys. Rev. B16, 5303 (1977).

W.E. Bron, W. Grill, "Phonon Spectroscopy II. Spectral, Spatial and Temporal Evolution of a Phonon Pulse", Phys. Rev. B16, 5315 (1977).

PHONON DETECTION BY THE FOUNTAIN PRESSURE

IN SUPERFLUID ^{4}HELIUM FILMS

Wolfgang Eisenmenger

Physikalisches Institut, Universität Stuttgart

Pfaffenwaldring 57, D-7000 Stuttgart 80, West Germany

INTRODUCTION

The importance of phonon focussing was first demonstrated by Taylor, Maris and Elbaum [1], [2], who calculated the phonon intensities for different modes in single crystals from the detailed form of the angle or k-dependent sound velocity surfaces. The results of these calculations are well-confirmed by experiment. It is quite surprising that phonon focussing leads in several crystals to very narrow beams of phonon energy propagation especially for transverse modes, as recently calculated by Rösch and Weis [3], who presented their computer results in a very instructive form, see Fig. 1. From these theoretical results it appears worthwhile to devise methods by which the intensity distribution of incoherent phonons propagating in single crystals can be measured in detail or directly imaged.

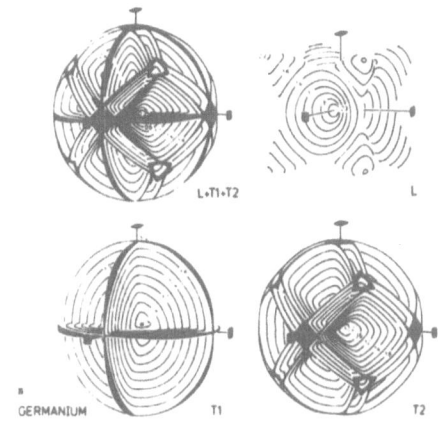

Fig. 1 Phonon Intensity Surfaces for Ge (After F. Rösch and O. Weis [3])

II EXPERIMENT

Whereas a distribution of phonon intensities in principle can be sampled by an array of bolometers or superconducting tunneling detectors, the possibility of an image converting system is also attractive for this purpose. In analogy to ultrasonic image converters based on the radiation pressure, it is the foun-

tain effect in liquid 4He below the λ-point that appears feasible
for image converting of incoherent phonon distributions at low
temperatures. The principle of this method is depicted in Fig. 2.

The single crys-
tal, e.g. Si |111|;
diameter 30 mm, thick-
ness 7 mm with a vacu-
um deposited constan-
tan heater on its bot-
tom side is immersed
in liquid 4He below
the λ-point, with the
liquid 4He level a few
mm below the upper
crystal surface. Pho-
non emission from the
heater increases the
local temperature or
phonon energy density
in the superfluid film
covering the crystal
surface in regions

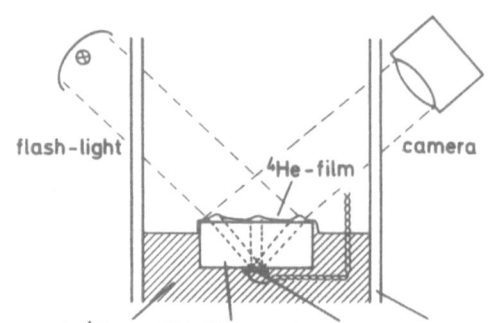

flash-light   4He-film   camera

superfl. 4He at 1.5K, Si-crystal, heater, cryostate-wall

Fig. 2 Experimental Arrangement for Imaging Phonon Distributions by the Fountain
Pressure

corresponding to directions of high phonon focussing. In these regions the fountain-pressure or the phonon radiation pressure,
which numerically equals one third of the phonon energy density,
leads to a locally increased film thickness. Under stationary conditions the film thickness is determined by the phonon energy density, whereas the inflow of superfluid 4He from surface regions
with lower temperature is balanced by the locally increased evaporation rate. The formation of a phonon intensity image by the
fountain-pressure, therefore, depends critically on the condition
that the evaporation rate from areas of increased film thickness
does not exceed the inflow of superfluid 4He limited by the critical film flow velocity. Otherwise the film disappears by evaporation at the spots of highest intensities. If the critical film
flow limit is increased by a somewhat rough crystal surface in addition to surface contaminations by condensed gases, higher phonon
intensities, i.e. thicker films, are possible and the phonon surface structure becomes directly visible under oblique light incidence and can be photographed, as shown in Fig. 3. This phonon in-

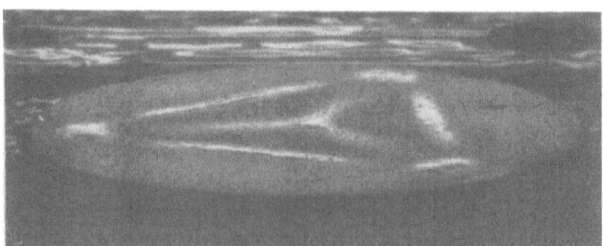

Fig. 3 Phonon Dis-
tribution Image at
the Surface of a
|111| Si crystal

tensity picture has been obtained using a Si |111| crystal with
the ^{4}He level about 5 mm below the crystal surface. The heater
power was 80 mWatts, and the bath temperature 1.5 K; surface contamination was accomplished by condensed air, 1 % air in the ^{4}He
gas of the cryostate before filling with liquid ^{4}He.

Fig. 3 has been obtained with the fountain pressure slightly
exceeding the pressure corresponding to the liquid level difference between crystal surface and ^{4}He bath. The heater power was
applied for a time of approximately 1 sec, sufficient for the collection of enough ^{4}He in the intensity maxima. The observed structure corresponds to the calculated distribution, c.f. Fig. 1 for
Ge, clearly indicating the strong |100|, |010|, |001| maxima, the
|111| maximum and the corresponding focussing planes of the slow
transverse modes in addition to the focussing planes of the fast
transverse modes. The splitting of the focussing planes of the
slow transverse modes and the conical structure of the |111| max-
imum in Fig. 1 are not resolved in the image Fig. 3. This possibly
reflects the difference between phonon focussing in Si and Ge.

With reduced liquid level difference pressure, the influence
of the inclination of the crystal surface against the liquid ^{4}He
bath level becomes visible, as shown in Fig. 4.

<u>Fig. 4</u> Phonon Distri-
bution Image as Fig.3
but with Reduced
Liquid Level Differ-
ence and Inclined
Surface.

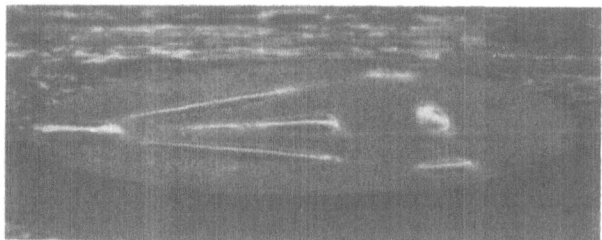

The |100| maximum on the left side of the picture with a smaller
liquid level difference has led to a strong drop-like collection
of ^{4}He spreading over a larger area than the phonon beam width.
This spreading corresponds to the finite lateral heat conduction
in a thick ^{4}He-layer, and reduces the phonon energy density in the
film.

In principle, it is also possible to use fountain pressures
below the liquid level difference. Under this condition, film
thickness changes are of the order of 1000 Å and it is necessary
to use schlieren-, interference- or ellipsometric techniques to
visualize phonon-surface structures.

III QUANTITATIVE DISCUSSION

Film thickness under the influence of the fountain pressure

The binding potential Φ of 4He atoms in the film surface can be described by c.f. |4|

$$\Phi = -\alpha/d^3 \qquad (1)$$

α = van der Waals constant for 4He atoms of the film bound to the substrate, d = film thickness.

Under equilibrium conditions Φ equals $-mg(h+d)$, (m = mass of He atoms, g = gravitational acceleration, h = liquid level difference with respect to crystal surface), the potential energy of He atoms at the liquid He level h below the crystal surface. Alternatively, the equilibrium condition can be described by an effective van der Waals pressure, which tends to increase the film thickness and is counterbalanced by the liquid level difference pressure $P_L = \rho_L g(h+d)$; (ρ_L = density of liquid 4He). This leads to the film thickness expression:

$$P_L = c/d^3 \text{ with } c = \rho_L \alpha/m \qquad (2)$$

Taking account of the fountain pressure and spatial variations of the film thickness, P_L has to be replaced by the total pressure P_t leading to the relation:

$$P_t = P_L - P_F + P_S = c/d^3 \qquad (3)$$

with P_F = fountain pressure and P_S = pressure by surface tension under the film surface curvature. In evaluating Equ. (3) with respect to the film thickness two limiting situations have to be considered:

i.) $|P_F| < |P_L|$: In this regime of "low" fountain pressure the film thickness increase Δd (neglecting P_S) results in the approximative relation:

$$\Delta d = d_o P_F/3P_L \qquad (4)$$

d_o = film thickness without fountain pressure.

ii.) $|P_F| > |P_L|$: In this region of "high" fountain pressure the film thickness d is approximated by

$$d = \{(P_F - P_S)/P_L - 1\} \cdot h \qquad (5)$$

Estimates of P_S for the experimentally observed film structures indicate that surface tension influences can be neglected.

Film temperature and fountain pressure

The experimental temperature dependence of the fountain pressure |4| can be described by:

$$P_F = 593 \ T^3 \cdot \Delta T \qquad (6)$$

P_F in cm liquid 4He, T = bath temperature and ΔT = film excess

temperature in K. The prefactor corresponds to a measurement at
1.5 K and exceeds the pure phonon contribution by a factor of 2 as
consequence of roton excitations and the curvature of the phonon
dispersion curve. Equ. (6) indicates that a temperature increase
in the film of $\Delta T = 10^{-4}$ K at a bath temperature of 1.5 K results
in a fountain pressure P_F = 2cm liquid ^{4}He. In calculating ΔT under phonon irradiation we use the balance between phonons transmitted from the crystal into the film against the energy transport
by excess ^{4}He atoms evaporating from the film and phonons backra-
diated from the film into the crystal. The detailed analysis shows
that direct heat conduction from the film into the ^{4}He gas can be
neglected in comparison to the energy transport according to the
latent heat of evaporation. Also lateral heat conduction within
the liquid ^{4}He film has little influence unless the film thickness
approaches the width of the structure. Using the rates of He atoms
incident from the vapor (sticking probability 1) which are in
equilibrium with ^{4}He atoms evaporated from the liquid film, the
excess evaporation rate at the elevated temperature T + ΔT of the
liquid film and the corresponding energy transport by the latent
heat of evaporation can be calculated. This, together with the phonon backradiation (assuming 20 % transmission within the angle limit of total reflection), leads to the total energy transport $\dot{q}$ as
function of ΔT:

$$\dot{q} = \{5.94 \cdot 10^4 \cdot T^{-2 \cdot 5} \cdot \exp\text{-}Q/RT + 1.39 \ T^3\} \cdot \Delta T \qquad (7)$$

$\dot{q}$ measured in Watt/cm^2, T and ΔT in K. R = 8.3 Joule/Mole·K, ideal
gas constant. Q = 82 Joule/Mole, latent heat of evaporation in the
1 K range.

In the calculation the latent heat of evaporation from the
film has been set equal to the latent heat of evaporation from the
free liquid surface. In thermodynamic equilibrium only the difference in potential energy mg(h+d) has to be taken into account,
which is small compared to the heat of evaporation under the conditions of the experiment. Furthermore, the energy contribution by
transport entropy, i.e. the energy necessary to heat the inflowing
superfluid component to the film temperature T + ΔT can be neglected.

Combining Equations (6) and (7) leads to the relation between
the fountain pressure and the phonon intensity transmitted into
the film. At the bath temperature of 1.5 K and the transmitted
phonon intensity of 1 mWatt/cm^2 this results in

$$P_F = 5.8 \text{x} 10^{-2} \text{cm liquid } ^4\text{He}$$

corresponding to 20 % phonon backradiation and 80 % evaporation.
Therefore, with a liquid level 0.5 cm below the crystal surface
the condition $P_F > P_L$ for a direct observation of the phonon intensity structures requires a transmitted phonon intensity of
I_{pht} = 8.6 mWatt/cm^2. For a phonon transmission factor of 20 % at
the crystal – film boundary this corresponds to an incident phonon

intensity of 43 mWatt/cm^2. In the experiment (c.f. Fig. 3) a total heater power of 80 mWatt has been applied. A rough estimate based on the crystal dimensions, using a phonon focussing factor of 10 and assuming that 2/3 of the heater power is directly emitted into the liquid He bath, results in the incident phonon intensity of 50 mWatt/cm^2. The comparison with the calculated minimum intensity of 43 mWatt/cm^2 indicates satisfactory agreement between experiment and theoretical model.

Intensity limit by the critical film flow

With increased transmitted phonon intensity and evaporation rate the liquid He mass transport becomes higher than allowed by the critical film flow. Considering a narrow maximum of the phonon intensity distribution in the film of 1 mm by 1 mm, the critical film flow of 8×10^{-5} cm ^{2} sec^{-1} |4| for clean polished surfaces corresponds to an evaporation heat flow of $\dot{q}$ = 9.5 mWatt/cm^2, which is close to the transmitted phonon intensity I_{pht} = 8.6mW/cm^2 considered before as necessary for obtaining a directly visible image of the phonon structure. For contaminated surfaces the critical film flow can be increased by one order of magnitude. In our experiments (c.f. Fig. 3) the surface contamination turned out to be very important. In experiments with a |100| oriented Si crystal the disappearance of the He film in the |100| phonon intensity maximum demonstrated that the critical film flow condition is easily reached with extreme differences in phonon focussing.

I gratefully acknowledge the technical assistance of Mrs. S. Döttinger, whose excellent experimental and photographic skill has led to the presented phonon distribution images.

REFERENCES:

1. B. Taylor, H. J. Maris and C. Elbaum, Phys. Rev. Lett. 23, 416, (1969).

2. B. Taylor, H. J. Maris and C. Elbaum, Phys. Rev. B 3, 1462, (1971)

3. F. Rösch and O. Weis, Z. Phys. B 25, 115, (1976)

4. J. Wilks, "The Properties of Liquid and Solid Helium", Clarendon Press, Oxford (1967)

BREMSSTRAHLUNG AND RECOMBINATION PHONONS STUDIED BY

Al-PbBi HETEROJUNCTION SPECTROMETER

W. Dietsche

Physik-Department E 10, TU München

8046 Garching, West Germany

In the past no practicable phonon spectrometer was available. Consequently the precise shape of many phonon sources was in doubt and the investigation of inelastic phonon scattering processes was difficult. Only very recently I reported that superconducting tunnel junctions made of Al and PbBi can be used as a spectrometer.[1] In this paper I will briefly review the technique, and present results on the spectra of bremsstrahlung and recombination phonons emitted by a Sn-Sn tunnel junction.

The inset of Fig.1 shows the tunneling density of states of the two superconductors. Phonons of energy hf impinge on the junction and generate a distribution of quasiparticles in the Al. These quasiparticles can tunnel into the PbBi only if the energy of the original phonons exceeds $\Delta_1 + \Delta_2 - eV$ where Δ_1 and Δ_2 are the energy gaps of the Al and the PbBi respectively and V the voltage across the junction. To have a large frequency range, Δ_2 should be much larger than Δ_1. To obtain a large tunnel current, i.e. high sensitivity, the life time of the quasiparticles should be long. In the case of Al and PbBi both conditions are satisfied. For an experimental test, an Al-PbBi junction was irradiated with bremsstrahlung phonons (0-290 GHz) emitted by modulation of a Sn-Sn junction.[3] This was placed on the opposite side of a Si substrate crystal. Fig.1 shows the detected phonon signal as function of frequency for a number of detector voltages. The expected detection thresholds are marked by the arrows. Clearly, each trace exhibits a distinct step at the expected frequency.

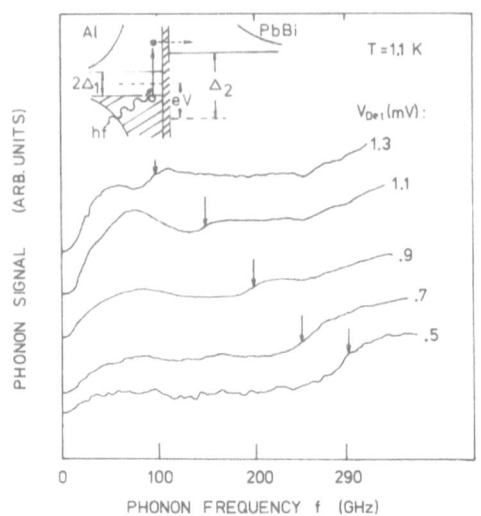

Fig. 1. Phonon signal vs. frequency with the Al-PbBi detector set
         at several frequency detection thresholds (dashed lines).
         Clearly a step like increase is observed when the generator
         frequency coincides with the threshold.   Insert shows the
         phonon detection process.

         However, an Al-PbBi junction is not only useful as
a detector with a voltage tunable frequency threshold but
also as a spectrometer. For this purpose a small ac-
modulation is applied to the detector together with the
bias voltage V. The phonon signal will then only be modulated if the frequencies of the incident phonons are
close to $\Delta_1 + \Delta_2 - eV$. Therefore by measuring the modulated
phonon signal and sweeping the detector bias one should
be able to determine the frequency distribution of an
unknown spectrum. In practice, the resolution is dependent on the modulation width which was generally about
10 GHz. Unfortunately, the sensitivity of the detector
is not independent of its bias. Therefore the modulated
signal also contains a term which depends on the derivative of the sensitivity with respect to the voltage.
The voltage dependence of the sensitivity can be determined from the height of the steps (Fig.1) as a function
of frequency and the correct phonon spectrum can be calculated. This method was applied to the phonons emitted
by a Sn-Sn tunnel junction

         If a generator junction is biased at $2\Delta/e$ the quasi-
particles can only tunnel into states just above the
energy gap. Phonons are then generated by recombination.
The measured spectrum of these phonons is shown in Fig.2a.

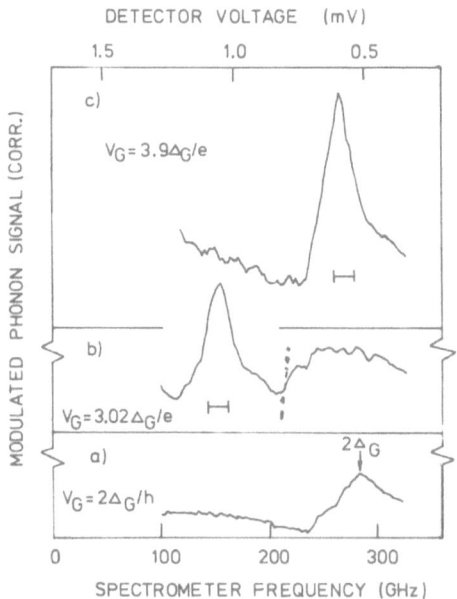

Fig. 2.   Measured spectra of phonons emitted by Sn-Sn junction at
         several biases as denoted in the figure.

A maximum shows up at the frequency corresponding to $2\Delta_G$,
as expected. The width of the spectrum, however, is [4]
slightly broader than that predicted theoretically. in
particular, no phonons are expected below $2\Delta_G/h$.

     At higher junction biases the injected quasipartic-
les relax first, emitting bremsstrahlung phonons before
they recombine. The generator tunnel current was modulated to produce a nearly monochromatic bremsstrahlung
phonon line.[3] Fig.2b) and c) shows spectra at generator
biases of $3.02\Delta_G/e$ (1.8mV) and $3.9\Delta_G/e$ (2.3mV), respectively. The horizontal bars under the traces denote the
width of the bremsstrahlung phonons as expected from the
generator modulation. Clearly, the measured spectral
width of the bremsstrahlung phonons agrees with the expected value. The band of recombination phonons, however,
became even broader than at the bias of $2\Delta_G/e$. This is
particularly visible in Fig.2b). The maximum even seems
to be shifted to a value below $2\Delta_G/h$.

     At present, I can not give a satisfactory explan-
ation for the unexpected broadness of the recombination
phonon spectrum. One might speculate that the energy gap
is locally reduced due to magnetic flux unintentionally

trapped in the generator junction. Recombination phonons
which happen to have frequencies below $2\Delta/h$ would escape
preferentially while those with frequencies corresponding to the average energy gap are strongly reabsorbed.
The spectrum of the bremsstrahlung phonons, on the other
hand, depends on the average gap and would therefore not
be much affected. More experiments with improved magnetic
shielding are necessary to clarify this point.

REFERENCES

1. W. Dietsche, Phys.Rev.Lett.40:786 (1978).
2. For another application of this detector see:
        W. Dietsche and H. Kinder, Inelastic Phonon
        Scattering in Glass, this volume.
3. H. Kinder, Phys.Rev.Lett.28:1564 (1972).
4. A. H. Dayem and J. J. Wiegand, Phys.Rev.B5:4390
        (1972).

QUANTITATIVE DETECTION OF INTENSE THERMAL PHONON PULSES USING SUPERCONDUCTING TUNNEL JUNCTIONS

M.Nover and O.Weis

Experimentelle Physik IV, Universität Ulm

79 Ulm, Germany

The momentary current/voltage characteristics of symmetric tin and lead junctions were measured during irradiation with intense longitudinal or transverse thermal phonon pulses. Using a chain of model assumptions the whole set of measured current/voltage characteristics can be reproduced theoretically astonishing well by fitting only two parameters for each phonon polarization and metal: i) the phonon absorption coefficient in the normal state and ii) a phonon cutoff-frequency for the detected phonons.

Fig.1 explains in the inset the radiator (1)/detector (2) geometry: the detector area was $A^{(2)}=0.3\times0.3$ mm^2. The area $A^{(1)}$ of a constantan radiator was evaporated just opposite onto a 6 mm thick a-cut

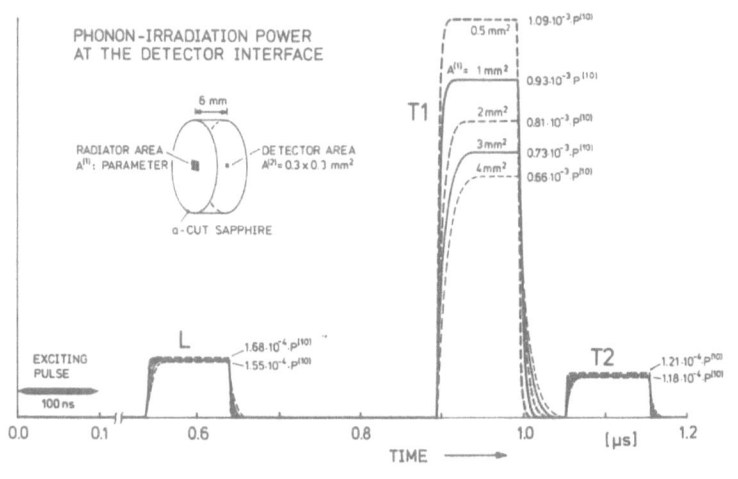

Fig.1: Theoretical phonon irradiation power

sapphire (= medium (0)). The main part of Fig.1 shows the calculated[1] momentary phonon irradiation power at the detector interface
(related to the whole radiation pulse power $P^{(10)}$ for different
radiation areas $A^{(1)}$ under the following model assumptions: i) the
constantan radiator is considered as a mechanically isotropic half
space possessing an isotropic phonon distribution, ii) the number
of phonons at each frequency in the radiator may change with frequency arbitrarily, only the ratio of longitudinal to transverse phonon
number is always taken equal to the ratio of the corresponding
phonon states, iii) the angular dependent phonon transition and
polarization-conversion probability at the radiator/sapphire interface is deduced from anisotropic continuum acoustics ('mismatch
model') assuming an isotropic/anisotropic interface, iv) anisotropic
ballistic phonon propagation in the sapphire crystal, starting with
the calculated $\vec{q}$-space distribution in the sapphire at the radiator
interface.

If we furthermore assume for the constantan radiator an one-tempera-
ture model, the emission power $P^{(10)}$ is determined[2,3] by the radiator temperature $T^{(1)}$ and the spectral phonon power is given by a
Planck distribution.

Earlier experiments[4] on 250 nm thick tin diodes revealed that the
momentary characteristic under phonon irradiation (taken with 150 ns
pulses at the end of each detection pulse where stationary conditions
are reached) do not deviate within experimental error from static
thermal characteristics. In lead diodes only small deviations at
low voltages are observable. So we are able to deduce an electronic
temperature $T^{(2)}$ from the observed gap voltages. In Fig.2 these
electronic temperatures $T^{(2)}$ of the diode are plotted over the
radiation temperature and a comparison is made with theoretical
curves resulting from the following model assumptions for the diode:
i) The phonon system of the diode is also described by $T^{(2)}$. Then
under stationary conditions the whole absorbed phonon power is
equivalent to the thermal radiation power of the heated diode, i.e.
$T^{(2)}$ is a measure of the absorbed phonon power, ii) the probability
that phonons (falling onto the detector area) are absorbed in the
film is determined by taking into account acoustic reflection at
the interface, multi-reflection effects in the film and finally the
bulk absorption according to the BCS-theory[5] which is related to
the absorption in the normal state, iii) the amplitude absorption
coefficient $\alpha_L^{(2)}$ of the normal state is taken for longitudinal
phonons equal to Pippard's theoretical value[6] $\alpha_{P,L}^{(2)}$ of the high-
frequency limit. Contrary, the observed absorption of transverse
phonons can not be described with Pippard's theory. We chose instead
$\alpha_T^{(2)} = \eta \cdot \alpha_{P,L}^{(2)}$ and determined $\eta=4$ as the best fit to experimental data
in tin as well as in lead at low radiation temperatures, where only
relatively low-frequency phonons are emitted. iv) In order to get
also an excellent fit at higher radiation temperatures a cutoff
frequency $\nu_{max,\sigma}$ was introduced for each polarization σ which may
be due to the cutoff frequency of the acoustic branches in the diode
and due to the onset of phonon scattering in sapphire.

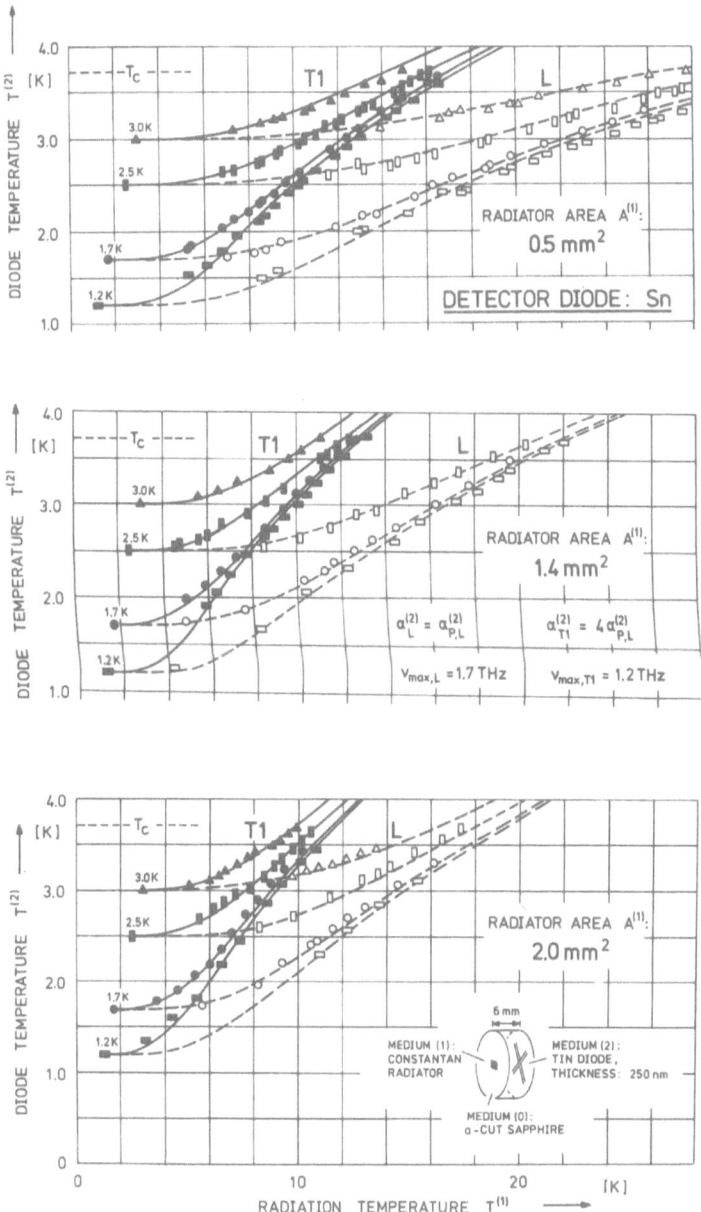

Fig. 2.   Comparison of measured tin-diode temperature $T^{(2)}$ with
          calculated temperatures as a function of radiation
          temperature $T^{(1)}$ for different radiator areas $A^{(1)}$ and
          for different sapphire temperatures $T^{(0)}$ (=starting
          point of each curve).

With this set of assumptions all observed diode temperatures $T^{(2)}$ are described astonishing well in tin (see Fig.2) and lead, even if one changes the bath temperature $T^{(0)}$ and the radiation area $A^{(1)}$ which produces significant changes in the thermal radiation spectrum.

REFERENCES

1. F.Rösch and O.Weis, 1977, Z.Physik B29, 71
2. O.Weis, 1969, Z.Angew.Phys.26, 325
3. P.Herth and O.Weis, 1970, Z.Angew.Phys.29, 101
4. M.Goetze, M.Nover and O.Weis, 1976, Z.Physik B25, 1
5. V.M.Bobetic, 1964, Phys.Rev.136 A, 1535
6. A.B.Pippard, 1955, Phil.Mag.46, 104

OBSERVATION OF A NEW TYPE OF ENERGY-SELECTIVE PHONON DETECTOR

Richard W. Cline* and Humphrey J. Maris

Department of Physics, Brown University,
Providence, RI 02912

Superconductors are commonly used as phonon detectors in either of two ways. Superconducting bolometers take advantage of the large change in resistance with temperature at the superconducting transition. Superconducting tunnel junctions, on the other hand, make use of the superconducting energy gap. These two mechanisms result in fundamentally different types of phonon detection. Bolometers are broadband in nature and respond to phonons of all energies whereas tunnel junctions respond only to phonons of energy greater than the superconducting energy gap. In this paper we report on experiments with a new form of superconducting detector. This has the unique property of energy-selectivity for phonons of energy significantly greater than the superconducting energy gap.

While carrying out heat pulse experiments on helium II at temperatures less than 0.8K using a thin film of zinc (nominally a superconducting bolometer) as a phonon detector, we observed that under some conditions the detector was more sensitive to phonons of one energy range than to those of another. Accordingly, we then used known properties of helium II to study the characteristics of this energy selectivity of the zinc thin-film detectors. The most quantitative information about the energy selective behavior was obtained from knowledge of the phonon part of the dispersion curve. Near the melting curve, the phonon group velocity continuously decreases as the phonon energy increases[1]. Experimentally we observed that the phonon velocity shifted as the magnetic field applied to the detector was changed. At higher magnetic fields the phonon velocity was faster. Knowledge of the dispersion curve then indicates that we have detected only high energy phonons in a low magnetic field, whereas in a large magnetic field all phonons are de-

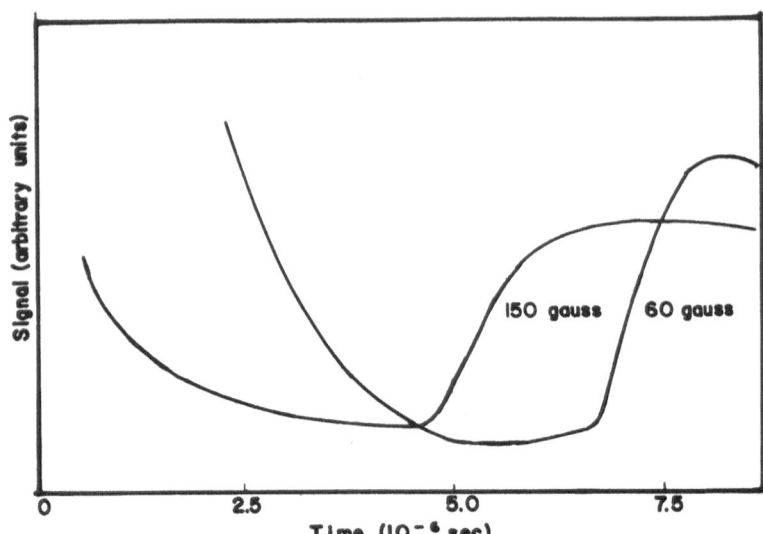

Fig. 1.   Received signal versus time for a zinc detector for two
          applied magnetic fields. The actual field at the detector
          is smaller than the applied field because of shielding.

tected. An example of this kind of behavior is shown in Fig. 1.
The data consist of a record of the signal as a function of time
obtained at different magnetic fields. The heater power was held
constant and the amplifier gain changed to give the signals roughly
the same magnitude. Above a certain magnetic field the velocity
ceased to increase. By taking this velocity to be the sound velocity we used the relative change in velocity at lower fields an an
indication of the phonon energy detected. At 24 bar the minimum observed phonon velocity corresponds to the detection of 7-10 K phonons, depending on the value of dispersion parmeter chosen[2]. This
was especially surprising as the energy of the detected phonons was
∿ 3x the superconducting energy gap of zinc (2.8 K). The highest
energy phonon signal is detected when the zinc film is resistanceless. In order to detect this signal, heat powers in the range 0.1
to 1 erg mm^{-2} are necessary, presumably because lower powers do not
generate sufficient numbers of high energy phonons.

    Three energy selective zinc film detectors were studied. The
two most extensively investigated were made by evaporating high
purity (99.999%) zinc in an alumina covered boat at 10^{-5} torr onto
a room temperature 3 MHz quartz ultrasonic transducer through a
stainless steel mask after first evaporating a layer of silver ∿ 5 A

thick to aid adhesion of the zinc. These detectors exhibited energy
selective behavior only after being thermally cycled between 300 and
4.2 K ∿ 5 times. The room temperature resistance, resistance ratio,
and transition temperature did not change with thermal cycling. The
third zinc film detector was made by etching away areas of a similarly evaporated zinc film that was not masked with photoresist.
This detector had characteristics similar to the others except that
the transition temperature was ∿ 0.2 K lower, and it exhibited energy selectivity without first being thermally cycled. In all experiments a bias current of about 1mA was used. A magnetic field
(0-300 gauss) was applied parallel to the film by a solenoid.
However as the sample cell was assembled with soft solder and contained an indium "O ring," the actual field seen by the superconductor was not precisely known.

By observing the signal as the helium II level rose between the
heater and detector we found that in a larger magnetic field (where
the detector acted essentially as a bolometer) the signal amplitude
changed smoothly as the cell filled, but in a smaller field (where
energy-selective behavior was apparent) the amplitude changed in two
steps as the level was raised. This suggests that the energy-
selective detection is localized. In another experiment we used a
geometry such that the phonons generated by the heater were pre-
dominately detected in the center of the zinc film. In this way we
showed that the contacts to the zinc film did not contribute to the
energy selective behavior. The V-I characteristics of the films
exhibit structure in the current-induced resistive state. This
structure was not exactly reproducible and depended somewhat on
parameters such as the rate at which the current was swept.

We feel that our observations of the energy-selective detection of phonons can best be explained in terms of the ability of
only high-energy phonons to drive trapped fluxoids over localized
energy barriers. If a fluxoid were trapped at some site in the
superconductor, and a phonon were able to provide enough energy to
drive the fluxoid over the barrier, the resultant movement of flux
would produce a voltage across the detector. In this way phonons
with sufficient energy to move fluxoids could be detected even
though the detector was resistanceless. The height of the energy
barriers could be 7-10 K as required to explain our observations,
and would not be limited by the superconducting energy gap. The
observation that the energy selective behavior is only seen after
thermally cycling the detectors several times suggests that the
energy barriers may be associated with defects in the film. This
mechanism would be tunable by increasing the applied magnetic field
because the force on fluxoids increases with magnetic field.

It was the intent of this work to show that the phonon detection we observed was actually energy selective. Further investi-

gation will be necessary to determine whether or not the energy selective behavior is produced through the mechanism we have proposed.

This work was supported in part by the National Science Foundation through grant number DMR 77-12249 and through the Materials Research Laboratory of Brown University.

*Present address: Department of Physics, Room 13-2061, Massachusetts Institute of Technology, Cambridge, MA 02139.

REFERENCES

1. H. J. Maris, Rev. Mod. Phys. 49, 341 (1977).
2. The dispersion parameter obtained from the neutron scattering measurements of E. C. Svensson, A. D. B. Woods, and P. Martel, Phys. Rev. Lett. 29, 1148 (1972) implies a phonon energy of 10 K. That from R. C. Dynes and V. Narayanamurti, Phys. Rev. B12, 1720 (1975) gives a phonon energy of 7 K. Dynes and Narayanamurti's measurements were made using a tunnel junction detector and are not all consistent with the same value of dispersion parameter. The 7 K result is obtained when the dispersion parameter obtained from the tunnel junction experiment in zero field is used.

FAST BOLOMETER AMPLIFIERS USING TUNNEL DIODES

F. Moss*

Department of Physics,
University of Lancaster,
Lancaster, U.K.

INTRODUCTION

The purpose of this paper is to describe two back diode amplifiers which have been operated at 1 K, and which deliver significant
amounts of power into 50 Ω cables terminated at room temperature.
The amplifiers are particularly suitable for use with moderately
fast resistance bolometers. Recently Lengeler[1] has discussed the
applicability and characteristics of various solid state devices,
including back diodes, for use at low temperatures. His measurements have inspired a number of recent low temperature amplifier
designs[2,3]. While back and tunnel diodes have long been used as
oscillators at low temperatures,[4,5] the possibilities for their use
as amplifiers seem to have been ignored.

AMPLIFIER DESIGN

The amplifier designs presented here have the following
advantages: simplicity, with only one coaxial cable necessary, low
power dissipation, stable and repeatable characteristics at low
temperature and insensitivity to thermal cycling. They have the
following disadvantages: nonlinear response, small dynamic range,
and the necessity of choosing a bolometer resistance comparable in
magnitude to the negative resistance of the diode.

The circuit used for gain and bandwidth measurements is shown
in Fig.1 (a). A signal generator v_i, provided the input to the
amplifier through a 50 Ω, terminated, low thermal conductivity,
microcoaxial cable UT2OSS.[6] The low temperature parts were
immersed in a bath of liquid helium whose temperature could be
varied between 1 and 4.2 K. The bolometer resistance R_B, was
simulated by a carbon resistor whose value was adjusted _in situ_ by
varying the bath temperature. The diode was biased by a constant

current entering the node marked V_b. Amplifiers were designed for
BD-2 and BD-4 diodes,[7] whose characteristics were measured at 1 K.

The frequency response can be calculated from the approximate
equivalent circuit shown in Fig.1 (b), where r_d is the negative
resistance of the diode at the operating point, C_d is the stray
capacitance and L the stray inductance.[8] C_c represents a stray
capacitance across the output cable due to imperfect termination.
The magnitude of the complex impedance for perfectly terminated
cables is given by

$$Z^2 = \left\{ R_L + r_d \Big/ \left[1 + (\omega\tau_d)^2 \right] \right\}^2 + \omega^2 \left\{ L - r_d^2 C_d \Big/ \left[1 + (\omega\tau_d)^2 \right] \right\}^2 ,$$

where $R_L = R_B + 2R_T$ is the total circuit resistance seen by the
diode and $\tau_d = r_d C_d$. The voltage and power gains are then given by
$G_v = R_T/Z$, and $G_p = (G_v/G_o)^2$, where G_o is the voltage gain of the
circuit with the diode replaced by a short circuit. In the low
frequency limit $G_v = R_T/(R_L + r_d)$; $G_o = R_T/R_L$ and $G_p =$
$\left[R_L/(R_L + r_d) \right]^2$. In these expressions, r_d is entered as a negative
number. The stability of the equivalent circuit is discussed in
Ref.8. For a damped response it is necessary that $R_L < |r_d|$.

(a)

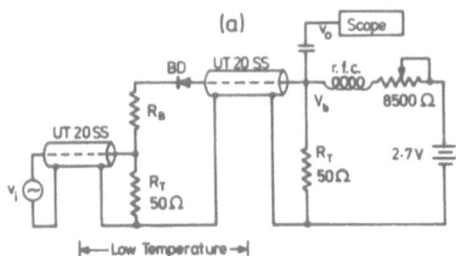

(b)

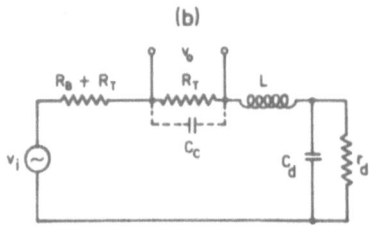

Fig. 1 (a) The actual circuit used to test the back diode amplifiers at liquid helium temperatures.   (b) An approximate equivalent circuit useful for frequency response calculations.

FREQUENCY RESPONSE AND GAIN MEASUREMENTS

     Measurements of the frequency response for the BD-2 amplifier
for two values of R_L are shown in Fig.2.   The low frequency voltage
gains are 1.4 and 0.5 with corresponding power gains of 87 and 7.3
respectively.   Curve A is the calculated response for perfectly
terminated cables.   Both cables were terminated with 50Ω ± 10%
resistors.   Curve B is the calculated response for the same parameters as Curve A but with a finite output cable time constant τ_c =
$R_T C_c$.   The response of the BD-4 amplifier is shown by the solid
triangles, with a low frequency voltage gain of 0.12, and a power
gain of 31.3, indicating that the diode is effective in coupling the
2.35 KΩ circuit resistance to the external 50Ω termination.

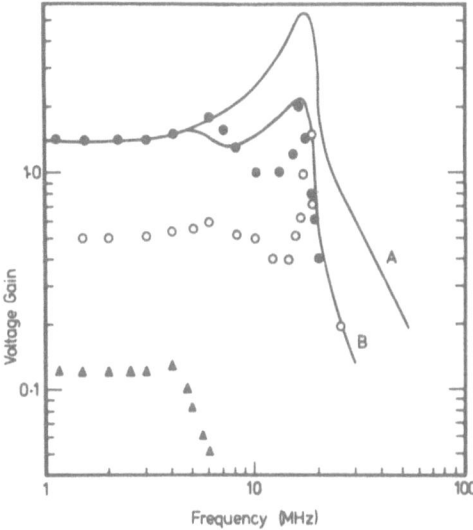

Fig. 2   The measured frequency response of the BD-2 amplifier
with R_L = 334Ω (solid circles) and R_L = 270Ω (open circles).
The power dissipated is 21 µW.   The curve A represents the
calculated frequency response for L = 1 µh, C_d = 8.5 pf and
R_L = 334Ω for perfectly terminated cables.   Curve B shows the
effect of including an output cable time constant τ_c = 28 ns.
The measured response of the BD-4 amplifier with R_L = 2.35 KΩ
(solid triangles), and power dissipation of 2.8 µW.

* Permanent address: Department of Physics, University of Missouri,
  St. Louis, MO 63131, U.S.A.

REFERENCES
1. Lengeler, B., Cryogenics, 14 439 (1974)
2. Forrest, S.R., Sanders, T.M., Rev. Sci. Instr. 49 1603 (1978)__
3. Ghola, H., Ehnholm, G.J., Ostman, P., and Rantala, B., Proc. XV
   Intern. Conf. on Low Temp. Phys.; J. De Phys. 39, C6 - 1184 (1978)
4. Heybey, O.W.C., Thesis, Cornell University (1962); Boghosian, C.,
   Meyer, H., Rives, J.E., Phys. Rev. 146 110 (1966)
5. Clover, R.B., Wolf, W.P., Rev. Sci. Instr. 41 617 (1970)
6. Uniform Tubes, Inc., Collegeville, Pennsylvania, U.S.A.
7. General Electric Co., Schenectady, New York, U.S.A.
8. see for example; Brophy, J.J., Basic Electronics for Scientists,
   McGraw-Hill, New York, pp 348-353, (1966)

STUDY OF ACOUSTOELECTRICALLY AMPLIFIED PHONONS IN InSb BY ANOMALOUS TRANSMISSION OF X-RAYS*

L. D. Chapman, R. Colella, and R. Bray

Physics Department
Purdue University
West Lafayette, IN 47907

Anomalous transmission of x-rays is a channelling phenomenon of photons in crystals. A convenient geometry for realizing anomalous transmission consists of using a crystal plate with the lattice diffracting planes perpendicular to the surface (see Fig. 1). This geometry is commonly referred to as the "Symmetric Laue Case". When the incident beam I_0 forms the correct Bragg angle θ_B given by Bragg's law, $\lambda = 2d \sin \theta_B$, with the lattice planes, a set of standing waves is formed in the crystal with nodes on the lattice planes and antinodes between lattice planes. In such a situation the electric field is maximum in regions of minimum electron density and photoelectric absorption is virtually absent. More quantitatively, the transmission observed on the T beam of Fig. 1 varies typically from 10^{-9} for $\theta \neq \theta_B$ to 10^{-2} for $\theta = \theta_B$. Clearly, the effect can be easily upset by small lattice perturbations. Any kind of atomic displacement with some component along the normal to the diffracting planes will greatly affect the anomalous transmission. By changing geometrical conditions in such a way that different atomic planes (hkl) are used for diffraction in the same crystal, it is possible to obtain information on the various components of vibrational atomic amplitudes along different directions.

We have previously reported on the effects of intense acoustoelectrically amplified phonons on anomalous transmission in InSb.[1] Acoustoelectric amplification is very strongly favored for [110] fast transverse (FT) phonons polarized along [001]; for the latter, we found, as expected, that anomalous transmission using the (004) planes was strongly quenched. We also expected to see no quenching

---
*Supported by National Science Foundation and NSF/MRL Grants.

Fig. 1. Geometrical ar-
         rangement for
         Laue case dif-
         fraction from
         (2$\bar{2}$0) planes
         parallel to the
         narrow side
         surface.

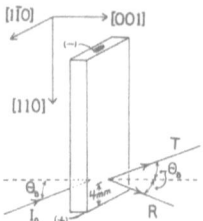

of anomalous transmission when planes parallel to [001] were used
for diffraction. Contrary to the latter expectations, we found
weaker quenching of the (2$\bar{2}$0) and (220) reflections. In those original experiments the effect of the amplified phonons on the anomalous
transmission was averaged over many round trips of the phonon flux
through the crystal. In order to discriminate between phonons generated within the [110] FT acoustic domain and phonons of possible
different modes generated by reflection at the side or end wall, we
have resorted to a time resolved technique. The time resolved
measurements were made for both the reflection geometry of Fig. 2
with the lattice planes parallel to the surface of the crystal
("Symmetric Bragg Case") in which only the [110] FT phonons can be
observed, and an anomalous transmission configuration using the
(2$\bar{2}$0) planes which is sensitive to phonons of any frequency which
have polarization components normal to the (2$\bar{2}$0) planes.

Fig. 2. Geometrical ar-
         rangement for
         Bragg case dif-
         fraction.

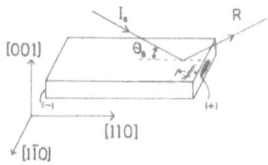

For reference, the applied voltage pulse and resulting current
profiles are shown in Fig. 3a. The current pulse shows the onset
of an acoustoelectric domain (current drop) and its arrival at the
anode (current rise). An x-ray beam (1.15 mm. wide) is incident on
the sample, 4 mm. from the anode end. Diffraction profiles were
simultaneously recorded during several scans in successive 0.5 μsec
time slots suitably delayed with respect to some reference t_0. All

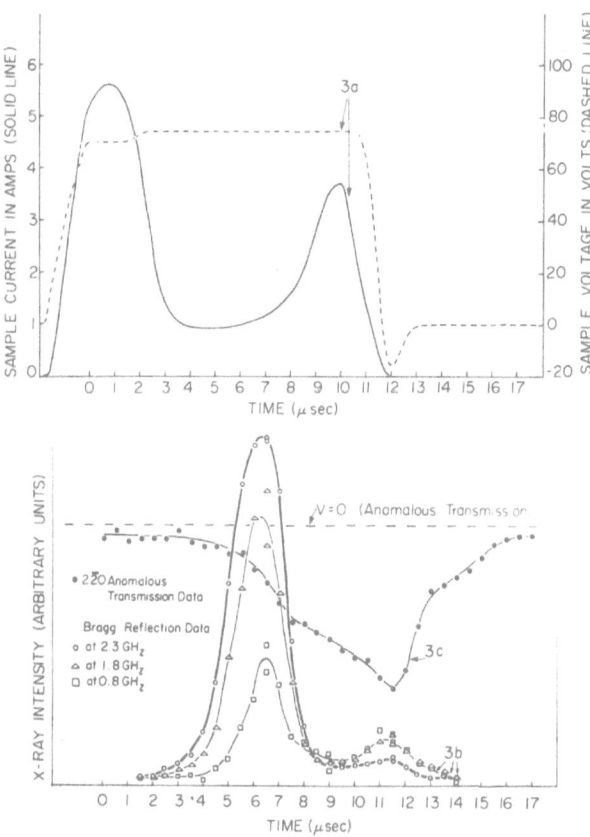

Fig. 3.   a) Applied voltage and current flow in crystal.   b) Bragg
          diffracted sideband intensity at three phonon frequencies.
          c) Intensity of $(2\bar{2}0)$ anomalous transmission.

measurements were made during the first round trip of the phonon
domain from the cathode to the anode and back again after reflection.   When the experiment is done using the Bragg reflection geometry of Fig. 2, the presence of amplified phonons is signalled by
two satellite peaks symmetrically located with respect to the main
diffraction peak.[2]   In this Bragg reflection arrangement the smallest observable frequency is ≈ 0.5 GHz for which the satellites merge
into the central Bragg peak.   The sequence of time resolved diffraction profiles in the Bragg case shows a maximum acoustic intensity at $t_1 = 6.5$ μsec (see Fig. 3b), consistent with the time of
flight of the $[110]$ FT phonon domain from the source end of the
crystal (cathode) to the region illuminated with x-rays.   A secon-

dary weak maximum is found at a later time due to phonons bouncing back from the end. Results are shown in Fig. 3b for a few acoustic frequencies. The reflectivity is obviously stronger at the lower acoustic frequencies.

The time evolution of the quenching of anomalous transmission is shown in Fig. 3c. In this situation the quenching is due to non-[110] FT phonons with a polarization component along the [1$\bar{1}$0] direction. The results for this case differ markedly from those obtained in the Bragg case for the [110] FT phonons. A decline in anomalous transmission appears near t_1 = 6.5 μsec coincident with the arrival of the phonon domain. However the absence of recovery after the domain passes suggests that there is continued growth in non-[110] FT flux at the position of the probing x-ray beam. Such continued growth suggests that non-[110] FT flux is generated and emanates backward from the growing, forward-propagating phonon domain even before the latter arrives at the end face. Note that a similar continued growth in backward travelling flux does not show up in the Bragg data (Fig. 3b), which seems to indicate that this flux must have such polarization and/or frequency as to be able to affect the anomalous transmission but not be observable in the Bragg reflection data. Further experiments will be necessary to determine the precise nature of the converted flux and its mechanism of generation in the domain.

Finally, we note the extended presence of the quenching signal appreciably past the 12 μsec mark. This may be due to off-axis phonons generated by scattering of the flux in the domain upon reflection at the end face.

REFERENCES

1. S. D. LeRoux, R. Colella, and R. Bray, PRL 37, 1056 (1976).
2. S. D. LeRoux, R. Colella, and R. Bray, PRL 35, 230 (1975).

TUNABLE OPTICAL DETECTOR AND GENERATOR FOR

TERAHERTZ PHONONS IN CaF_2 AND SrF_2

W. Eisfeld and K.F. Renk

Institut für Angewandte Physik
Universität Regensburg
8400 Regensburg, W-Germany

The lowest excited electronic state $4f^6 5d$ (Γ_8^+) of Eu^{2+} ions in CaF_2 and SrF_2 can be split in two sublevels by applying uniaxial stress along the [001] crystalline axis.[1] This splitting is used for phonon detection up to a frequency of 2.4 THz. The phonons are detected by phonon-induced fluorescence radiation due to their resonance interaction with the stress-split sublevels.

Our experimental arrangement is shown in the inset of Fig.1. Phonons were generated by the heat-pulse technique by injecting current pulses (duration 200 ns, repetition rate 50 kHz, peak power 20 W/mm^2) in a 1 mm^2 constantan heater. Continuous optical pumping with a Hg-Xe lamp created in the crystal a small volume containing excited Eu^{2+} ions in the lower sublevel. The phonon-induced fluorescence radiation W_2 from the upper sublevel was analyzed with a fast multichannel analyzer triggered by the current pulses.

An experimental result is shown in Fig.1. The signals were measured for a stress of P = 1.5 kbar corresponding to a phonon frequency of 600 GHz. The signal curves from a detector volume located in two different distances x between heater and detector, indicate that phonons at 600 GHz propagate nearly ballistically in a CaF_2 crystal with 0.01 mole% Eu^{2+} ions. The broadening of the pulses towards longer times is due to phonon scattering at Eu^{2+} ions. The time-independent background is mainly caused by an imperfect suppression of the fluorescence radiation W_1 from the lower sublevel.

A first application of this detector principle for the measurements of phonon spectra of heat pulses within a CaF_2 crystal

329

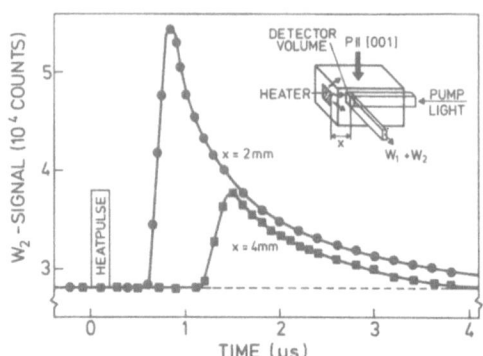

Fig. 1. Phonon-induced fluorescence signals by 600 GHz-phonons
         in $CaF_2:Eu^{2+}$.

has already been published.[2] In this paper we report measurements
of the defect-mass scattering of the phonons by the Eu^{2+} ions. For
a first study we have chosen $SrF_2:Eu^{2+}$ in order to compare with
results obtained from the vibronic sideband spectroscopy.[3]

    We have measured the spatial distribution of phonons at dif-
ferent frequencies some time after heat-pulse injection in a SrF_2
crystal containing .01 mole% Eu^{2+} ions. Fig.2 shows two experimental distributions for phonons at 1.4 THz and 2.4 THz obtained at a
time t= 1 μs after phonon generation. The shape of the curves is
typical for a diffusive propagation where the high frequency phonons diffuse much slower than the low frequency phonons. We should
note that the occupation number n which is obtained from the ratio
of the fluorescence intensities of the two sublevels (W_2/W_1) is
much smaller for the high frequency phonons than for the low frequency phonons as expected for a Planck phonon spectrum.[2]

    From spatial phonon distributions we have determined the mean
free path l_D of the phonons in SrF_2. The results are shown in Fig.3
and compared with the results obtained by the vibronic sideband
spectroscopy for a crystal containing .1 % Eu^{2+} ions.[3] Regarding
the concentration differences the experimental results are in
reasonable agreement. We think, however, that the new results indicate that l_D has a stronger frequency dependence than the $1/\nu^2$

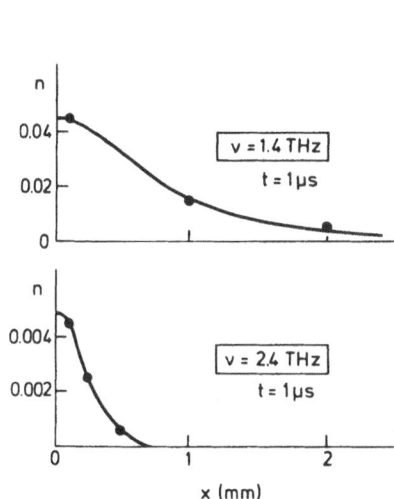

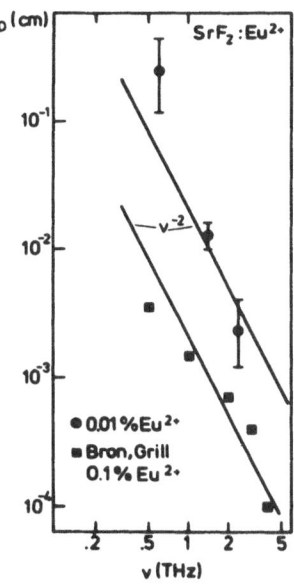

Fig. 2. Spatial phonon   Fig. 3. Phonon mean free
         distributions.   path for defect-mass
                                          scattering.

dependence which has been concluded by Bron and Grill.[3] It will be very interesting to determine the frequency dependence more accurately in order to understand the phonon-impurity interaction in $SrF_2:Eu^{2+}$.

The stress-split sublevels of Eu^{2+} ions in CaF_2 and SrF_2 are also suitable for the tunable generation of terahertz phonons. The crystal is optically pumped by a nitrogen laser. Fast relaxation leads to equal population of the two sublevels. By the subsequent one-phonon relaxation of the upper sublevel monochromatic phonon pulses are generated.

Financial support by the Deutsche Forschungsgemeinschaft is acknowledged.

REFERENCES

1. A. A. Kaplyanskii and A. K. Przhevuskii, Opt.Spectrosc. 19, 331 (1965).
2. W. Eisfeld and K.F. Renk, Appl.Phys.Lett. 34, 481 (1979).
3. W. E. Bron and W. Grill, Phys.Rev. B 16, 5303 (1977); Phys.Rev. B 16, 5315 (1977).

****

DISCUSSION

R. S. Meltzer: Do you have any idea what the spin-lattice relaxation times are in this crystal?

R. F. Renk: We have measured them. It is about 3 nsecs at 1 THz.

P. Taborek: Can this method generate transverse phonons also?

K. F. Renk: Yes. We have done time-of-flight experiments also with heat pulses. We could only see the transverse phonons.

P. Taborek: Is there some selection rule?

K. F. Renk: We are not yet clear about the selection rules.

W. E. Bron: The lowest state is a well-known and complicated Jahn-Teller state. In fact there are several states in it. Don't you have to worry about what straining such a state will do on the transition probabilities?

K. F. Renk: They are the same because the absorption to both states is the same. So we have the same transition probabilities to the ground state.

ELECTROMECHANICAL CONVERSION IN MS-CONTACTS; DETERMINATION

OF FREQUENCY RESPONSE, DEPLETION DEPTHS AND ENERGY LEVELS

K. Fossheim and A.M. Raaen

Department of Physics, The Norwegian Institute

of Technology, N-7034 Trondheim - NTH

Piezoelectricity is absent in centrosymmetric crystal classes. Yet, surface acoustic wave generation and detection can be performed with high efficiency in centrosymmetric dielectrics by deposition of interdigital metal electrodes on the dielectric surface[1]. In this paper we discuss results of a similar nature for volume waves. We outline a model theory for acoustic volume wave generation and detection by metal-dielectric (or metal-semiconductor) contacts deposited on centrosymmetric, cubic crystals, and present experimental verification of the main results.

As shown below an effective piezoelectricity $d_{eff} \sim mE_o$ exists in the depletion region of a metal-dielectric contact. Here m is an electrostrictive constant and E_o is the built-in electric field. In the simplest case, when a space charge ρ_o is taken to exist down to a depth d (see Fig.1), the built-in field is given by[2] $E_o = \rho_o z/\varepsilon_s \varepsilon_o$ for $0 < z < d$ where z is the thickness coordinate, as in Fig.1, ε_o is the dielectric permittivity of vacuum, and ε_s is the relative dielectric constant in the depletion region.

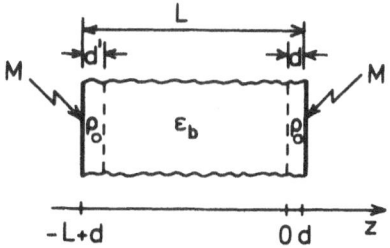

Fig. 1. Crystal with metal electrodes (M) and depletion layers, d. Also, the coordinate system is defined.

Since ρ_o and/or ε_s can be affected very strongly by temperature, illumination, and dc or rf electric fields d_{eff} displays a number of interesting properties, in far richer variety than ordinary piezoelectricity. An important property of the field E_o is that it builds up under externally applied rf fields. This is caused by the rectification of electric current across the barrier.

Taking strain S_{ij} and electric field E_i as independent variables in the free energy[3] of a cubic, centrosymmetric crystal, with electrostriction included, the wave equation for generation of longitudinal waves along [100] is

$$\frac{\partial^2 u_z}{\partial z^2} - \frac{1}{v_l^2}\frac{\partial^2 u_z}{\partial t^2} = \frac{\partial}{\partial z}[\frac{1}{2}\frac{m_{11}}{C_{11}}E_z^2 + \frac{1}{2}\frac{m_{12}}{C_{11}}(E_x^2 + E_y^2)] \tag{1}$$

In Eq.(1) u_z is the displacement, $v_l = (C_{11}/\rho_m)^{\frac{1}{2}}$ and ρ_m is the density of the material. C_{11} and m_{11}, m_{12} are elastic and electrostrictive constants, respectively. Source terms also exist for transverse waves. Hereafter we specify the electric fields E_o and E_{rf} to lie in the z-direction. Taking $E = E_o + E_{rf}$ in Eq.(1), using the explicit expression for E_o given above, and introducing an image source in the region $d < z < 2d$ to ensure stress free surface, we find a source term $[m_{11}\rho_o/C_{11}\varepsilon_s\varepsilon_o]E_{rf}[1-2d\delta(2-d)]$ for $0 \le z \le d$, and zero otherwise. Here we have dropped terms of frequency 0 and 2ω since we shall be investigating solutions at frequency ω, i.e. the frequency of the applied rf field E_{rf}. Introducing $u_z = u_o e^{-i\omega t}$ and $E_{rf} = E_{rf}^o e^{-i\omega t}$ the wave equation takes the form

$$\frac{\partial^2 u_o}{\partial z^2} + k^2 u_o = \frac{m_{11}}{C_{11}}\frac{\rho_o}{\varepsilon_s\varepsilon_o}E_{rf}[1-2d\delta(z-d)] \tag{2}$$

The equation is solved by the Green's function method. In the region $z < 0$ we obtain for the real displacement

$$u_z = -\frac{m_{11}}{C_{11}}\frac{\rho_o d^2}{\varepsilon_s\varepsilon_o}\frac{\varepsilon_b}{\varepsilon_s} \cdot \frac{V_{rf}}{L}\frac{1}{kd}(\frac{\sin kd}{kd}-1)\sin\{k(z-d)+\omega t\} \tag{3}$$

after introducing the rf voltage through the relation $E_{rf} = \varepsilon_b V_{rf}/\varepsilon_s L$. The expression may be written in a form which resembles precisely that found for piezoelectric generation. A full derivation of the detected voltage is given elsewhere[3].

$$V = 2\frac{m_{11}^2}{C_{11}}\frac{\rho_o^2 d^3}{(\varepsilon_s\varepsilon_o)^3}\frac{\varepsilon_b}{\varepsilon_s}\frac{V_{rf}^o}{L}\frac{1}{\kappa}(1-\frac{\sin\kappa}{\kappa})^2 \tag{4}$$

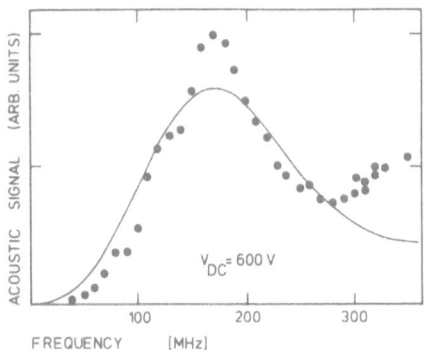

Fig. 2. Measured ($\bullet$) frequency dependence of volume wave generation
and detection in KTaO$_3$. Fully drawn: Eq. (4) with d = 28μm^2.

with $\kappa = kd = \omega d/v_l$. This function is plotted in Fig.2.

Experiments were carried out on KTaO$_3$ with $\vec{k} \| [100]$. The
first step in any experiment was to build up sufficiently high space
charge ρ_0 by means of the rf transmitter voltage and/or a dc
voltage. After this is done, and a stable saturation value is
reached, the generated acoustic amplitude u_z is found to be a
linear function[3] of V_{rf} , in accord with Eq.(4).

In order to test Eq.(4) further the frequency dependence of the
generated and detected signal was measured, and comparison made[4] with
the predictions, as shown in Fig.2. In these cases a biasing field
was applied across the crystal in order to enhance ρ_0 and make
possible accurate measurements at low frequency where the signal
approaches zero according to Eq.(4). In fitting the theory to the
experimental points (Fig.2) the depletion depth d is the adjustable parameter. In the present case its size is found to be
d = 28 μm . This high value of d is caused by the high bias field.
We conclude that the method described here provides a rather direct
way of measuring depletion depths by acoustic means, in addition to
its applicability as an electromechanical transducer. Methods for
measuring band gaps and donor levels are discussed in Ref.3.

REFERENCES

1. K. Fossheim and L. Bjerkan, Appl.Phys.Letters 32, 199
   (1978); L. Bjerkan, J.O. Fossum and K. Fossheim,
   J.Appl.Phys. (in press)
2. H.K. Henisch, Rectifying semiconductor contacts, Oxford,
   at the Clarendon Press (1957)
3. A.M. Raaen and K. Fossheim,J.Phys.C. (to be published)
4. K. Fossheim, R.M. Holt and A.M. Raaen, J.Appl.Phys.
   (in press)

****

DISCUSSION

W. Arnold: Have you tried this in any other material yet?

K. Fossheim: Yes, we have observed it in $BaTiO_3$ and $SrTiO_3$. In $SrTiO_3$ we had to precharge the sample by connecting it to a DC field. Once you have done that you can remove it and put it into a microwave cavity or use it as a piezoelectric material, and it's perfectly stable once you have charged it up. In $BaTiO_3$ the RF field itself was enough to generate the space-charge which was required.

PHONON MAGNIFICATION IN GaAs

M. Lax* and V. Narayanamurti

Bell Laboratories
600 Mountain Avenue
Murray Hill, NJ 07974

Our experiments[1] on phonon generation by hot electrons in epitaxial layers on GaAs have demonstrated a strong dependence of phonon
intensity on phonon propagation direction, phonon polarization and
detector shape. The analysis of these experiments is strongly influenced by focussing and defocussing effects whose importance in
heat pulse experiments is well known.[2,3] In this paper we present
a new expression for the phonon magnification for infinitesimal
apertures. We also present graphic illustrations of the transformation of the detector shape from real space to wave vector space for
the particular case of GaAs and emphasize the significant difference
between finite aperture and infinitesimal aperture results (to which
previous calculations have largely been confined).

As a byproduct of our work in optics[4] we can express the infinitesimal aperture magnification A expressible in terms of an intrinsic, geometric property -- the Gaussian curvature K -- of the
slowness surface

$$\lambda(\vec{q}) - \rho\omega^2 \equiv \rho\omega^2(\vec{q}) - \rho\omega^2 = 0 \qquad (1)$$

where $\vec{q}$ is the phonon wave vector, $\omega(\vec{q})$ is the frequency of one
polarization, and ρ is the density. The remarkably simple relation
is

$$A^{-1} = q^2 K/\cos \delta \qquad (2)$$

where δ is the angle between the phase velocity (along $\vec{q}$) and the
group velocity [along $\nabla_q\omega(\vec{q})$]. An expression has been found for the
Gaussian curvature K of an arbitrary surface that does not require
the introduction of a surface coordinate system. Previous calcula-

tions of the magnification use such a coordinate system.[3,5]

The magnification is found to depend on the detector shape even for infinitesimal aperature for degenerate modes. Maris[3] recognized this point and calculated the magnification for a circular detector in the [100] direction. This shape dependence does not occur at nondegenerate points where a unique curvature tensor exists. Such shape independence is valid even when the two principal curvatures and hence the two principal magnifications are unequal as in the [110] direction. In the case of finite aperture, however, the magnifications are shape dependent even along directions possessing no degeneracy. Thus care has to be exercised in taking over the results from infinitesimal aperture to finite aperture. We will present calculations for rectangular and square detectors typical of experiments.[1]

[100] Results: Because of the transverse mode degeneracy in the [100] direction, the transverse magnifications A+, A- depend on the azimuthal angle ϕ and hence on the detector shape. In Fig. 1 we present the dramatic variation of A+, A- with ratio r (of z to y sides) for rectangular detectors. In Fig. 2 we display (for one quadrant) the limits of the region in q space which after focussing arrives at a detector with ratio r=0.1. The nondegenerate L mode shows shape preservation and strong demagnification. The T+ and T- modes show appreciable distortion. For r<.118, the T+ mode in addition has a notch cut out caused by the z limit. This notch deepens as r→0 causing A+ to level off before A- (see Fig. 1).

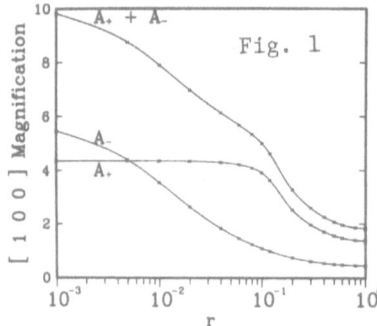
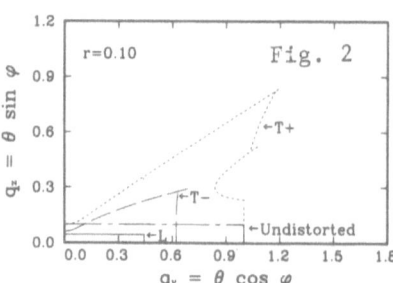

Fig. 1. Infinitesimal magnifications A+ (A-) for a rectangualr detector of ratio r. Fig. 2. The deviations q_y, q_z from the [100] direction that are permitted to enter a detector with r=0.1.

For r=1/8 and finite aperture the T- mode has an A-=8.4 and the startling "question mark" boundary of acceptable q̂'s (Fig. 3). This boundary is a highly distorted version of the corresponding region for a square detector (Fig. 4). These modes are the off-axis piezoelectrically generated phonons seen in the [100] direction in our experiments.[1]

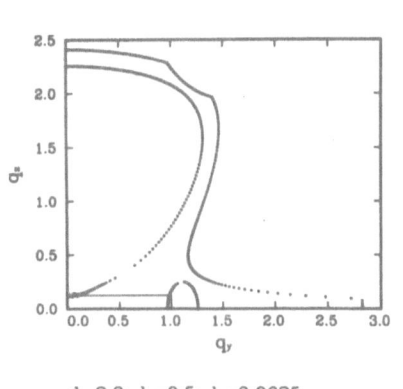

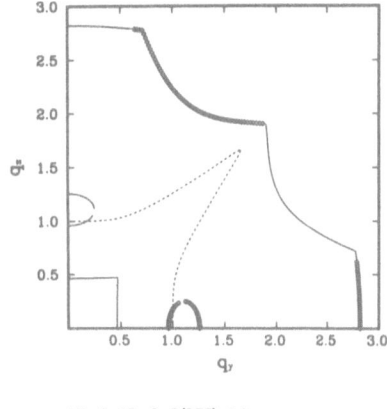

d=2.3; l_y=0.5; l_x=0.0625

LY=.5, LZ=.5, D(DET)=2.3
SQUARE DETECTOR [1 0 0]

Fig. 3.   The allowed region for the T_ mode in the [100] direction is the interior of the question mark.   The rectangle is the undistorted region.

Fig. 4.   The square is the boundary for the L mode, the dashed curve applies to the T+ mode and the T_ mode has a scalloped outer boundary and two bumps on an inner boundary.

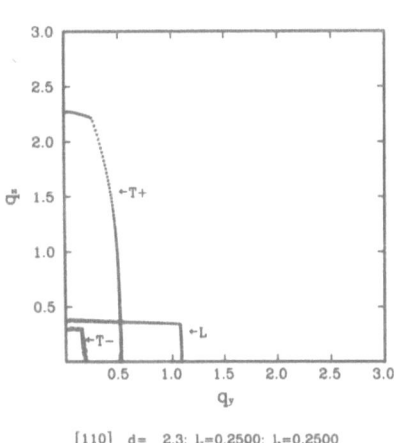

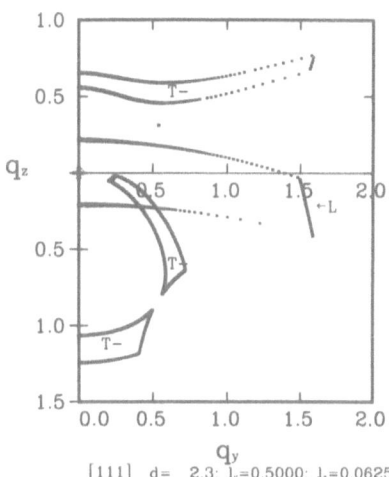

[110] d= 2.3; l_y=0.2500; l_x=0.2500

[111] d= 2.3; l_x=0.5000; l_z=0.0625

Figs. 5 and 6. Boundaries for finite aperture [110] and [111] cases.

[110] Results: For the [110] direction, which possesses no degeneracy, the infinitesimal aperture results are not shape dependent although the magnification is different in the z direction than in the [1$\bar{1}$0] direction. The magnifications in a cubic crystal are given by

$$A^{-1} = A_z^{-1} \cdot A_L^{-1} = \frac{\lambda_{33}}{\lambda_{11} + \lambda_{12}} \cdot \frac{\lambda_{11} - \lambda_{12}}{\lambda_{11} + \lambda_{12}} \text{ where } \lambda_{ij} = \frac{\partial^2 \lambda(\vec{q})}{\partial q_i \partial q_j} \qquad (3)$$

These derivatives are expressible in terms of the elastic constants, and the values of the overall magnification A agree with Maris. The [110] boundaries are fairly conventional, see Fig. 5. Thus the [110] direction is a particularly useful one, for comparison of theory with experiment.

[111] Results: Because there is a conical refraction angle of about 17° in GaAs in the [111] direction only finite aperture results are relevant for transverse modes. A sample of our results is displayed in Fig. 6, in which no T+ modes are detected, and the T_ modes come from three disjoint regions. Note also the necessity to plot results for negative as well as positive displacements "q_z" in the [1$\bar{1}$2] direction because of the absence of a symmetry plane perpendicular to this direction.

We thank R. C. Fulton for extensive computational aid.

*Also at Phys. Dept., City College of the City University of New York, New York, NY 10031. Work at CCNY supported by DOE and ARO.

REFERENCES
1. V. Narayanamurti, R. A. Logan, M. A. Chin and M. Lax, Sol. St. Elec. 21, 1295 (1978) and Phys. Rev. Lett. 40, 63 (1978).
2. B. Taylor, H. J. Maris and C. Elbaum, Phys. Rev. Lett. 23, 416 (1969).
3. H. J. Maris, J. Acoust. Soc. Am. 50, 812 (1971).
4. M. Lax and D. F. Nelson, J. Opt. Soc. Am. 65, 668 (1975), ibid 66, 694 (1976).
5. J. Philips and K. S. Viswanathan, Phys. Rev. B 17, 4969 (1978).

IONON FOCUSING AND PHONON CONDUCTION

I ELASTICALLY ANISOTROPIC CRYSTALS

A.K. McCurdy

Worcester Polytechnic Institute

Worcester, Mass. 01609

ITRODUCTION

Heat-pulse measurements have shown striking differences (up to ictors of 100) in the intensity of phonons propagating ballistically ι an elastically anisotropic crystal.[1,2] These results were shown ɔ arise from phonon focusing due to elastic anisotropy.[3] Phonon ɔcusing occurs when the direction of the group velocity varies more .owly with wave vector than in an elastically isotropic solid, as ɔr example, along cuspidal edges in the group-velocity surface.[4] ιbsequent measurements of the thermal conductivity of silicon and ιlcium fluoride in the boundary-scattering regime demonstrated ιisotropies of up to 50% for silicon and 40% for calcium fluoride. ιe predictions of Casimir's theory,[5] end-corrected for finite ιermal length,[6] and generalized to allow for phonon focusing gave ιantitative agreement with experimental results.[7,8] Similar ani- ɔtropies in the thermal conductivity have been predicted in ιfficiently defect-free superconducting lead and niobium at 'T$_c$<<1.[9] Phonon-focusing effects have also been predicted in .astically anisotropic hexagonal,[10] tetragonal and orthorhombic ·ystals.[11]

:SULTS AND DISCUSSION

Calculations of phonon intensity are described elsewhere.[2,11] ιgle θ is measured with respect to the [001] axis and φ is measured ι the (001) plane with respect to the [100] axis. Corresponding ιtensities for an elastically isotropic solid are unity. Similar .ots to those in Fig. 1a,b have been generated for other materials. ιonon conductivity in the boundary-scattering regime is enhanced ιen the sample rod axis is aligned along highly focused directions.

Table 1. Highest phonon intensities and their direction for selected
         crystals. The designation L, Tl or T2 refers to longi-
         tudinal, first or second transverse modes, respectively.

Material	Lattice	L	θ	ϕ	Tl	θ	ϕ	T2	θ	ϕ
SnF$_2$	mono.	10.3	52	100	13.1	70	82	16.2	66	70
SiO$_2$	trig.	11.7	56	92	10.3	0	..	13.2	22	150
KB$_5$O$_8$	ortho.	12.0	90	0	19.1	60	0	19.7	48	60
TeO$_2$	tetra.	40.4	88	46	17.0	90	42	5.9	36	30
MgO	cubic	2.7	50	38	24.7	90	4	11.9	90	44

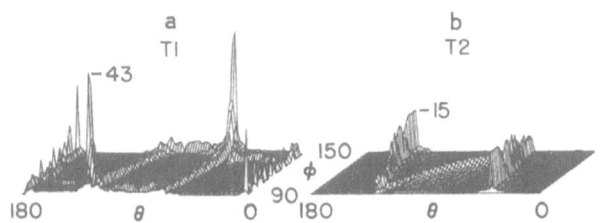

Fig. 1a,b. Computer plotted profiles of phonon intensity for the Tl
           (slower) and T2 (faster) transverse modes of Al$_2$O$_3$ using
           the elastic constants of J.H. Geiske and G.R. Barsch,
           [Physica Status Solidi 29, 121 (1961)].

        Calculations of the boundary-scattered phonon mean-free path,
Λ_{eff}, for cubic crystals as a function of the elastic anisotropy
factor, A = $2c_{44}/(c_{11}-c_{12})$, are performed using Eq. 1:

$$\Lambda_{eff} = 3\kappa/(C_v v_{eff}).\qquad\qquad(1)$$

κ is the end-corrected thermal conductivity for samples of square
cross-section with side dimension, D, and thermal length, L, [7,8]
and C$_v$ is the specific heat per unit volume. The effective
velocity, v_{eff}, is defined as:

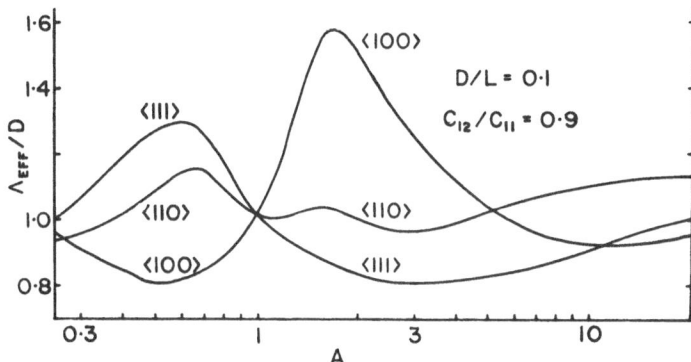

Fig. 2. Boundary-scattered phonon mean-free path for cubic crystals
as a function of the elastic anisotropy factor, A, for heat
flow axis along principal directions.

$$v_{eff} = <|v|s^{-3}>/<s^{-3}> \qquad\qquad (2)$$

where $<|v|s^{-3}>$ is the average of the phonon group velocity times
the inverse cube phase velocity. Note that for c_{12}/c_{11} = 0.9 and
D/L = 0.1 the value of Λ_{eff}/D for the <100> direction increases from
0.8 to almost 1.6 as A increases from 0.5 to 1.7, but for the <111>
direction decreases from 1.3 to 0.8 as A increases from 0.6 to 3.
For elastic isotropy A = 1 and Λ_{eff} is 1.016D. The large maximum in
Λ_{eff}/D along the <100> direction is due to strong phonon focusing
about the <100> axes arising from cuspidal edges about the <100>
direction in the {100} and {110} planes. The maximum along the <111>
direction is due to the strong focusing near the <111> directions
in the {110} planes and the cuspidal edges about the <110> axes in
the {100} planes which, because of symmetry, extend toward the <100>
directions.   Curves of Λ_{eff}/D and κ/T^3 calculated as a function of
c_{12}/c_{11} and D/L will be presented elsewhere.

REFERENCES

1.   B. Taylor, H.J. Maris, and C. Elbaum, Phys. Rev. Lett. 23,
      416 (1969).
2.   B. Taylor, H.J. Maris, and C. Elbaum, Phys. Rev. B3, 1462
      (1971).

3.   H.J. Maris, J. Acoust. Soc. Am. $\underline{50}$, 812 (1971).
4.   M.J.P. Musgrave, Proc. Camb. Philos. Soc. $\underline{53}$, 897 (1957).
5.   H.B.G. Casimir, Physica (Utr.) $\underline{5}$, 495 (1938).
6.   R. Berman, F.E. Simon, and J.M. Ziman, Proc. R. Soc. A$\underline{220}$, 171 (1953).
7.   A.K. McCurdy, H.J. Maris, and C. Elbaum, Phys. Rev. B$\underline{2}$, 4077 (1970).
8.   A.K. McCurdy, Ph.D. thesis (Brown University, 1971) (University Microfilms, Inc., Ann Arbor, Michigan, Order No. 72-12,047).
9.   C.G. Winternheimer and A.K. McCurdy, Solid State Commun. $\underline{14}$, 919 (1974).
10.   A.K. McCurdy, Phys. Rev. B$\underline{9}$, 466 (1974).
11.   C.G. Winternheimer and A.K. McCurdy, Phys. Rev. B$\underline{18}$ 6576 (1978).

PHONON INTERACTIONS AT CRYSTAL BOUNDARIES BELOW 1 K

M. LOCATELLI

L.C.P.,S.B.T., C.E.N.G. 85X 38041 GRENOBLE Cédex FRANCE

Low temperature thermal conductivity is a very useful technique for phonon interaction studies. However, the analysis of the data implies a knowledge of all the scattering processes involved.

At very low temperature boundary scattering is important (1) and depends on the surface roughness (2).

In order to test the theoretical models (3), (4), we have mesured the thermal conductivity between .05 and 1.5 K using a standard method (5), of a pure Al_2O_3 sample for different surface roughnesses. The surface roughness was characterized by observation with a surface profile measuring system which has a sensitivity of ten nanometers but a stylus diameter of a few microns.

Results and Analysis

The results are shown in figure 1 for three different grades of surface roughness.

These results have been analysed according to two models

a) Ziman's model (3), Ziman introduces a mean free path

$$1/\ell_B = (1 - p) / ((1 + p) (\ell_c)+ 1/L_s \qquad [1]$$

ℓ_c is the Casimir value, (corresponding to the sample diameter in

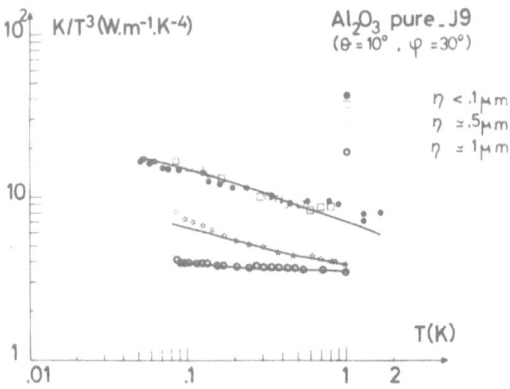

Fig. 1 : K/T^3 versus T, experimental points and calculated curves

our case); L_s is the sample length and p is given by the relation

$$p(\omega) = \exp \left(- \left(\frac{2\omega}{v}\right) \eta\right)^2)$$

   where v is the sound velocity, ω the phonon frequency and η represents an average value of the surface roughness considering a gaussian distribution.

   The thermal conductivity is calculated with the expression [1] above using the Callaway model (6).

b) Nonnenmacher's model, Nonnenmacher introduces the size effect on the phonon mode density and the influence of two roughness parameters η and L, where η corresponds to η in Ziman's model and L is the correlation length.

   The thermal conductivity is given, using Callaway's model, by a sum

$$K = K_I + K_{II}$$

where K_I corresponds to the standard bulk conductivity including Casimir term, and K_{II} introduces the dependence of the thermal conductivity on the two roughness parameters η and L and of the size effect on the density of phonon states.

Attemps to fit our experimental results on the basis of these models were not satisfactory. Nevertheless the Ziman's model gives the better fit.

The disagreement may have two origins:

a) The surface observation seems to show that the roughness distribution is not gaussian

b) In the two models, we do not consider phonon focusing (7) which can exclude practically   all possibilities of specular reflexions.

For the assumption a), we propose a phenomenological expression for the boundary scattering, introducing a frequency dependence into the Ziman's model which modifies the roughness distribution.

The best fits were obtained with the following expression for p

$$p = \exp \left(\left(- a\omega^{-.8} \ \frac{2\omega\eta}{v}\right)^2 \right)$$

a is a dimensional constant which takes account the $\omega^{-.8}$ dependence. The best fits are shown in Fig. 1.

In conclusion, it seems that theoretical models do not adequately account for phonon boundary scattering and probably phonon focusing has also to be considered.

References

1) H.B.G. CASIMIR, Physica (Utrecht) 5, 495 (1938)

2) G.J. CAMPISI and D.R. FRANKL, Phys. Rev. B, 10-6, 2644 (1974)

3) J.M. ZIMAN, Electrons and Phonons, Clarendon Press, Oxford (1960)

4) T.F. NONNENMACHER and R. WUNDERLE, Phys. Stat. Sol. (b) 91, 147 (1979)

5) M. LOCATELLI, D. ARNAUD, Phys. Let. 42 A-2, 181 (1972)

6) J. CALLAWAY, Quantum Theory of the Solid State, Part A, Academic Press, New York (1974)

7) J. DOULAT, M. LOCATELLI, J. RIVALLIN, Phonon Scattering in Solids, Edited by L.J. Challis, V.W. Rampton, and A.F.G. Wyatt (Plenum Press, New York and London) 383 (1976)

PHONON EMISSION INTO DIFFUSIVE MEDIA*

W.E. Bron

Department of Physics
Indiana University
Bloomington, IN 47405

It is a basic assumption of the acoustic mismatch (AM) model
of heat pulse transport through solid-solid interfaces that a
steady state is reached between an electrically driven heater film
and the phonon flux into the substrate within a time, τ, comparable
to the electron-phonon interaction time and the transit time of
phonons across the heater film.[1] A necessary consequence of this
assumption is that none of the phonons which enter the substrate
return to the heater film. Furthermore, since typically $\tau \sim 10^{-10}$
sec, and since normally the temporal resolution is $\geq 10^{-9}$, it
follows that no time dependent change in the film's temperature
should be observed during or after the electrical excitation
pulse.

In Figs. 1 and 2 we show experimental results of the instantaneous temperature, during and after a Joule heating pulse, of
$\sim$ 3000 Å x 5 mm x .2 mm nickel films deposited either on a "pure"
crystal of SrF_2 or on one containing 0.1 mole % Eu^{2+}. The ambient
temperature is 4.2°K. It is seen that the film temperature is not
a constant of time. Both the "pure" and the doped sample lead to
similar effects, although it takes more Joule power, P_A, to reach
the same terminal temperature in the "pure", compared to the
doped, sample. No such time-dependent temperature changes were
observed under the same experimental conditions in a nickel film
deposited on a Al_2O_3 crystal.

Since both the pure and Eu doped SrF_2 exhibit a time-dependent
temperature it is clear that the Eu probe ions are not the sole
source of the observed effects. Moreover, since the time scale of

---

*Work supported by ONR Contract N14-78-C-0249.

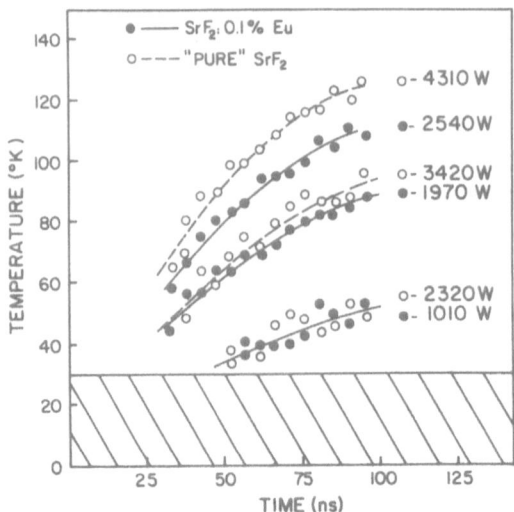

Figure 1.

the effect is of the order of tens of nanoseconds, it is unlikely
that the source is in the metal film or the film-substrate interface. The main difference between "pure" Al_2O_3 and "pure" SrF_2
appears to be their relative natural isotopic impurity content
which is a total of ~ 0.24% of the oxygen in sapphire and 17.26%
of the Sr in SrF_2. Both foreign and isotopic impurities lead to
elastic phonon scattering and to diffusive flow of high frequency
phonons[2,3] part of which can re-enter the heater film. The
scattering strength of the isotopes can be compared to that of the
Eu probe ions by assuming only the mass defect and Rayleigh-like
scattering as proposed by Klemens.[4] Using handbook values for the
various isotopic impurities, we obtain relative scattering rates
for $Al_2O_3/SrF_2/SrF_2$:Eu to be roughly 1/23/173, which is in qualitative agreement with our results.

We have also applied a new model[5] of phonon transport into
diffusive media. Good overall agreement is obtained with the
observed heating in the doped SrF_2 for the case of P_A = 2540 W.
Some disagreement arises, however, between observed and calculated
temperatures at lower P_A with, for example, the calculated temperatures for P_A = 1010 W being some 10°K higher than the observed
temperatures, although the rates of change are quite similar. The
ratio of the scattering probabilities of Eu versus isotopic impurities in SrF_2 is determined by the model to be 5, which compares
well with the value of 7.5 cited above. The decay of the film
temperature after the voltage pulse is also in qualitative agreement with the model treatment. The model also predicts steady
state (AM) behavior if the scattering probability is 20 times less

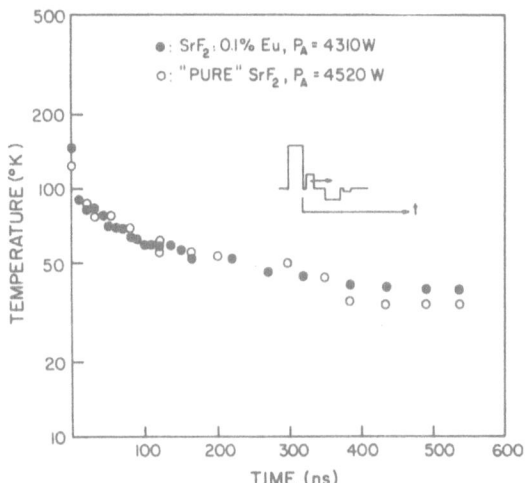

Figure 2.

than "pure" SrF$_2$; again in good agreement with Al$_2$O$_3$ results.

We have now demonstrated that sufficiently diffusive scattering of phonons in the substrate can cause observable time – dependent temperature changes in thin-film Joule heaters. The theoretical analysis[5] shows that the frequency dependence of the scattering can lead to the previously observed distortions of the phonon spectral distribution in the substrate when the latter is determined over a macroscopic volume near the interface[2,3]. It is an important consequence of these results that theoretical models of the heat pulse method which neglect diffusive scattering in and the return of energy from the substrate, as does the acoustic mismatch model, are not applicable to most crystal systems except in the limit of low heating, i.e. low effective film temperatures (of order $\leq$ 10°K, see Ref.2), or in crystals which contain only a few parts per million of foreign or isotopic impurities.

The author acknowledges the collaboration of J. Patel and W. Schaich.

References

1. N. Perrin and C. Budd, Phys. Rev. Lett. 28, 1701 (1972); J. Phys. (Paris) 33, C4-33 (1972).
2. W.E. Bron and W. Grill, Phys. Rev. B16, 5303 (1977).
3. W.E. Bron and W. Grill, Phys. Rev. B16, 5315 (1977).
4. P.G. Klemens, Solid State Physics, Vol. 7, ed. by F. Seitz and D. Turnbull, Academic Press, NY 1958, pg. 1.
5. W.L. Schaich, J. Phys. C:Solid State Phys. 11, 4341 (1978).
****

DISCUSSION

A. C. Anderson: In support of your conclusion I'll just mention
that similar insertions of phonons into dispersive media have been
done with very dispersive media - and that is amorphous materials
where you get exactly the same results, you apply essentially the
same model, the same analysis, and you get very good agreement.

J. P. Maneval: What was the temperature coefficient of your nickel
films, the temperature range you used them?

W. E. Bron: We measured all the way up to room temperature. But
the temperature coefficient is essentially zero until you get out
of the impurity range. That's about 30°K.

J. P. Harrison: What would be the effect of any damage to the substrate?

W. E. Bron: That's a good question. That has to do with the
aluminum oxide. We repeated the experiments done by Herth and Weis
some time ago. Although we see no temperature dependent change in
the resistivity we do not get an acoustic-mismatch temperature of
the kind that they do. That probably has to do with the fact that
we use somewhat different cleaning techniques and polishing techniques so of course there is an underlying effect having to do with
the surface. But the surface cannot be the source of this problem
because the equilibrating time of anything that occurs on the surface is of the order of nonseconds, whereas the equilibration time
of the heater you might have noted is of the order of hundreds of a
nanosecond. You have time enough to get into the crystal and come
back. Any surface scattering or surface effect would be over much
earlier.

THE STORAGE ECHO IN PIEZOELECTRIC POWDERS

Neil Thomas, W. Arnold and K. Dransfeld

Max-Planck-Institut für Festkörperforschung,

D-7000 Stuttgart 80, FRG

ABSTRACT

A model is proposed to account for the long-lived "storage" echo observed in piezoelectric powders. We suggest that the two RF writing pulses burn holes in the frequency distribution of the powder grains such that the powder remembers the spectral density of the writing sequence. When the powder is subsequently excited with a weak reading pulse, the linear ringing response of the previously uniform population, now modified by the hole burning, produces the storage echo. At weak writing powers, the information is only written into grains preferentially oriented relative to the RF field, producing a storage echo which is sensitive to tapping and which has a $\cos^2\theta$ angular dependence. At higher writing powers, the echo should become more isotropic and persist even after sieving the powder.

Piezoelectric powders subjected to intense RF pulses produce polarization echoes rather similar to spin echoes. Fossheim et al. /1/ studied the two-pulse and three-pulse echoes for short time delays and showed that these echoes are due to non-linear oscillations of the powder particles; this important role played by non-linearity in echo effects has recently been reviewed by Shiren and Melcher /2/. An additional echo effect has been reported by Popov et al. /3/: When two intense RF pulses separated by a delay τ of a few microseconds are followed by a third pulse at time T, an echo at time $T + \tau$ is produced even though T is very much longer than the decay time of the mechanical oscillations. Information about the first two pulses is somehow stored in the powder, and such storage can apparently persist for days or even weeks.

353

The distribution of particle sizes in the powder produces a distribution of resonant frequencies for the particles, and furthermore
each particle is piezoelectrically anisotropic. On the basis of
this, Melcher and Shiren /4/ proposed that the second RF pulse
rotates those powder particles whose oscillating dipole moment due
to the first pulse is suitably phased. The static frequency-
dependent distribution of piezoactive axes is probed by the third,
reading, pulse, and this leads to the storage echo. They found that
the echo is removed upon sharp tapping of the powder samples, a
result which is consistent with this rotation model.

The model does not, however, explain the results of Berezov
et al. /5/ and Billmann et al. /6/, who showed that the storage
echo persists, albeit somewhat reduced in amplitude, after sieving
the powder. Furthermore, Cheeke et al. /7/ found a $\cos^2 2\theta$ angular
dependence of the echo amplitude, θ being the angle between the
reading and writing RF electric fields; this is also at variance
with the rotation model.

To take account of the persistence of the storage echo after
sieving, we have developed a more general model, incorporating the
idea of hole burning. We consider the sample to be a lossy capaci-
tor, the loss at a frequency ω being proportional to the frequency
distribution of the grains $g(\omega)$. Denoting the two-pulse writing
sequence by $V_1(t)$, with a spectral density $I_1(\omega)$, we assume that
the writing pulses burn a hole in the previously uniform population
g_o such that

$$g(\omega) = g_o - \alpha \, I_1(\omega), \tag{1}$$

where α is a constant. Since the losses in the sample are quite
small, the response to the reading pulse $V_2(\omega)$ is given approximately by

$$V(\omega) = V_2(\omega) \, (1 - \beta \, g(\omega)), \tag{2}$$

where β is another constant, depending on the electrical parameters
of the system and also on the piezoelectric coupling to the powder
grains: in the rotation model it is β which is made frequency dependent, rather than $g(\omega)$, to produce the storage echo. Considering only hole burning, the term responsible for the echo is

$$\Delta V(\omega) = \alpha \, \beta \, I_1(\omega) \, V_2(\omega). \tag{3}$$

Fourier transforming this and using the Wiener-Khintchine theorem
yields

$$\Delta V(t) = \alpha \, \beta \, \gamma \, \Gamma_1(t) * V_2(t), \tag{4}$$

where γ is an additional constant and $\Gamma_1(t)$ is the autocorrelation function of the writing sequence. (The star denotes convolution.) For a two-pulse writing sequence, $\Gamma_1(t)$ has peaks at $t = 0$ and $t = \tau$, so that a reading pulse applied at $t = T$ produces a storage echo at $t = T + \tau$. The rotation mechanism modifies the coupling ß to produce an additional term in (4) of the same form. It is probable that both storage mechanisms occur simultaneously in some powders, only the hole burning persisting after sieving.

Since the quartz grains are piezoelectrically anisotropic, we expect at low powers only to burn holes in the frequency spectrum of those grains which are favourably oriented. This establishes an axis in the powder and produces the $\cos^2\theta$ angular dependence of the storage echo, as observed by Cheeke et al. /7/. At higher writing powers, the holes should be burnt into the whole population, leading to an isotropic echo which persists even after sieving, as observed by Billmann et al. /6/. The storage echo at low writing powers should virtually disappear on tapping, but after writing and tapping many times the echo should persist.

The most direct form of hole burning is to drive the grains so hard that they break. The static electric field to break a grain is about 2×10^5 Vmm^{-1}, but with resonant excitation at 50 MHz by a 5 µs pulse this is reduced to 200 Vmm^{-1}. In practice, brittle fracture due to crack propagation from the surface may occur at somewhat lower fields, although the non-linear response of a grain will reduce the enhancement due to resonant excitation. The quartz grains used in our experiments have much smaller particles on their surfaces, so that hole burning could easily occur by moving these particles to shift the resonant frequency.

We acknowledge helpful discussions with R.L. Melcher. One of us (N.T.) is grateful to the Royal Society for a research fellowship under the European Exchange Programme and wishes to thank MPI-FKF and the Humboldt-Stiftung for their hospitality.

REFERENCES

1. K. Fossheim, K. Kajimura, T.G. Kazyaka, R.L. Melcher, N.S. Shiren, Phys. Rev. B17, 964 (1978)
2. N.S. Shiren and R.L. Melcher in "Internal Friction and Ultrasonic Attenuation in Solids" pp. 11-19 (Univ. Tokyo, 1977).
3. S.N. Popov, N.N. Krainik, and G.A. Smolenskii, Sov. Phys.-JETP 42, 494 (1976).
4. R.L. Melcher and N.S. Shiren, Phys. Rev. Lett. 36, 888 (1976).
5. V.M. Berezov and V.S. Romanov, JETP Lett. 25, 151 (1977).
6. A. Billmann, Ch. Frénois, G. Guillot, and A. Levelut, J. de Phys. Lett. 39, 407 (1978).
7. D. Cheeke, A.A. Lakhani, and H. Ettinger, Solid State Commun. 25, 289 (1978).

DISCUSSION

H. Stormer: How hard can you read the sample?

W. Arnold: When your reading pulse gets too high, then you destroy
even this more persistent echo. So this is also a non-linear
process - you can't read it out with an infinite amount of power
because then you would destroy the picture right to begin with.

P. G. Klemens: When you further space this powder, you don't fill
it to the optimum density. When you subsequently apply vibration
provided the acceleration is more than 1.5 g, you raise the density.
The actual process is not well understood, but remember that the
contact points are very small and that the yield stress is usually
exceeded at the contact points so that you have complicated and
irreversible effects.

K. Fossheim: I have participated in this work with Shiren and
Melcher, and I would like to mention that we studied several
powders and in fact in all the cases we studied the echoes would go
away with tapping. So that at least this shows there must probably
be two different mechanisms involved.

W. Arnold: Yes. This echo, which stays even after sieving, does
not occur in all powders. Not even in all powders of one chemical
composition. So the choice is when the powder is sticky when it's
really very dry and powderlike you don't get it, it's completely
gone after tapping. But some powders really show there's a storage
mechanism even after sieving.

A. C. Anderson: I thought part of the reason for the holographic
scheme was to explain the large amplitudes that were observed in
the echoes. Will your scheme provide the required amplitudes?

W. Arnold: I don't know that in detail. I believe in that holographic picture too and that this explains most of the things.
But this would not explain what one sees after sieving and there is
on top of it an additional mechanism. For example, the French
group reports that after tapping their echo goes down of the order
of 10 dB, so they also have something which goes away with tapping
which could be holographic.

ACOUSTIC IMPULSE GENERATION BY PIEZOELECTRIC SHOCK IN ACOUSTO-ELECTRIC PHONON DOMAINS*

W. C. Chang, S. Mishra, and R. Bray

Physics Department
Purdue University
West Lafayette, IN 47907

In the course of Brillouin scattering studies of acoustoelectrically amplified phonon domains,[1] at scattering angles very close to and within the Gaussian width of the reflected light beam, we observed very strong, unexpected modulation of the reflected light by the domains. This led to the discovery of a new phenomenon, - the emanation from the domains of mechanical impulses propagating both in the forward and backward directions. This effect was observed most easily in CdS. The impulses manifested themselves optically by inducing a rocking motion of the reflected light beam with respect to a narrow slit placed in front of the detector [Fig. 1(a)]. Examples of the optical signals are shown in Fig. 1(b). The shapes of the signals depend on the initial position of the light beam with respect to the slit. From Fig. 1(c) it is evident that a simple rocking of the reflected light beam with respect to the slit (in the sequence of positions $0 \to 1 \to 0 \to 2 \to 0$) is capable of generating signals (as shown encircled), closely matching the observed signals. From the shape and phase of the signals shown, it can be readily deduced that the mechanical disturbance has the form of a ridge-like bulge propagating to the positive end of the sample.

The signals are quite large, corresponding to as much as 30% modulation of the reflected light. From a study of the beam deflection as a function of slit position, the width and height of the ridge could be determined: ridge width $\approx$ domain width $\approx$ 0.5 mm, and height $\leqslant$ 200 Å in CdS.

That the mechanical impulses propagate in both the forward and backward directions from the domain is shown in Fig. 2, by the sequence of signals at a series of positions of the focussed light spot along the sample. The forward travelling ridge (F) can be seen at

357

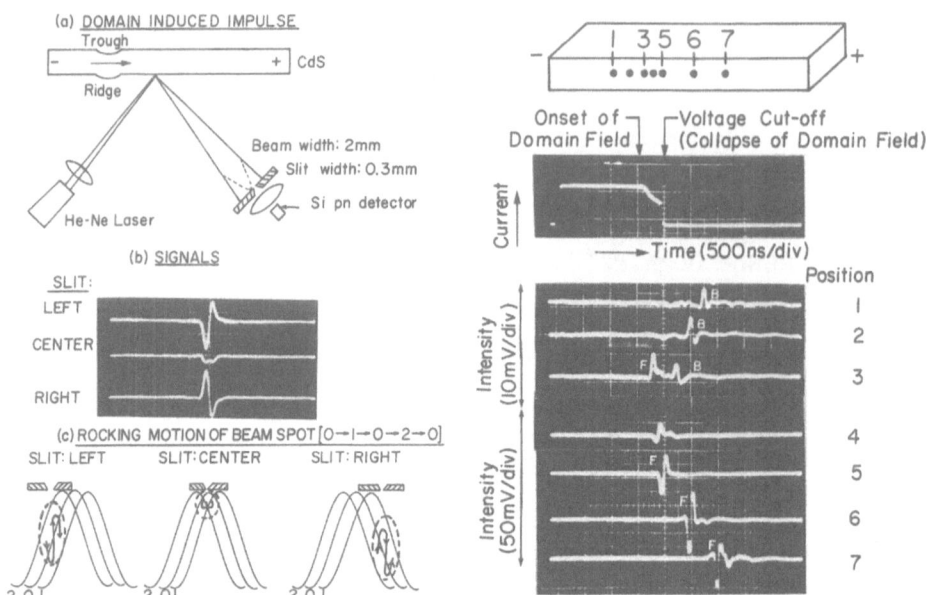

Fig. 1. Experimental setup and form of photodiode signals generated by rocking of reflected light beam by domain-induced ridge-shaped impulse.

Fig. 2. Si photodiode signals at indicated position along sample. Forward (F) and Backward (B) propagating impulses originate in propagating acoustoelectrically amplified phonon domains.

positions 3 to 7. It travels with the domain, and it is seen to originate when the non-ohmic drop in the current pulse occurs, signifying the onset of the domain. It is continuously augmented as the current continues to drop. The backward propagating ridge (B) is seen at positions 1 to 3. The strongest portion of this signal originates when the voltage pulse terminates and the domain field collapses. However, earlier, weak signals are generated at the onset of the domain field and during the whole current drop interval. Note that continuous generation can augment the signal for the forward propagating impulse which travels along with the domain, but not for the backward travelling signal which is running away from the generation source within the domain.

Two facts identify the piezoelectric shear nature of the acoustic modes associated with the mechanical impulse. (1) The velocity of the propagating impulses can easily be measured from their time of flight (Fig. 2). For both forward and backward propagating signals, the speed is the same as that of the piezoelectric shear waves constituting the domain, 1.76×10^5 cm/sec. (2) The mechanical impulses are transverse disturbances, always forming a ridge on one face of the sample and a trough on the opposite face. This asymmetrical, uni-directional displacement is along the c-axis in CdS samples, in the same direction as the polarization of piezoelectric shear waves propagating in the basal plane.[2]

The determination of the mechanism of generation of impulses relies on the following facts: (1) the piezoelectric shear wave identification of the impulse, (2) the generation of both forward and backward propagating signals, and (3) the association of the generation process with a rapid onset or collapse of the high electric field in the domain (as noted in Fig. 2). These three observations suggest a piezoelectric "shock" generation mechanism. It is well-known that a time-varying gradient in piezoelectric stress is the source of generation of forward and backward travelling waves.[3] The time-varying aspect is inherent in domain formation, in propagation, and in the cut-off of the applied voltage pulse. The concentration and distribution of the applied voltage in the domain provides the requisite gradients in a high electric field.[1,4] This high electric field is dc and longitudinal, with spatial extent determined by the domain width. These characteristics provide the necessary coupling of a longitudinal piezoelectric field with a transverse displacement, and are consistent with the width of the impulse being correlated with the domain width. We note that there is no direct correlation of the very low frequency waves forming the mechanical impulses with the GHz frequency phonons (and their ac longitudinal piezoelectric fields) in the domain.

This phenomenon has been exploited to produce two-pulse "phonon echoes" with the unique feature of optical detection. First, a high voltage pulse is applied which generates a propagating domain and produces a forward propagating signal (F) at the site of the focussed light spot. A second voltage pulse is applied some time after the first pulse, such that the high frequency phonons in the propagating domain have not yet attenuated appreciably. The second pulse re-establishes the high field in the domain and also generates a backward travelling impulse (B) which arrives back at the probing light spot, as an "echo", i.e., a time-reversed version of the first signal. We note that the forms of the signal and the echo are predetermined by the domain shape and are not subject to external control.

*Supported by National Science Foundation and NSF/MRL Grants.

REFERENCES
1. D. L. Spears, Phys. Rev. B2, 1931 (1970).
2. A. R. Hutson and D. L. White, J. Appl. Phys. 33, 40 (1962).
3. E. H. Jacobsen, J. Acoustical Soc. of America 32, 949 (1960).
4. M. B. N. Butler, Rep. Prog. Phys. 37, 421-479 (1974).

****

DISCUSSION

J. P. Maneval: Why is this phenomenon peculiar to piezoelectric materials?

W. C. Chang: You need a piezoelectric stress gradient to provide the shock mechanism.

W. Arnold: How does this relate to the backward-wave echoes observed in cadmium sulphide a couple of years ago by the Russian group, IBM, and so on?

W. C. Chang: This type of a crystal is of a completely different nature. We use DC electric voltages and a backward travelling impulse constitutes our low-frequency waves.

CAPTURE AND DELAY OF PHONONS IN RESONANT REGIONS

Max Wagner and Hans Nusser

Institute of Theoretical Physics, University of Stuttgart,

Pfaffenwaldring 57, D-7000 Stuttgart 80, Germany

In disturbed crystalline regions phonons lose their particle properties to a large degree. New effects, such as capture, delay and spectral redistribution, occur. In view of these it seems worthwhile to handle phonon transport in a non-scattering type of approach. An additional motivation for such an investigation is the strong progress in the spectral, spatial and temporal discrimination of lattice modes which recently has been made.

As a first attempt we consider a linear chain with singular non-resonant (mass and spring defects) and resonant (foreign harmonic oscillators, 2-level systems) perturbations. In an earlier paper[1] a method has been given to decouple singular oscillators from lattice modes. The spirit of this method has been to remove the foreign degrees of freedom with the help of a local Green function ("layered" GF's). In this manner the foreign modes formally play the role of a disturbance for the regular lattice modes. A slight modification of this method can be used here too.

We will further assume that the number of lattice coordinates ξ_μ which are involved in either of the disturbed Hamiltonian parts is small ("small space"). We will denote these by μ, ν ($\mu, \nu = 1, 2, \ldots p$). It then turns out that the Kubo response to an external stimulus can be written in the form

$$\langle \xi_m \rangle_E = \sum_{\mu\nu} G^{(o)}_{m\mu}(E) \left[i + \lambda g \right]^{-1}_{\mu\nu} \left[F_\nu(E) - \sum_s \alpha_{\nu s} \gamma_s \Phi_s(E) \right] \tag{1}$$

$$E = \omega + i\varepsilon$$

where $\langle \xi_m \rangle_E$ is the Fourier transform of the time dependent displacement expectation value $\langle\xi_m\rangle_t$ and $F_\nu(E)$ and $\Phi_s(E)$ are the Fourier transforms of the respective stimulating forces onto the lattice sites $(F_\nu(t))$ and the foreign modes s $(\Phi_s(t))$. Further

$$G_{m\mu}^{(0)}(E) = \frac{1}{N}\sum_k \frac{e^{ik(m-\mu)}}{\omega_k^2-E^2} \; ; \qquad k=\frac{2\pi}{N}K, \; K=0,\pm1,\pm2,\dots \qquad (2)$$
$$m,\mu=0,\pm1,\pm2,\dots\pm(N/2)$$

$$g_{\mu\nu}(E)=G_{\mu\nu}^{(0)}(E), \quad i_{\mu\nu}=\delta_{\mu\nu} \quad \text{(unity matrix in "small space"), and}$$

$$(3) \quad \lambda_{\mu\nu}(E) = (V-\kappa_\mu E^2 i+W)_{\mu\nu}, \qquad W_{\mu\nu}= -\sum_s \alpha_{\mu s}\gamma_s\alpha_{\nu s} \qquad (4)$$

V is the deviation of the spring constant matrix from the ideal form and κ_μ is the relative mass defect, $\kappa_\mu=(M_\mu-M)/M$. W represents the effective disturbance due to the foreign degrees of freedom, which are coupled to the lattice (coupling constants $\alpha_{\mu s}$). We further have

$$(5) \quad \gamma_s^{(1)}= \frac{1}{\Omega_s^2-E^2}, \qquad\qquad \gamma_s^{(2)} = -\frac{\Omega_s}{\Omega_s^2-E^2}(\frac{\overline{n}_1-\overline{n}_0}{2}) \qquad (6)$$

where $\gamma_s^{(1)}$ pertains to foreign harmonic oscillators and $\gamma_s^{(2)}$ to foreign 2-level systems. Ω_s respectively is the oscillator frequency or the level splitting and $\overline{n}_1$, $\overline{n}_0$ the respective thermal occupation of the 2 levels.

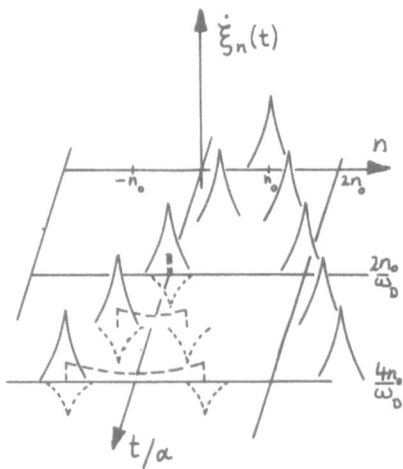

<u>Fig. 1.</u>   Pulse excitation at $n=n_0$, $t=0$.   Mass defect at $n=o$.

We will discuss some of the relevant properties of the solutions by a selection of 3 single models.

I. If the lattice has only a mass deviation at $n=0$ and experiences a Lorentzian pulse at $n=n_0$, $t=0$, we encounter the situation given in Fig. 1. There the velocity evolution $\dot{\xi}_n(t)$ is shown. The curves represent convolutions of the different contributions to $\dot{\xi}_n(t)$. After time $\frac{2n_0 a}{\omega_D}$ the left pulse "reaches" the disturbed mass: a reflected pulse is created. But in addition to the reflected and transmitted pulses there is a delayed part with a relaxation time given by $\tau = \kappa_0/\omega_D$ ($\kappa_0 = M_0 - M/M$).

II. Two foreign harmonic oscillators ($\mu = \pm n_0$). Fig. 2 depicts the spectral behaviour of $\xi_m(\omega)\omega$ (Fourier transform of $\dot{\xi}_m$),

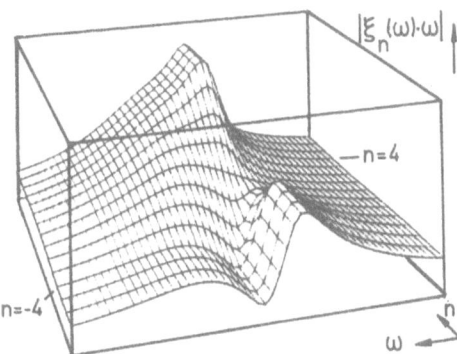

Fig. 2. Lattice excitation, if foreign oscillators are coupled to it at $n=\pm n_0$ and the one at $n=\pm n_0$ experiences a pulse excitation at $t=0$.

if at t=0 a δ-pulse is exerted at $n=n_o$. The main relevant features here are the strong spectral changes in the region $-n_o<n<n_o$ and the spectral asymmetry of the regions $|n|>n_o$.

In view of the preceding study it seems promising to continue this kind of approach to phonon transport. In 3-dimensional lattices the spectral redistribution effect may be less spectacular, but it still may be appreciable. In particular for boundary regions as well as for glass-like regions one may gain some new insight, which may also shed some further light on the Kapitza resistance problem.

REFERENCES
1. M. Wagner, Phys. Rev. 131, 2520 (1963)

DYNAMICS OF 29 CM^{-1} PHONONS IN RUBY USING OPTICAL GENERATION AND DETECTION

W. C. Egbert, R. S. Meltzer and J. E. Rives

Department of Physics and Astronomy
University of Georgia
Athens, GA 30602, USA

In the past few years there have been several attempts to understand the decay of 29 cm^{-1} phonons resonantly trapped in the $E \rightarrow 2\bar{A}$ excited state resonance of the Cr^{3+} ions in ruby. Several mechanisms have been proposed to explain the decay of the resonant phonons including spectral[1-3] and spatial diffusion,[3,4] and anharmonic decay.[3] We suggest here another process, resonant phonon assisted energy transfer, which we feel is active under conditions of strong resonant scattering in ruby. It is probable that all of these mechanisms are important under certain conditions.

In this paper we will discuss the results of experiments on 0.08% ruby in which the relaxation of 29 cm^{-1} phonons is studied as a function of excited Cr^{3+} ion concentration, N^*, and spot diameter. The sample is excited with a pulsed tunable dye laser beam directed 25° from the c axis at a wavelength of 5820Å, near the peak of the $^4A_2 \rightarrow {}^4T_2$ Cr^{3+} broadband absorption (see insert Fig. 1). Rapid relaxation to the two components of ^2E occurs (<<1ns), with about 30% of the ions passing to the $2\bar{A}$ state.[5] Further relaxation (T_1=1.1ns) from $2\bar{A}$ to $\bar{E}$ produces a monoenergetic pulse of 29 cm^{-1} phonons. The phonons are resonantly reabsorbed leading to a non-thermal population of $2\bar{A}$ and enhanced R_2 luminescence which is proportional to the resonant phonon population. For the 500µm spot a coaxial flash lamp-pumped tunable dye laser (400ns pulse width) was used for the excitation, while for the 100µm and 250µm spots, a N_2 laser pumped dye laser (3ns pulse width) was focused on the sample to define the excited cylindrical volume. Excited ion concentrations were estimated (±30%) from the laser pulse energy and known absorption coefficients.

Under most conditions, the decay is nearly exponential,

although for small pulse energies and small spot sizes the decay
often departs significantly from exponential behavior. For the
largest pulse energies, it appears that the high energy phonons
produced in the initial non-radiative relaxation from the broad
bands to the ^2E state feed the 29 cm^{-1} phonon modes during the
sample thermalization producing an additional enhanced R_2 emission.
For non-exponential behavior we approximated the early part of the
decay after termination of the laser pulse by an exponential decay,
thereby defining τ.

     The relaxation time as a function of N* is shown in Fig. 1
for three spot sizes. For small N*, τ increases with increasing
N* and is strongly dependent on the spot size, a result observed
previously.[1-4] This implies that at least part of the phonon loss
occurs through spatial migration out of the excited volume due to
spatial and/or spectral diffusion. For large N* however, τ appears
to asymptotically approach a limiting value of about 1μs and becomes almost independent of spot size, indicating an additional
spot size-independent loss mechanism.

     An additional mechanism which can explain the saturation of τ
and its spot size independence is the resonant phonon assisted
energy transfer process described by Holstein, Lyo, and Orbach[6]
where the 29 cm^{-1} phonons induce an energy transfer to non-reso-

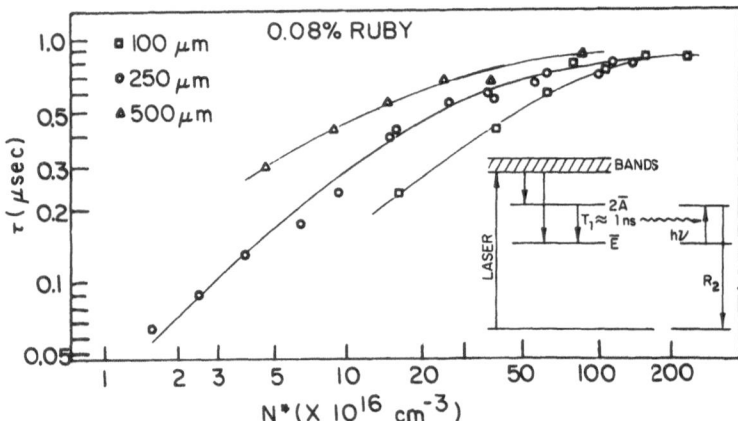

Fig. 1. Decay time, τ, of the R_2 luminescence after laser excita-
         tion into the broadband absorption of 0.08% ruby as a
         function of N* and spot diameter. The insert shows the
         phonon generation and detection scheme. The 29 cm^{-1}
         phonons are produced by $2\bar{A} \rightarrow \bar{E}$ relaxation and detected from
         the enhanced R_2 luminescence whose intensity is propor-
         tional to the resonant phonon population.

nant neighbors and take up or supply the energy required to con-
serve energy. The non-resonance may occur either due to the 0.2
cm^{-1} inhomogeneous broadening of the ^2E states as well as the 0.38
cm^{-1} ground state splitting. In either case the phonons are
shifted well outside the excited state resonance width of 0.01
cm^{-1}. The model contributes a rate proportional to $N*(1+N*/\Sigma)^{-1}$
which becomes nearly independent of N* when $N*>>\Sigma$ where Σ is the
number of phonon modes on speaking terms with the excited state
resonance ($\sim 3.5 \times 10^{16}$ cm^{-3}). From the estimated dependence of
the exchange on Cr^{3+} ion separation,[7] we calculate a rate of
10^6-10^7 sec^{-1}, consistent with the data. The model further predicts a rate proportional to the ground state Cr^{3+} concentration
from which we expect the saturated value of τ to vary inversely
with Cr^{3+} ion concentration as has been observed in our laboratory
with other Cr^{3+} concentrations.

Unfortunately, it is not possible to ascertain from the data
the anharmonic decay time nor to decide upon a model for the spectral diffusion because of the presence of the competing decay
mechanisms. We can say however that it seems possible under conditions of sufficiently large N* and spot size to spatially trap
the resonant phonons so that decay must then proceed by mechanisms
which do not involve spatial migration from the excited volume.

Supported in part by the Army Research Office, Durham.

REFERENCES

1. J. I. Dijkhuis, A. van der Pol, and H. W. de Wijn, Phy. Rev.
    Lett. 37:1554(1976).
2. G. Pauli and K. F. Renk, Phys. Lett. 67A:410 (1978).
3. R. S. Meltzer and J. E. Rives, Phys. Rev. Lett. 38:421 (1977).
4. J. I. Dijkhuis and H. W. de Wijn, Phys. Rev. B, September 1,
    1979 (to be published).
5. J. E. Rives and R. S. Meltzer, Phys. Rev. B16:1808 (1977).
6. T. Holstein, S. K. Lyo and R. Orbach, Phys. Rev. Lett. 36:891
    (1976).
7. R. J. Birgeneau, J. Chem. Phys. 50:4282 (1969).
                        ****
DISCUSSION

K. R. Renk: I have the impression that the saturation occurs at a
lower value in the more strongly doped crystals than for the other
crystals.

J. E. Rives: The saturation value is ground-state concentration
dependent, and this was for 0.08% ruby. For the work you did at

0.03% ruby one would expect a saturation value of 3.5 or 4 μsecs.
So in fact it is in agreement with your results at lower concentrations.

W. E. Bron: What is the lifetime of 29 cm^{-1} phonons in ruby - the
anharmonic decay time?

J. E. Rives: In a lot of the earlier experiments we both thought
that one could measure anharmonic decay times. Now I think we both
have to agree that the anharmonic time which was the thing we wanted
to get out of this to begin with is certainly longer than 1 μsec,
and probably several μs.

K. F. Renk: I would like to make a comment. We have also measured
with strong pulses, and we found lifetimes of the order of 100 μs.
Then when the crystal cooled down, we have sent a second pulse and
this has a lifetime of less than 10 μs. So these are really not
phonons - that is some kind of heat.

J. E. Rives: If one goes to even larger spot sizes, one fills up
the entire crystal to about 40°K, and this lasts for about 1ms.
So yes, there are problems, but one can decrease the power, go to
small spot sizes and get the initial decay out.

ULTRASONIC SPECTROSCOPY OF THE

ACCEPTOR GROUND STATE IN CUBIC SEMICONDUCTORS

Kurt Lassmann and Heinrich Zeile

Physikalisches Institut, Universität Stuttgart

Pfaffenwaldring 57, D-7000 Stuttgart 80, FRG

INTRODUCTION

Phonons are effectively scattered by electrons and holes bound to donors and acceptors in semiconductors at low temperatures corresponding to the relatively large deformation potential constants involved. This has been found over twenty years ago in measurements of the thermal resistivity and the ultrasonic attenuation of bound donor states in Germanium. Since then many phonon scattering experiments on both donors and acceptors have been carried out and a lot of results have been obtained[1]. Much of the progress in understanding is due to the progress of industry and research in manufacturing pure and well-defined crystals. This applies particularly to the acceptor ground state which is very sensitive to crystal imperfection. In the following we will discuss ultrasonic measurements on p-type semiconductors reflecting spontaneous and defect sensitive modifications of the acceptor ground state.

For the bound states of shallow impurities in semiconductors a hydrogenic model is justified where the wavefunctions are constructed from the neighbouring band states leading to relatively small ionization energies $E_A \simeq 10$ to 100 meV and relatively large Bohr radii $r_B \propto 1/\sqrt{E_A} \sim 100$ to 10 Å. This means that these states are extended point defects. The ground state of acceptors in cubic semiconductors is a fourfold degenerate Γ_8 state that may be split by elastic or electric fields into two Kramers doublets. A magnetic field will lift the degeneracy completely. Such a Γ_8 state should, in principle, show an instability against spontaneous deformation, viz. a Jahn-Teller effect. However, because of the extended wavefunction, a Jahn-Teller effect is expected to be weak for shallow acceptors, it should be more pronounced for deeper acceptors with their smaller Bohr-radii.

In fact, for two acceptors, namely GaAs(Mn) (E_A=110 meV, r_B=10.4 Å) and Si(In) (E_A=156 meV, r_B=7.3 Å) we have found resonance energies at 3.1 meV and 4.2 meV respectively, in reasonable accordance with first calculations of the problem[2]. Experimentally, these values have been obtained from an analysis of the ultrasonic relaxation attenuation[3][4], from thermal conductivity measurements[5], and, in the case of Si(In), also from experiments with quasimonochromatic phonons[6] and from a satellite in luminescence[7].

The vibronic ground state of the system will exhibit full symmetry, that is, it will also be a Γ_8 state. Internal elastic and electric fields originating from crystal defects will lead to a distribution of splittings which,in turn, may be regarded as an indicator of residual defects. Such a distribution of two-level systems is quite analogous to that encountered in glasses, and we are faced with the same problems as to the form of the distribution, the relaxation times and the interaction between defects.

THE DISTRIBUTION OF SPLITTINGS

The form of the distribution function may be obtained by measuring the resonant attenuation in a wide frequency range: Since the attenuation α is proportional to the spectral density $N(\delta)$ of the splittings δ at a certain frequency $\nu=\delta/h$, the distribution is thus probed in the corresponding frequency range.

For the resonant absorption we have

$$\alpha(\nu,T) = \int N(\delta)\alpha(\delta,\nu,T)d\delta$$

$$\alpha(\nu,T) \propto \int\left[\tanh(\delta/2kT)D_{eff}^2(\delta)N(\delta)\nu\Gamma(\delta,T)/\left[(h\nu-\delta)^2+\Gamma^2(\delta,T)\right]\right]d\delta$$

If the linewidth $\Gamma(\delta,T)$ of the resonance line is small enough compared to the variation of $N(\delta)$, $D_{eff}(\delta)$ we can write:

$$\alpha(\nu,T)\propto\left[N(\delta=h\nu)D_{eff}^2(\delta=h\nu)\right](\nu^2/T)$$

for $h\nu\ll kT$. The effective coupling constant D_{eff} is a combination of the two well-known deformation potential constants of the ground state corresponding to the "average" symmetry of the disturbing field. Assuming that this "average" symmetry does not vary with δ we may write $\alpha(\nu)/\nu^2\propto N(\delta=h\nu)$ at a fixed temperature. In Fig.1 we have plotted the variation of $\alpha(\nu)/\nu^2$ in the frequency range 0.5 GHz$\leq\nu\leq$4.2 GHz for a boron doped high quality silicon crystal. Under the assumption that D_{eff} is constant and that $\alpha(\nu)/\nu^2$ is negligibly small outside our measuring range we can integrate the

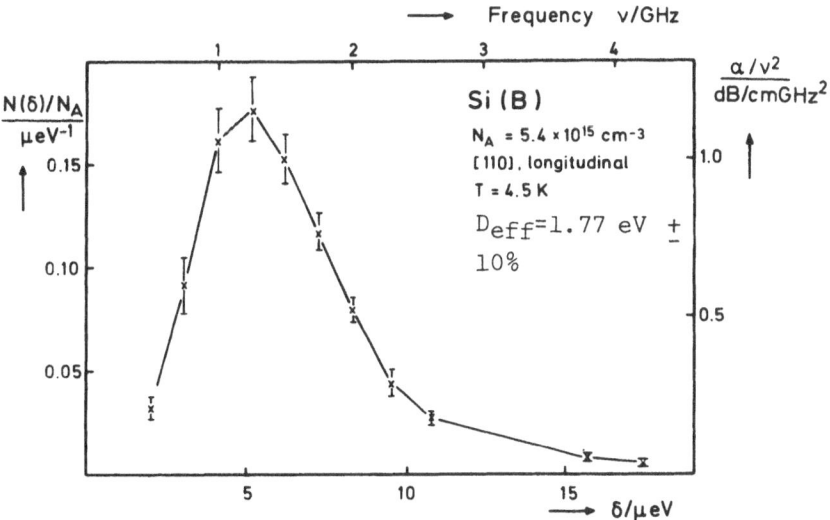

Fig. 1: Distribution of splittings of the boron acceptor ground
        state for a high quality silicon crystal.

area to obtain $N_A \cdot D_{eff}^2$ where N_A is the known acceptor concentration.
The value $D_{eff}=1.8$ eV seems reasonable in comparison with the two
deformation potential constants $D_u^a = 2.4$ eV and $D_u^{a'} = 3.9$ eV. In
applying uniaxial pressure we have verified for a similar crystal
that there is no peak in the distribution function at $\delta=0$. Measuring
the attenuation for other directions and polarizations one may separate the different elastic constants involved to get additional
information on D_{eff}. In Fig. 2 the distribution functions are shown
for a series of high quality silicon crystals of varying boron concentrations. They are all similar with maxima between 1 GHz and 2
GHz, falling rapidly off at low frequencies. With growing boron
concentration there is a tendency to flatten the maximum and to
broaden the distribution to higher energies. All these crystals
(supplier Wacker Chemitronic, Burghausen, F.R.G.) are floating
zone-pulled, dislocation-free, with low carbon and oxygen
contents (below 10^{15} cm^{-3} to 10^{16} cm^{-3} typically). The
pulling velocity was high enough to avoid detectable clusters of
selfinterstitials ("swirl-free" crystals), however, the concentration of selfdefects is unknown. It may be as high as 10^{17} cm^{-3}.
(The concentration of selfdefects may be reduced in pulling the
crystal at extremely low velocity, where the defects have time to
diffuse out of the crystal during growth[8]). The crystal of Fig. 1
differed from those of Fig. 2 in that the pulling conditions were
such as to produce an even higher homogeneity of the crystal.

   In less pure crystals (Czochralski-grown Si(B) and Si(In) with

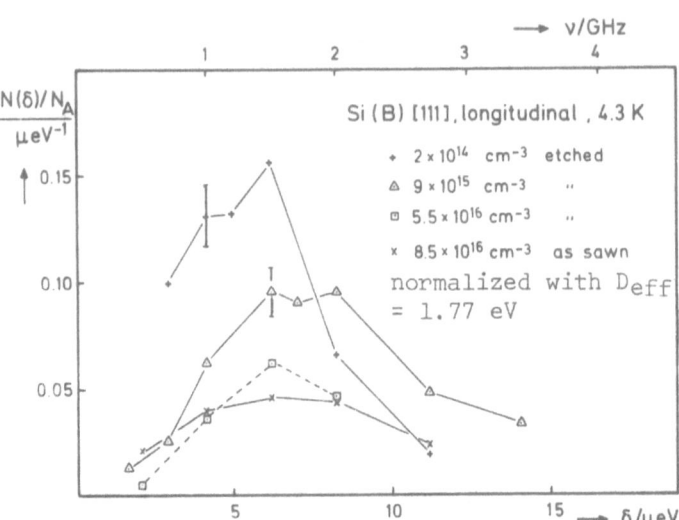

Fig. 2: Distribution of splittings for various boron concentrations.

high carbon and oxygen content) we find only a rise in α/ν^2 indicating a maximum beyond our limit of about 4 GHz. Earlier we have found
for GaAs($\sim 10^{18}$ cm^{-3} Mn)[3] α/ν^2 to be practically constant from 0.4 to
2.0 GHz and also for Ge of various doping[9] between 8×10^{15} cm^{-3} and
3×10^{16} cm^{-3} α/ν^2 constant between 1.2 GHz and 2.5 GHz indicating
a width of the distribution of about 50 µeV and 100 µeV respectively.
(In contrast, for a very pure Ge(2×10^{14} cm-3 Ga) crystal
grown and kindly supplied by E.E. Haller of Lawrence Berkeley Laboratory, California, we find a distribution even narrower than that
of Fig. 1 with maximum near 1 GHz.)

     So far, we don't know the reason for the residual linewidth of
the purest crystals. We shall shortly discuss three possibilities:
(i) Though there is some correlation with boron concentration visible
in our results, ground state splitting due to wave function overlap
seems improbable, since we would expect a monotonous exponential
dependence of the splitting on mean acceptor distance for this range
of small energies[10] . (ii) Electric fields from residual compensa-
ting donors should be too small for the low concentrations of residual donors ($\sim 10^{12}$ cm^{-3} in the purest crystal). In Ref. 11 we over-
estimated the electric field in our calculation. Furthermore, at low
temperatures the ionized donors and acceptors should not be uncorrelated, as assumed[12] in that calculation, because a nearest neighbour
correlation will be energetically more favourable. Such a correlation
will lead to a further reduction of the electric field, as we have
found by a computer simulation. (iii) Strain fields from point defects would be expected at relatively high concentrations. In Fig. 3

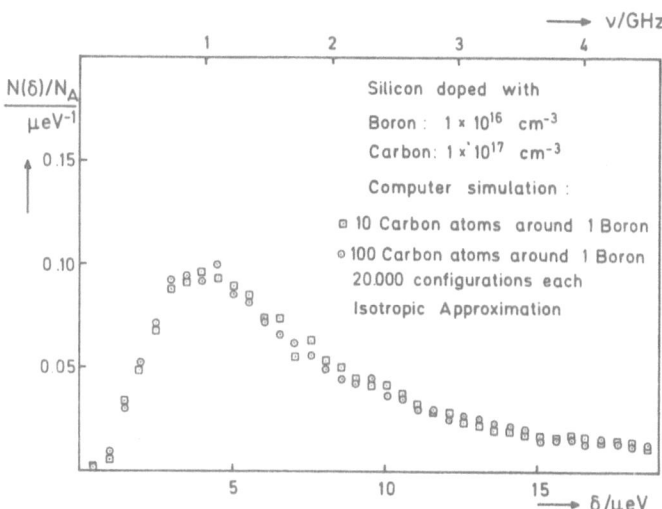

Fig. 3: Computer simulation of splitting distribution due to substitutional carbon.

we have plotted the distribution of splitting resulting from a computer simulation for substitutional carbon with an elastic strength $A = -0.85$ $Å^3$ corresponding to its covalent radius in an isotropic approximation for the crystal. Since the elastic strength of boron is much smaller, we see that neither the boron nor the carbon or oxygen content should be responsible for the residual splitting. Since the concentration of intrinsic defects or their agglomeration is unknown, at the present state of our investigation we cannot say whether selfdefects or some other defects or perhaps some effects intrinsic to the isolated acceptor are responsible for the residual splitting.

THE CRITICAL INTENSITY

To separate the resonant attenuation α_{res} from relaxation attenuation and geometrical effects we make use of the possibility of saturation at high acoustic intensities. From these intensity dependent measurements we obtain the critical intensity J_c.

In a Bloch equation formalism we obtain from J_c the product $\tau_1 \cdot \tau_2$ of the "spin"-lattice or energy relaxation time τ_1 and the transverse or dephasing relaxation time τ_2

$$J_c \propto 1/(\tau_1 \cdot \tau_2)$$

If only "spin"-lattice relaxation were effective, one might assume

$\tau_1 \cdot \tau_2 = 2\tau^2$ [13] ; i.e. I_c should not depend on N_A. A dynamical interaction between the acceptors should shorten τ_2, leading to a larger
I_c for smaller mean distances between the acceptors. In Fig. 4 we
have plotted the critical intensity for various acceptor concentrations. We see that below about 10^{16} cm^{-3} the critical intensity does
not depend on N_A. From $I_c = 10^{-4}$ W/cm^2 we estimate $\tau_1 \cdot \tau_2/\sqrt{2} = \tau_1 =$
1.4×10^{-8} sec, whilst from our measurement of the relaxation attenua-

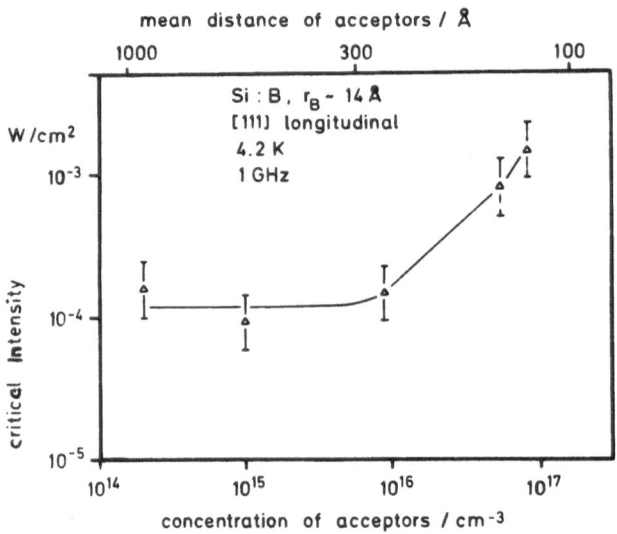

Fig. 4: Critical intensity as a function of average acceptor
        distance.

tion we have $\tau_1 = 1.3 \times 10^{-8}$ sec, which means, that for average distances larger than about 300 Å the critical intensity is determined
by "spin"-lattice relaxation alone. For smaller distances d the
critical intensity rises roughly as $1/d^3$, as one might expect for
a dipolar interaction[13 14]. However, applying the formula for spectral diffusion by elastic dipolar interaction given in 14 for our
experimental conditions (Raman/Orbach relaxation, narrow distribu-
tion), we arrive at $\Delta\omega/Hz = 5.6 \cdot 10^{-8}$ (N_A/cm^{-3}) (independent on temperature) which leads to τ_2 being too small by an order of magni-
tude. Also the temperature dependence of J_c varies more slowly than
expected. We are, therefore, preparing experiments for a direct
determination of τ_2.

    Applying uniaxial pressure will "shift" the distribution and
reduce the concentration of acceptors on speaking terms. In our
first experiments we have seen no effect on I_c at 4.2 K reducing

$\alpha(\nu)/\nu^2$ by about a factor of five. This is as expected since the energy-independent Raman/Orbach relaxation is dominant in this range so that spectral diffusion should remain unchanged and on the other hand because the acceptors on speaking terms are at any rate relatively minor in number to have an influence on spectral diffusion.

For the narrow distribution in a magnetic field acoustic paramagnetic resonance can be seen and lines be resolved at 4 GHz. Also EPR at 10 GHz is possible in these clean crystals, as was recently shown[15]. By detailed investigation of the APR at somewhat higher frequencies we should get additional information on the symmetry and distribution of the random splittings of the acceptor ground state.

We hope that investigations of this kind may help to some practical informations concerning residual crystal defects as well as to a .better understanding of the dynamical processes connected with the two-level systems.

This work has been supported by the Deutsche Forschungsgemein-schaft (SFB 67), Stuttgart.

REFERENCES
1. For reference to earlier work see
    H. Tokumoto and T. Ishiguro, Phys. Rev. B 15:2099 (1977)
2. E. Sigmund and K. Lassmann, These Proceedings
3. K. Lassmann and Hp. Schad, Solid State Commun. 19:599 (1976)
4. Hp. Schad and K. Lassmann, Phys. Lett. 56A:409 (1976)
5. A. de Combarieu and K. Lassmann, Phonon Scattering in Solids,
    Plenum Press, N.Y. (1976)
6. H. Schenk, W. Forkel and W. Eisenmenger, Verhandl. DPG 1/1978;
    page 328
7. R. Sauer, W. Schmid and J. Weber, Solid State Commun. 27:705
    (1978)
8. P.J. Roksnoer, W.J. Bartels and C.W.T. Bulle, J. Cryst. Growth
    35:245 (1976)
9. E. Ortlieb, Hp. Schad and K. Lassmann, Solid State Commun.
    19:599 (1976)
10. E. Kaczmarek and Z.W. Gortel, Phys. Rev. B10:2535 (1974)
11. H. Zeile, O. Mathuni and K. Lassmann, J. de Physique L40:53
    (1979) (I_c of Fig.2 has to be increased for the overlap of the
    ultrasonic pulse train by a factor of 2.)
12. D. Larsen, Phys. Rev. B13:1681 (1976)
13. J. Joffrin and A. Levelut, J. Physique 36:811 (1975)
14. J.L. Black and B.J. Halperin, Phys. Rev. B16:2879 (1977)
15. H. Neubrand, phys. stat. sol. (b) 86:269 (1978)
                                     90:301 (1978)
                        ****

DISCUSSION

R. Collella: These impurities are on the average inhomogeneously distributed in these crystals. I was wondering whether you have considered using crystals like silicon in which the doping is achieved by neutron irradiation, and you can have the same kind of doping with the same percentages, but you are sure that it is much more homogeneous. It would be interesting to compare if you get the same results.

K. Lassmann: By this method you can obtain donors by transportation of silicon into phosphorous, and this we used already in one case. For acceptors I don't know of a possibility to make it more homogeneous. What we tried to do was to look with an infrared microscope and in the case of some silicon-indium crystals where we also have some broader distribution there we saw some striations. With these silicon crystals, especially the first one I showed, it would be very inhomogeneous, the suppliers told me but they didn't tell me how they told it. There we didn't see with the microscope any striations, but we cannot tell any more about that. What we want to do is to see if there is any inhomogeneity correlated say with the pulling direction so we might work in different directions if we are allowed to do it. For instance, under pressure we could see whether there are three types of strain. This is really a problem. Also association between defects might be a problem.

J. P. Harrison: Could you measure the strain dipole moment by applying a stress and then repeating your measurements?

K. Lassmann: It is potentially given by the deformation potential constants which have been measured.

IMAGES OF ELECTRON-HOLE DROPLETS AND BALLISTIC PHONONS IN Ge

J. P. Wolfe, M. Greenstein, G. A. Northrop and M. Tamor

Physics Department and Materials Research Laboratory
University of Illinois at Urbana-Champaign
Urbana, IL   61801

INTRODUCTION

Since L. V. Keldysh's original 1968 postulate of electron-hole droplets (EHD) in semiconductors, the spatial distribution of such droplets has provided an intriguing and sometimes controversial subject. Only recently has its resolution appeared imminent, precipitated by another remarkable insight by Keldysh:[1] droplets are pushed into the crystal by a "phonon wind."

In this paper two distinct imaging experiments in Ge are discussed: a) luminescence imaging of the electron-hole droplet cloud which reveals the anisotropic effect of the phonon wind, and b) ballistic phonon imaging which provides a new look at the general phenomenon of phonon focusing in solids.

BACKGROUND--ELECTRON-HOLE DROPLETS

When light impinges on an ultrapure Ge crystal at low temperatures, the photo-produced electron-hole (e-h) pairs bind by their coulomb attraction into excitons. The exciton is a highly mobile hydrogen-like species with a binding energy of about 3.5 meV, a Bohr radius of 177Å, and a lifetime of 1-10 μs. At sufficiently high density, say $n_x \simeq 10^{12} cm^{-3}$ at 2 K, the exciton gas condenses into droplets of a rather high density plasma with $n_{eh} = 2.3 \times 10^{17} cm^{-3}$. This unique electron-hole liquid phase has a metallic conductivity like copper, a surface energy 10^{-6} that of water, a Fermi energy of 6.4 meV, a critical temperature of 6.5 K, and a lifetime due to pair recombination of 40 μs. Such internal properties of the electron-hole liquid have been well-predicted by theory and accurately determined by a remarkable variety of experiments.[2] This paper on the other hand will deal with an external aspect of the liquid: how does it move within the crystal?

   In a typical experiment e-h pairs are produced by a continuous, focused Ar^+ laser beam which is absorbed within 0.1 μm of the surface. At 2 K and moderate excitation levels ($P_{abs} \gtrsim 1$ mW) almost all the excitons condense into liquid within a nanosecond. Consequently, liquid is produced very near the excitation region. Yet, what was observed was a rather large cloud of small droplets, each 1-10 μm in size (from Rayleigh scattering), extending up to several millimeters inside the crystal. Neither thermal diffusion nor ballistic motion of these microscopically-massive droplets (containing 10^8 pairs) could explain this phenomenon. In particular, it was argued that a droplet in motion should quickly come to rest within a carrier-phonon momentum relaxation time, $\tau_{ph} \simeq 10^{-9}$ sec. Thus a continual force F is needed to push droplets macroscopic distances, via a drift relation $Mv = F\tau_{ph}$, where M and v are the mass and velocity of a single drop.

   Convincing experimental evidence has recently emerged indicating that F is due to a non-thermal flux of phonons. By applying external sources of phonons, Bagaev et al.[3] and Hensel and Dynes[4] have succeeded in transporting droplets macroscopic distances. Doehler and Worlock[5] measured the drift velocities of droplets by laser scattering, obtaining $v \sim 10^3$ cm/sec. The experiments described below strongly support the phonon wind hypothesis by showing that the electron-hole droplet distribution is highly anisotropic, with features explained by anisotropic carrier-phonon absorption and phonon focusing.[6]

IMAGING THE CLOUD OF DROPLETS

   The distribution of droplets is experimentally determined by the recombination luminescence they continuously emit at $\lambda = 1.75$ μm. A snapshot of the cloud of droplets presents something of a challenge, requiring both spatial and spectral resolution at this infrared wavelength. The unavailability of a sensitive image device in this region, however, precipitated our development of scanning technique based on a single-element, cooled Ge photodetector. While this system lacks the multiplex advantage of an array, it permits spectral- and time-resolved imaging. Basically a sharply-focused image of the cloud is raster-scanned across a small aperature at the entrance plane of a spectrometer, as shown in Fig. 1. The x-y scanning by 2 galvanometer-driven mirrors can be controlled by a minicomputer which also records the luminescence intensity in a 256 × 256 array. The data may be replayed on a storage oscilloscope and photographed, or processed digitally for detailed information.

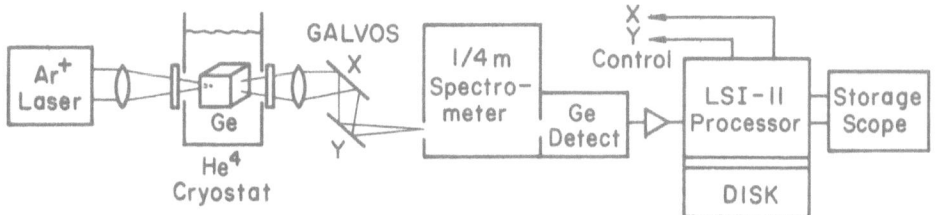

Fig. 1.   Computer-controlled infrared imaging system. The two
           galvos rotate mirrors which translate the image of the Ge
           crystal.

     Figure 2a shows a luminescent image of the cloud of droplets
looking through the <001> face of the crystal. The laser strikes
the crystal at the center of the pattern on the opposite face. The
crystal is transparent to the luminescence but completely absorbs
the laser light at $\lambda = 0.515$ μm. A digitally-produced contour map
of the luminescence intensity is displayed in Fig. 2b. Here light
and dark bands separate curves of constant intensity. The cloud
of droplets is seen to be highly anisotropic with broad <111> lobes
and sharp <100> flares radiating outward from the excitation point.

     The basic features of this droplet distribution can be semi-
quantitively understood in terms of a phonon wind which emanates
(at least in part) from the excitation region. Two likely sources

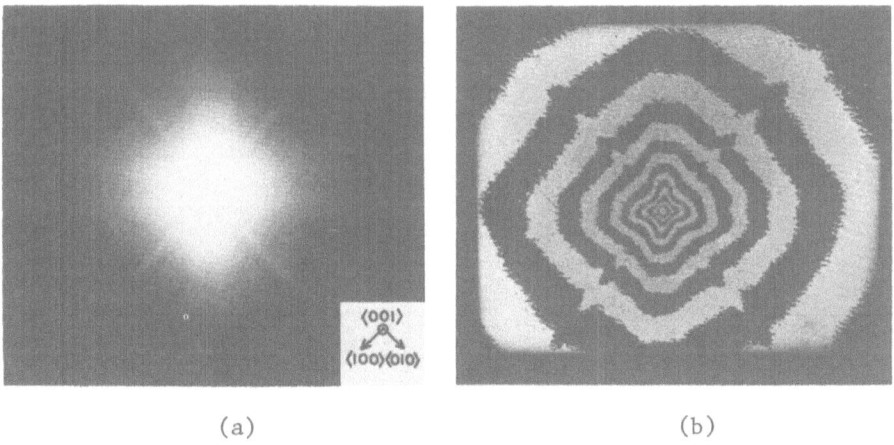

(a)   (b)

Fig. 2.   a) Image of the cloud of droplets at $\lambda = 1.75$ μm. Width of
           photo equals 5 mm on the crystal. b) Contours of constant
           intensity. Vertical and horizontal lobes point into the
           crystal along <111> directions.

of non-equilibrium phonons are those originating from the thermalization of carriers (T-phonons) and those resulting from recombination processes (R-phonons). A localized source of T-phonons near
the laser excitation point is needed to explain the sharp cloud
structures but the overall size may be affected by mutual respulsion
of droplets due to R-phonons. A quantitative model along these
lines has been developed by Markiewicz.[7]

The force on a droplet depends upon the phonon absorption
probability, the phonon momentum $\hbar\vec{q}$ which is transferred to the
droplet, and the phonon flux. The absorption probability depends
linearly upon wavevector q up to a cutoff $2k_{Fermi}$ of the liquid,
and for a given q, depends quadratically upon deformation potential
E_i for electrons in the ith valley. Fig. 3 illustrates how these
factors explain the broad <111> lobes of the cloud in terms of a
flux of LA phonons. Fig. 3a is a sum of E_i^2 over all four valleys
for LA phonons, showing a pronounced peak in absorption probability
along <111>. The LA phonon flux also peaks in the <111> direction,
illustrated in Fig. 3b, due to phonon focusing, as discussed in the
next section. In addition the four conduction valleys are ellipsoids with their long axes oriented along <111>, implying $2k_F$ is a
maximum in this direction.

Thus the droplets should be pushed strongly along <111> due
to LA phonon absorption, explaining the broad lobes. The contour
map shows that the pronounced <111> lobes dissipate with increasing
distance from the excitation point. We believe that this represents
an attenuation in the LA phonon flux as phonons are absorbed by
droplets. An explanation of the sharp flares lies in the sharp
focusing of TA phonon flux, which is now examined in
detail from quite a different point of view.

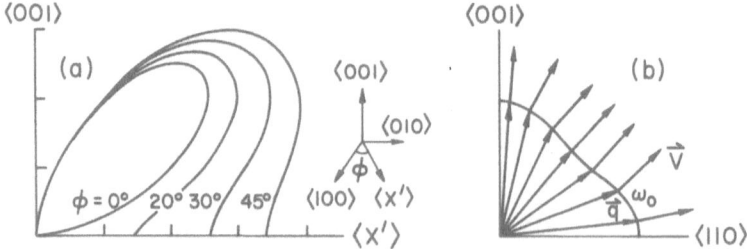

Fig. 3.   a) Polar plot of LA-phonon screened deformation potentials
          E_i^2 summed over the electron valleys, by the method of
          Ref. 8. b) Schematic of phonon focusing effect for LA
          phonons, showing constant $\omega(\vec{q})$ surface in q-space which pro-
          duces an energy-flux maximum near <111>.

BALLISTIC PHONON IMAGING IN CRYSTALS

    At low temperatures where phonon mean free paths are comparable
to crystal dimensions, an anisotropic channeling of phonon flux
results from the elastic anisotropy of the crystal. Taylor, Maris
and Elbaum[9] first demonstrated this phonon focusing effect using
heat pulses[10], and recent experiments by Hensel and Dynes[11] have
shown sharp focusing of fast TA phonons in Ge. The phonon focusing
effect arises because the phonon energy flux corresponds to the
group velocity $\vec{V}$ which is not colinear with the wavevector $\vec{q}$, as
shown schematically in Fig. 3b.[12] Thus a uniform distribution over
wavevectors, such as that produced by an ideal heat pulse, corresponds to a distinctly non-spherical distribution of phonon flux.

    A new technique is introduced here which produces a two-
dimensional map of the phonon flux from a localized heat pulse in
Ge. In our experiments a heat pulse is generated with a Q-switched
YAℓG:Nd laser and detected with a small superconducting Al bolometer
(220×180 μm^2), as illustrated in Fig. 4a. Typical phonon signals
are shown in Fig. 4b. The propagation direction is changed by
raster-scanning the laser beam in x-y fashion with two galvanometer-
driven mirrors. As in Fig. 1 the scanning is controlled by a minicomputer and the phonon intensities are stored digitally in a
256×256 array. To determine the phonon flux for a given (θ_V, ϕ_V)
a boxcar integrates the total bolometer signal within the gate
period of Fig. 4b.

    Figure 5a shows the resulting ballistic phonon image as viewed
into a <001> face of the Ge crystal. One can imagine the heat source

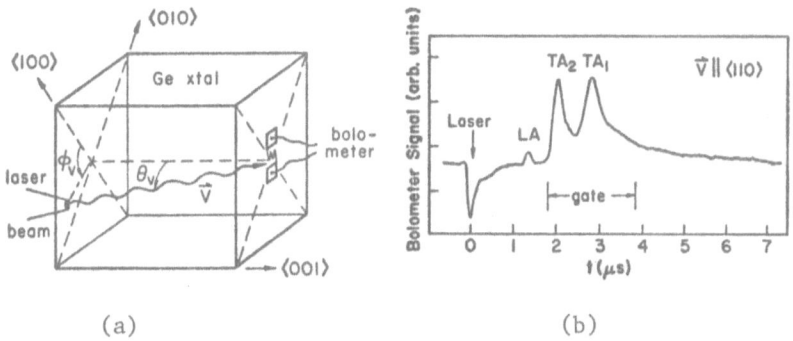

(a)   (b)

Fig. 4. a) Schematic of ballistic phonon imaging experiment. A
         pulsed laser beam (10 W, 150 ns) is continuously scanned
         across a metalized or polished surface while phonon inten-
         sities are recorded for each propagation direction.
         b) Typical heat pulses at 2 K for one laser position.

fixed at the center of the pattern but on the opposite face. The
bright areas represent directions of high TA phonon intensities
with the <001> direction at the center of the pattern. By changing
the boxcar delay and gate times, different features of this pattern
are emphasized and can easily by correlated with different phonon
modes. The central square and horizontal and vertical ramps are
identified with the slow TA phonons. The diagonal ramps are due to
the fast TA modes.[13] A quantitative output is shown in Fig. 5b by
plotting the intensity of successive horizontal scans. Figure 6a
is a blowup of the central region, obtained by reducing the laser
scanning range. The bright square spans $\pm$ 7.4° in propagation direction from the <001> axis. A smaller, less intense square is discernible. Figure 6b is a contour map in which boundaries between
black and white regions are curves of constant intensity.

An understanding of these complex patterns can be gained by
extending some of the basic ideas of Maris.[14] We have found that
the bright regions in Fig. 5a are propagation directions that result
from more than one $\vec{q}$ vector. The still brighter boundaries of these
regions correspond to integrable infinites in phonon flux. Mathema-
tically these infinities can be seen as lines of inflection in the
constant $\omega(\vec{q})$ surfaces, and occur when the Jacobian J of the transformation from $\vec{q}$ direction (θ_q, ϕ_q) to $\vec{V}$ direction (θ_V, ϕ_V) is zero.

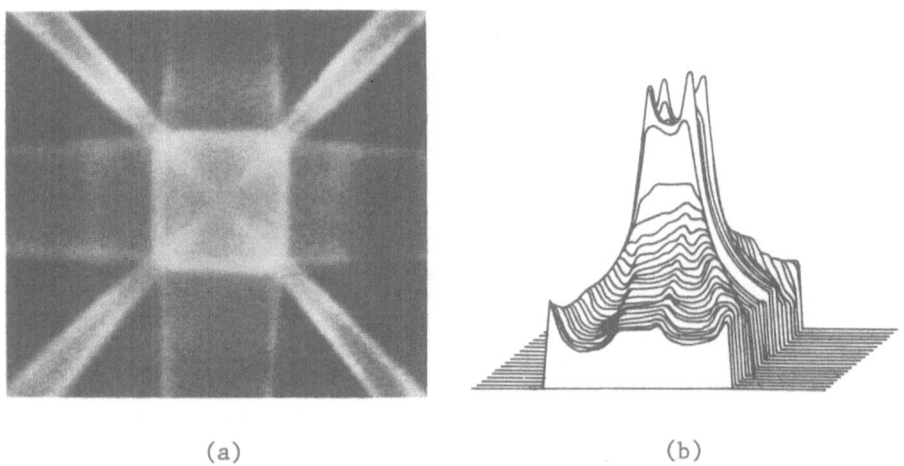

(a)   (b)

Fig. 5.   a) Image of ballistic TA phonon intensities as the direc-
          tion of $\vec{V}$ is scanned about the <001> axis at center. Center
          to left edge is 25° in propagation direction. b) Linear
          intensity plot for a number of horizontal scan lines.

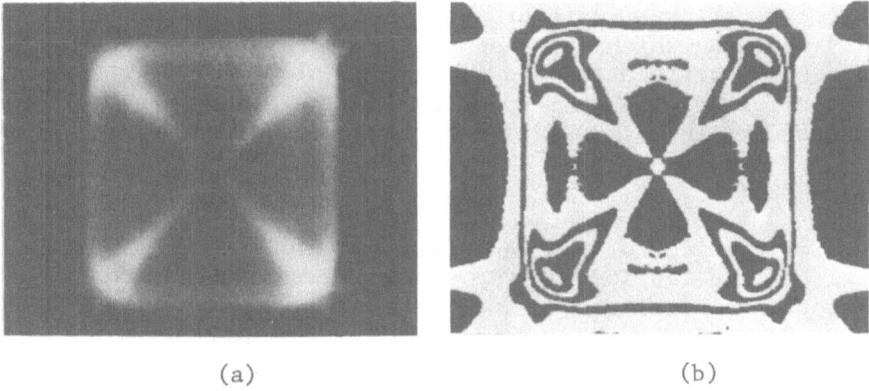

(a)   (b)

Fig. 6.   a) Magnification of the central region in Fig. 5.
           b) Contours of constant phonon flux.

These lines separate regions of convex or concave constant $\omega(\vec{q})$
surface from saddle regions, as shown for the slow TA mode in
Fig. 7a. By mapping these lines into $\vec{V}$ space we have determined
the spatial positions of the infinities in phonon flux corresponding to our experimental geometry, as plotted in Fig. 7b. These
lines are based on previously determined elastic constants
$C_{11}:C_{12}:C_{44}::1:0.38:0.52$ and agree within a few percent with the
photographic image of Fig. 6.

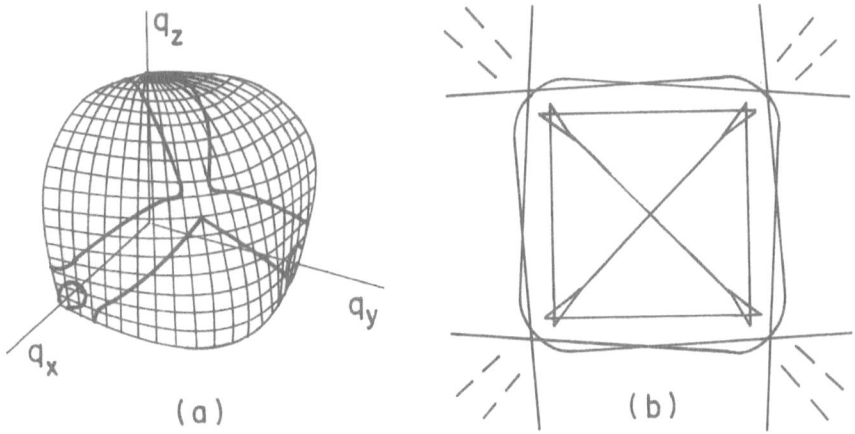

(a)   (b)

Fig. 7.   a) Constant $\omega(\vec{q})$ surface for the slow TA phonon in Ge.
           Dark lines of $J = 0$ separate convex, saddle and concave
           regions, corresponding to $\vec{V}$ directions with infinite
           focusing. b) $J = 0$ lines mapped into V space on the scale
           of Fig. 6. Dashed lines are due to fast TA.

This type of experiment produces detailed ballistic phonon
intensity patterns which are a sensitive probe of the fundamental
constant-energy surface for elastic waves. In addition the detailed description of phonon focusing is a necessary step for un-
raveling the complex distribution of electron-hole droplets in Ge.

ACKNOWLEDGEMENTS

We are indebted to A. C. Anderson for his encouragement, and
essential help in fabricating the superconducting bolometer. E. E.
Haller supplied the ultrapure Ge. This work was supported by NSF
grants DMR 77-11672 and MRL-DMR-77-23999 and a Cottrell Grant.

REFERENCES

1. L. V. Keldysh, JEPT Lett. $\underline{23}$, 86 (1976).
2. See for example C. D. Jeffries, Science $\underline{189}$, 955 (1975); T. M.
     Rice, J. C. Hensel, G. A. Thomas, T. G. Phillips, Solid
     State Physics $\underline{32}$ (1977).
3. V. S. Bagaev, L. V. Keldysh, N. N. Sibel'din and V. A.
     Tsvetkov, Sov. Phys. JETP $\underline{43}$, 362 (1976).
4. J. C. Hensel and R. C. Dynes, Phys. Rev. Lett. $\underline{39}$, 969 (1977).
5. J. Doehler, J.V.C. Mattos, and J. M. Worlock, Phys. Rev. Lett.
     $\underline{38}$, 726 (1977); Sol. St. Comman. $\underline{27}$, 229 (1978).
6. M. Greenstein and J. P. Wolfe, Phys. Rev. Lett. $\underline{41}$, 715 (1978).
7. R. S. Markiewicz, preprint.
8. R. S. Markiewicz, Phys. Stat. Sol. (b) $\underline{83}$, 659 (1977).
9. B. Taylor, H. J. Maris and C. Elbaum, Phys. Rev. Lett. $\underline{23}$,
     415 (1969).
10. R. J. von Gutfeld, Physical Acoustics (Academic Press, New
     York, 1968) Vol. 5, 223.
11. J. C. Hensel and R. C. Dynes, Proceedings of Int. Conf. on
     Physics of Semicon., 1978 (Inst. of Physics, Conference
     Series, no. 43), p. 371.
12. See also M.J.P. Musgrave, Crystal Acoustics (Holden-Day,
     San Francisco, 1970); F. I. Federov, Theory of Elastic Waves
     in Crystals (Plenum, New York, 1968).
13. The larger features of this data were in fact predicted by
     numerical calculations of F. Rosch and O. Weis, Z. Physik
     B$\underline{25}$, 115 (1976).
14. H. J. Maris, J. Ac. Soc. Amer. $\underline{50}$, 812 (1971).

PHONON OPTICS, CARRIER RELAXATION AND RECOMBINATION IN SEMICONDUC-

TORS: CASE OF GaAs EPITAXIAL LAYERS

V. Narayanamurti

Bell Laboratories
600 Mountain Avenue
Murray Hill, NJ 07974

INTRODUCTION: The phenomena of carrier-capture and energy transfer processes in semiconductor epitaxial layers are of considerable fundamental and technological importance. The luminous output of semiconductor pn junctions for example, depends on the relative strength of radiative (photon) and non-radiative (phonon or Auger) processes. As another example, the mobility at low temperatures of electrons in an n-epitaxial layer of a semiconductor such as GaAs is determined by ionized impurity scattering but the energy relaxation of the carriers under the application of a field takes place via phonon emission. Until the present time, because of the dominance of impurity scattering, one has not been able to determine the strength of the intrinsic electron-phonon scattering processes from transport measurements.

In this paper we present a brief review of some recent experiments[1] using superconducting bolometers and/or tunnel junctions to study directly phonon processes in epitaxial layers and pn junctions (also grown epitaxially) of GaAs and GaAℓAs. As a by-product of our work we have also studied defects in such layers using more conventional heat pulse type techniques.

In Fig. 1 we show a schematic of the sample and the variety of transducers which we have used. The thick (typically 0.5 to 5 mm) insulating GaAs substrate serves as a "lossless" transmitting medium for the phonons which are generated by electrical pulses applied to the epitaxial layers or across a pn junction. The phonons are detected at the other end, after a ballistic time of flight, by either a bolometer or a tunnel junction or a low concentration epilayer. The epilayers had a variety of dopants and in addition controlled photoexcitation could be used to vary the charge state of

385

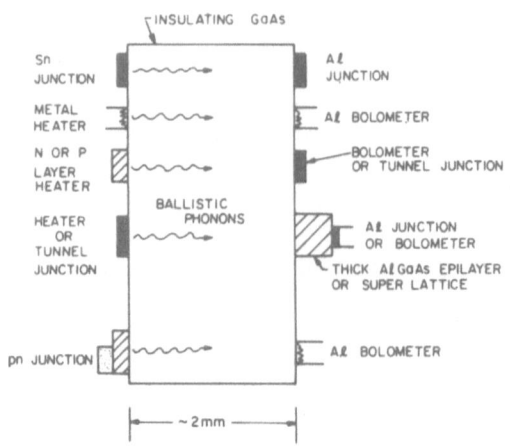

Fig. 1. Schematic of variety of transducers used to study phonon interactions in GaAs.

donor centers in the layers or in the bulk. We would like to em-phasize the epitaxial nature of the transducers. Such transducers have a highly uniform doping (unlike diffused material) and a very low density of interface states which makes them ideal for phonon studies. Because of the large number of examples of heteroepitaxy in semiconductors, such experiments should prove to be quite useful in the understanding of transport processes in such materials.[2] In addition, the use of modern developments in lithography combine with the growth of layered structures by techniques such as molecular beam epitaxy, should make possible the manipulation of phonon beams in technologically important semiconductors such as GaAs. Such manipulation of phonons with novel transducer materials and the resulting physical studies may be best described as "phonon optics".

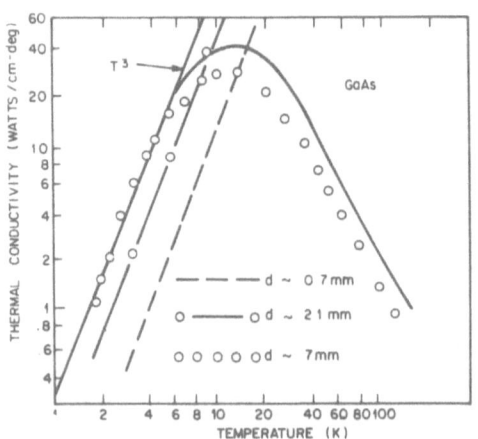

Fig. 2. Thermal conductivity as a function of temperature for "pure" GaAs after Holland.[3]

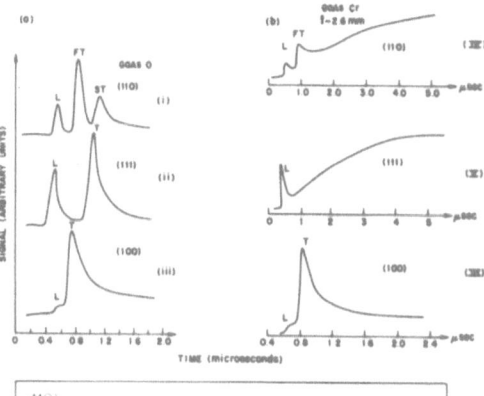

Fig. 3. Heat pulses in GaAs:
and GaAs:Cr as a function of
orientation. Propagation
length $\sim$ 2.5mm.

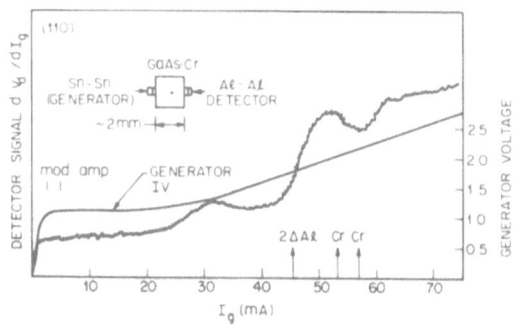

Fig. 4. Tunnel junction
spectroscopy on GaAs with
$\sim$ 7 x 10^{16} cm^{-3} Cr ions.

Substrates: In Fig. 2 we show the thermal conductivity as a function of temperature for "pure" GaAs as measured by Holland.[3] Also
shown are three T^3 lines for samples of different Casimir Lengths
(7, 2.3 and 0.7 mm). These lines define the temperature region for
ballistic phonon transmission. This is true only for substrates
which are semi-insulating and free of transition metal impurities[4]
such as Cr. In Fig. 3 we show the effect of such impurities on
heat pulse propagation in GaAs of d $\sim$ 2.5 mm. The Cr doped samples
suffer from considerable diffusive propagation and show selective
scattering of certain modes consistent with the tetragonal symmetry[5] of the defects. In Fig. 4 we show data taken on the frequency dependence of the absorption for a [110] Cr doped ($\sim$ 7×10^{16}
cm^{-3}) sample using a Sn tunnel generator and an Aℓ tunnel detector.
Absorption in the range 0.7 meV to 1 meV is observed which is the
expected region for the spin splitting under the influence of a
tetragonal Jahn-Teller distortion.[6] Many details, however, of the
GaAs:Cr system are still incompletely understood and in the rest of
this paper, for simplicity, we will confine ourselves to substrates
free of such impurities.

Epilayers: In Fig. 5 we show some typical phonon generation data
at low fields ($\sim$ 5 V/cm) for n-layers (n $\sim$ 2×10^{17} cm^{-3}) as a function of orientation. The data show that the generation rate of
longitudinal (L) and transverse (T) phonons is highly anisotropic.
For example in the [110] direction we observe only the L and fast

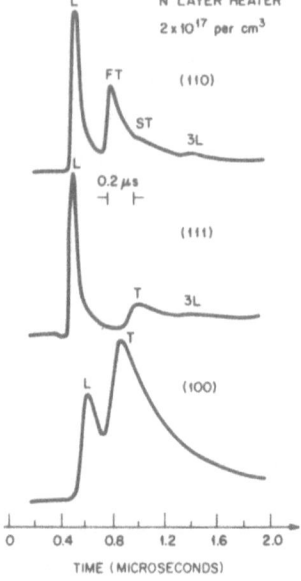

Fig. 5. Orientation dependence of
ballistic phonons with n layer heater.

transverse (FT) phonons but not the slow transverse (ST) phonons.
In the [111] direction the degenerate T modes are extremely weak
compared to L but are of the same order in the [100] direction.
These data are explained as follows: The spherical symmetry of the
conduction band of GaAs causes the deformation potential scattering
to be isotropic and allows only the generation of L phonons. Piezoelectric scattering is, however, highly anisotropic and depending
on orientation allows the generation of both L and T phonons. The
relative importance of the two mechanisms depends strongly on the
carrier concentration and the phonon momentum q. For degenerate
layers and small electron temperatures (T_e), $q \sim 2$ k_F (where k_F is
the Fermi momentum). For low concentration layers ($n \lesssim 10^{16}$), $q \sim$
$2(3mkT_e/\hbar^2)^{1/2}$. The measurements of the phonon generation when
combined with calculations of the effects of screening[7] on these

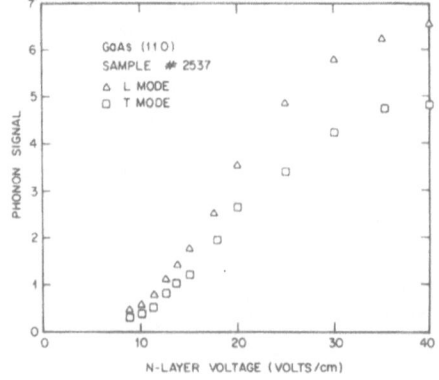

Fig. 6. Amplitude dependence
for a low mobility [110] sample.

scattering processes and corrections due to the finite aperture of
the detector (which allows detection of some off-symmetry phonons)
allows us to quantitatively describe electron-acoustic phonon
scattering in n-GaAs.

This quantitative description is valid only in low fields. At
high fields the phonon signals show unusual behavior. For low mobility ($\mu \sim$ 1000 cm^2/Vsec) samples the signals show a saturation
with field (see Fig. 6). No such saturation (at comparable field
strengths) is observed with metal heaters and appears to be a property of the n-layer heater. Figure 7 shows the current density J

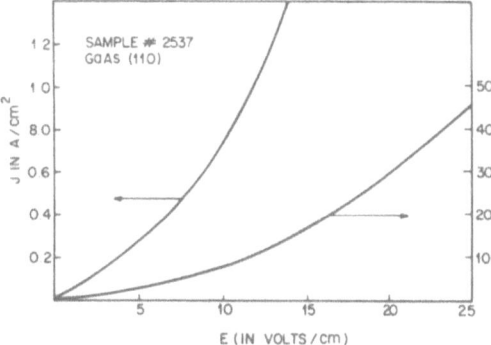

Fig. 7. Current density J vs.
electric field (E) for same
sample as in Fig. 6.

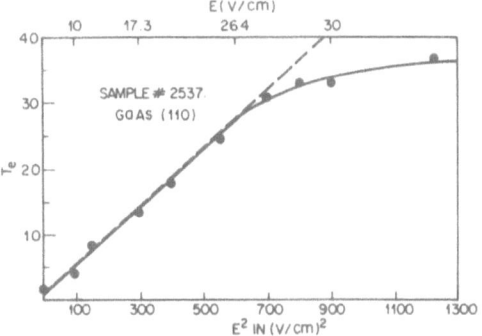

Fig. 8. Calculated electron
temperature (T_e) as a function
of E^2.

as a function of field (E) which yields the field dependence of the
mobility and, after some analysis, T_e as a function of E as shown
in Fig. 8. It is clear that T_e also saturates in the range 30 to
50 K. In this temperature range higher order phonon processes and
optic mode scattering become important and the rate of transfer of
energy to long wavelength, long lived acoustic phonons is no longer
effective. In higher mobility samples ($\mu(0) \gtrsim 5 \times 10^3$ cm^2 volts/
sec), however, growth of the T and L phonons occurs once $\mu E \gtrsim V_T$,
V_L where V_T and V_L are the transverse and longitudinal sound

velocities. Figure 9 shows some data in the [111] direction at a
low and two intermediate fields where $V_T < \mu E \lesssim V_L$. With increasing
field an initial increase of background "diffusive" heating occurs,
but for values of $\mu E \sim V_T$ the T pulse picks up and overtakes the L
mode. Fig. 10 shows a plot of the field dependence of the step
heights which shows that the L mode also grows once the field is
high enough. We believe, this phomonenon is related to "phonon
bunching effects" in the quantum regime ($q \sim 2 k_F$) once the carrier
velocity exceeds the velocity of sound. The quantum nature is
proved directly through data on a high mobility, high concentration
($\sim 5 \times 10^{17}$ per cm^3) layer in the [100] direction and a Sn tunnel
detector. For such a layer $2 k_F \sim 5 \times 10^6$ cm^{-1} and the phonon energy is very close to the Sn energy gap. Figures 11 and 12 show
the ballistic pulses and their amplitude dependence. The non-
linear growth of the T pulse in this direction is evident. The
tunnel junction data as a rule show no "diffusive" background
heating when compared to the bolometer data even on the same sample.

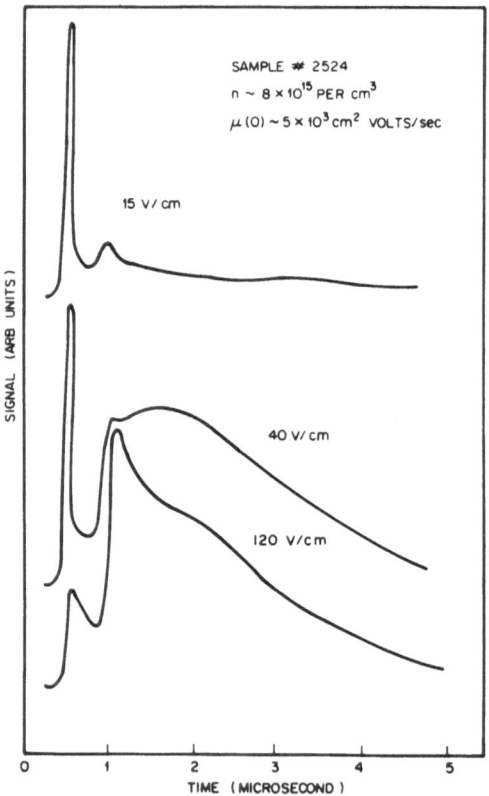

Fig. 9. Ballistic phonons for 3 fields. Direction [111]. Bolometer detector.

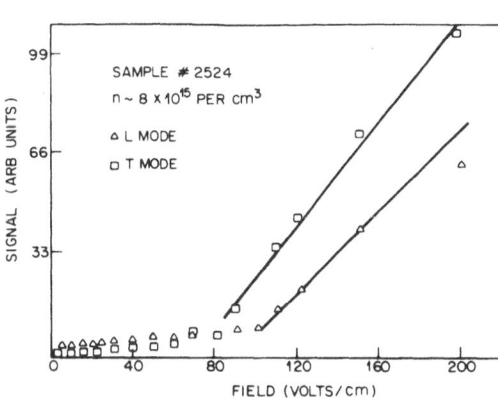

Fig. 10. Signal amplitude
vs. field for sample of Fig.
9.

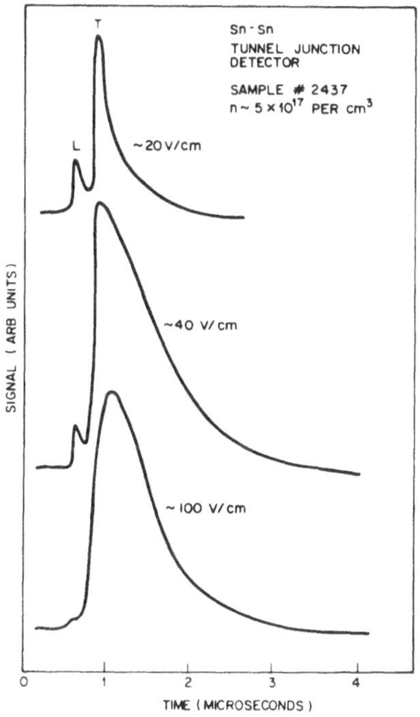

Fig. 11. Ballistic phonons in
[100]. Sn junction detector.

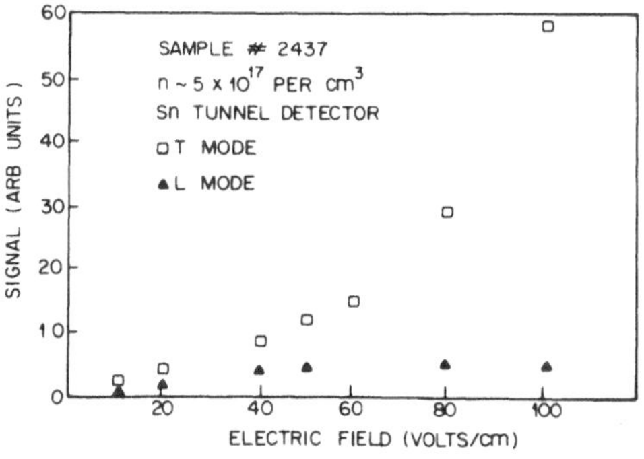

Fig. 12. Signal amplitude
vs. field for sample of Fig.
11.

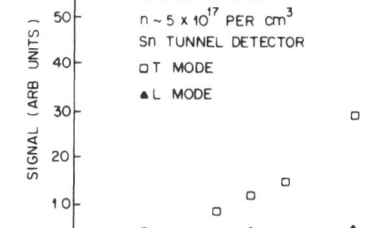

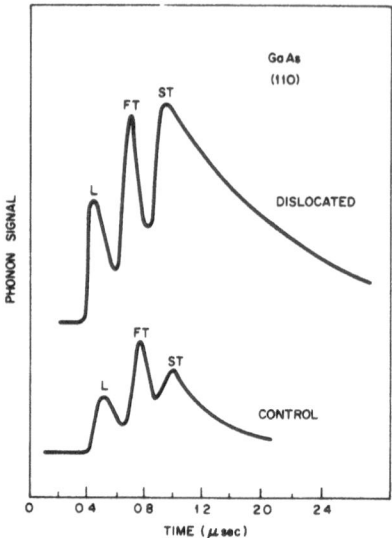

Fig. 13. Phonon emission from control and dislocated pn junctions.

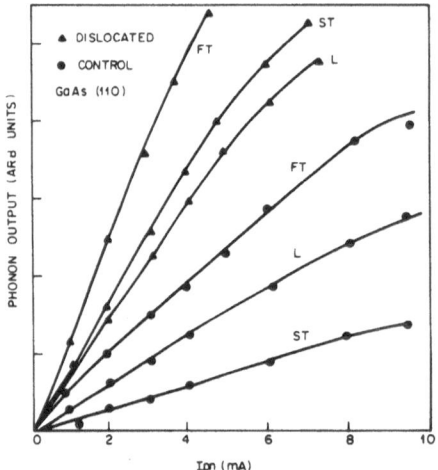

Fig. 14. Signal amplitude vs. junction current.

Besides the phonon generation work with n-epilayers, we have studied the transmission through layers of $Al_xGa_{1-x}As$ of varying Al composition and doped with Sn and Te donors. These measurements shed new light on the phenomenon of persistent photoconductivity centers[8] in the mixed crystal. The details are presented else-where[9] in the conference proceedings.

pn Junctions: We have made extensive studies of non-radiative
recombination in pn junctions at misfit dislocations, introduced
by the addition of small amounts of phosphorous to the growth melt
at 850°C. In Fig. 13 we present data for a control (no phosphorous
added) and a dislocated (GaAs$_{.95}$P$_{.05}$) junction grown simultaneously
on the semi-insulating GaAs substrate by the growth of bi-composi-
tional layers. It is clear from Fig. 13 that the phonon emission
is considerably enhanced in the dislocated junction, particularly
for the ST mode as shown quantitatively in Fig. 14 where we have
plotted the step heights. Measurements of the radiative output
show that it anticorrelates with the phonon generation. Measurements in other orientations ([111] and [100]) shows enhancement of
T modes in the dislocated junctions also. The overall effect is,
however, largest in the [110] presumably because the misfit dislocations occur in this direction.

Summary and Future Directions: In summary, I have briefly reviewed
recent experiments on phonon generation by and propagation through
Al$_x$Ga$_{1-x}$As epilayers and pn junctions. When combined with optical
and transport measurements, such studies can reveal detailed information on a variety of interesting scattering processes, the
symmetry and structure of anisotropic defect centers and the nature
of energy transfer in pn junctions. Many variants of the techniques
described and the nature of the physical system are possible and
many unanswered questions remain. The area of defects in epitaxial
layers itself is an extremely rich field in which phonon techniques
can play an extremely useful role, particularly in the absence of
spin resonance data. Phonon studies to probe the layered two dimensional electron gas[10] could also be profitably pursued. Detailed
understanding of "phonon bunching" phenomena in epitaxial layers and
possible Bragg diffraction of high frequency phonons by semiconductor
superlattices[11] are other examples of current interest. Clearly
much remains to be done in a developing field such as "phonon optics".

Acknowledgements: Finally, I would like to thank several of my
collaborators. I would expecially like to thank Maurice Chin,
Ralph Logan and Melvin Lax with whom most of the original work was
done and Harry White for the epilayer growth and sample preparation
in various ways. Recent close interactions with Horst Störmer,
Art Gossard and Willie Wiegmann is also acknowledged. I would also
like to thank Rosemarie Fulton for extensive computational aid.

REFERENCES

1. V. Narayanamurti, R. A. Logan, M. A. Chin and M. Lax, Sol. St.
    Elec. 21, 1295 (1978); Phys. Rev. Lett. 40, 63 (1978) and work
    to be submitted to Physical Review.
2. The earliest work in this area is due to K. Hubner and W.

Shockley, Phys. Rev. Lett. 4, 504 (1960). For a summary of subsequent DC work mainly in Si and Ge, see R. W. Keyes, Comments Sol. State. Phys. 6, 63 (1975). For pulsed experiments in Ge, see A. Zylberstejn, Phys. Rev. Lett. 19, 838 (1967); see also W. Reupert, K. Lassmann and P. deGroot in Proceedings of Nottingham Conference on Phonon Scattering in Solids (Plenum Press, New York) p. 315, 1976. For DC work in GaAs see R. S. Crandall, Sol. St. Commun. 1, 1109 (1968). We know of no work on time of flight spectra using epitaxial (n, p or pn) trans-ducers.

3.   M. G. Hollana, Phys. Rev. 134, A471 (1964).

4.   L. J. Challis and A. Ramdane, Proceedings of this conference.

5.   V. Narayanamurti, M. A. Chin and R. A. Logan, Appl. Phys. Lett. 33, 481 (1978).

6.   J. J. Krebs and G. H. Staus, Phys. Rev. B 15, 17 (1977) and 16, 971 (1979).

7.   M. Lax, private communication. Care has to be taken in calcu-lating the effects of finite wave vector ($q \sim 2$ k_F) and the effects of exchange and correlation.

8.   D. V. Lang and R. A. Logan, Phys. Rev. Lett. 39, 635 (1977).

9.   V. Narayanamurti, M. A. Chin and R. A. Logan, Proceedings of this conference.

10. R. Dingle, H. L. Stormer, A. C. Gossard and W. Wiegmann, Appl. Phys. Lett. 33, 665 (1978).

11. H. L. Stormer, M. A. Chin, A. C. Gossard and W. Wiegmann, private communication of work with V. Narayanamurti.

INTERACTION OF NON-EQUILIBRIUM, BALLISTIC PHONONS IN A HEAT PULSE

WITH THE ELECTRON-HOLE LIQUID IN Ge: THE PHONON WIND

J. C. Hensel and R. C. Dynes

Bell Laboratories
Murray Hill, New Jersey 07974

The discovery of the electron-hole (e-h) drop in Ge and subsequent investigation of its unique properties[1] have stimulated exciting new directions in phonon research. One remarkable property of the e-h drop is its mobility. In the past several years a number of experiments[2-4] have demonstrated that mobile e-h drops can be propelled by a flux of non-equilibrium phonons; this novel concept,[5] dubbed the "phonon wind", has been instrumental in explaining how the e-h droplet cloud, typically a mm in size, is produced. The mechanism underlying the phonon wind entails interesting phonon physics: (1) absorption of non-equilibrium phonons by the metallic e-h liquid; (2) the characteristics of ballistic phonon propagation in the Ge lattice, i.e. phonon focusing and low temperature phonon scattering. We have made a comprehensive investigation[4,6] of all aspects of the problem by means of a phonon transmission technique utilizing sharply collimated beams of ballistic phonons in a heat pulse. The following is a short synopsis.

ABSORPTION OF NON-EQUILIBRIUM PHONONS BY THE e-h LIQUID

The e-h liquid in Ge is a metal (a two component metal, in fact) and being a metal absorbs phonons by the scattering of quasi-particles, the scattering rate[7] $P(q)$ being proportional to the magnitude of the phonon wave vector $\vec{q}$. The phonon wind force derives from the crystal momentum $\hbar\vec{q}$ transferred to the e-h drop. We note several salient features of the scattering process: (1) the process cuts off at $|q| \simeq 2k_F$ where k_F is the Fermi wavevector (k_F for the e-h liquid spans only a small fraction of the Brillouin zone); (2) for electrons the $2k_F$ cutoff and, hence, the maximum value of $P(q)$ depends strongly upon the orientation of $\vec{q}$ with respect to the spheroidal electron Fermi surface (scattering by holes, having

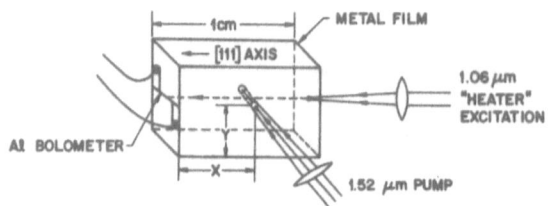

Fig. 1.   Experimental geometry.

smaller deformation potentials than electrons, can be ignored);
(3) also P(q) obeys certain selection rules - it vanishes for transverse acoustic (TA) phonons [but not for longitudinal acoustic (LA)
phonons] for $\vec{q}$ along any symmetry axis of the spheroidal Fermi surface. It follows then that the absorption of phonons will be highly
anisotropic with respect to the orientation of the Fermi surface and
will exhibit a strong dependence on $|\vec{q}|$, effects that we directly
observe in heat pulse experiments.

The experiments were done in liquid HeII at $\sim$ 1.8 K; the geometry is shown in Fig. 1. Phonons are generated by a Q-switched
Nd:YAG laser beam focused on an evaporated constantan film (on the
right) and detected by a granular Al superconducting bolometer (on
the left). Temporal and spatial resolution are defined by the pulse
width ($\sim$ 150 nsec) and bolometer dimensions (a stripe 0.1 mm ×
1.0 mm). A cloud of e-h drops was interposed in the phonon beam by
means of volume excitation with a focused beam of a He-Ne laser
operating at 1.52 µm wavelength.

The large anisotropy in absorption for LA phonons is illustrated
in Fig. 2. The data represent a spatial absorption profile of the
e-h drop cloud obtained by translating the 1.52 µm beam in the y-
direction (see Fig. 1). The absorption profile contains some information about the cloud size and shape, but the main point we wish to
make is the marked difference in LA absorption - approximately one
order of magnitude - for the case $\vec{q}\|[111]$ versus $\vec{q}\|[110]$ (and
$\vec{q}\|[001]$). For TA phonons, on the other hand, absorption is essen-
tially zero for $\vec{q}\|[111]$, but finite for other directions. Qualita-
tively, the explanation resides with the Ge band structure, i.e.
electron ellipsoids of an extremely elongated shape oriented along
the star of [111]. Owing to the extreme anisotropy of the Fermi
surface the LA absorption peaks strongly when $2k_F$ is along [111].
For TA modes absorption for $\vec{q}\|[111]$ is essentially forbidden by
selection rules.

An example of the dependence of phonon absorption on $|q|$ is shown
in Fig. 3. Inasmuch as the phonons in the heat pulse are distributed
according to a Planck distribution, the energy of the peak is propor-
tional to the heater temperature T_h, i.e. $\hbar\omega_{max} = \hbar v_s q_{max} \approx 2.82 k_B T_h$,
so that the overlap of the distribution function with P(q) is

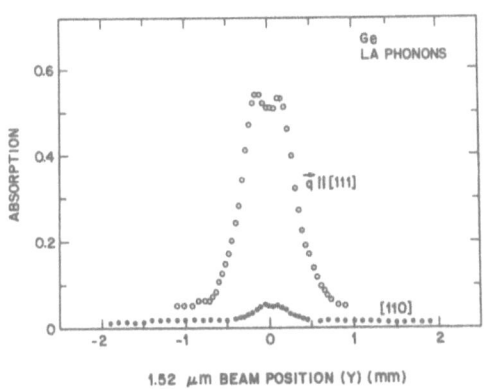

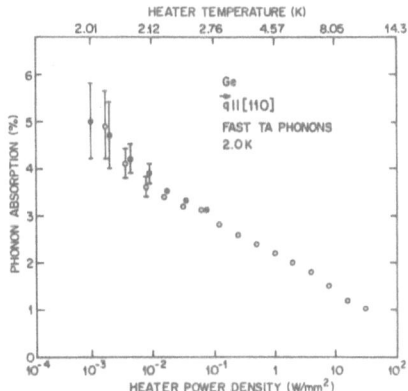

Fig. 2. Absorption profiles.   Fig. 3. Absorption q-dependence.

adjustable by varying the heater excitation power. The decrease in absorption in Fig. 3 as T_h goes from 2 K (ambient) to $\sim$ 14 K demonstrates the existence of the $2k_F$ cutoff in $P(q)$. For this temperature range the peak of the distribution actually lies beyond the $2k_F$ cutoff; a maximum overlap would occur at $T_h \sim 1.4$ K.

The foregoing measurements were done at moderate to very low phonon flux densities; if the heater excitation (referring to Fig. 1) is raised to the highest levels available ($\sim$ 200 watt optical pulse excitation) it is possible to observe movement of drops, a direct manifestation of the phonon wind force. By varying the beam position coordinate "x" and registering the time delay of the droplets arriving at the bolometer, we made a time-of-flight measurement of the droplet drift velocity v_d. From the value obtained, $v_d \approx 4 \times 10^4$ cm/sec, we estimate a droplet damping time $\tau_p \approx 1.2 \times 10^{-9}$ sec at 1.8 K in reasonable accord with theory.[8] This agreement leaves little doubt that the electron-phonon interaction is the predominant mechanism for damping of droplet motion.

To recapitulate we see that the phonon wind force will depend, via the absorption mechanism, upon both the magnitude and direction of $\vec{q}$. In addition, the phonon wind force must also depend upon the characteristics of the phonon flux. This we discuss next.

PHONON FOCUSING

It is well known that elastic anisotropy of a solid imparts directional characteristics to ballistic phonon propagation - an effect called phonon focusing.[9,10] Such directional behavior will obviously also be present in the phonon wind.

We have directly measured the anisotropy of phonon focusing in a transmission experiment employing the geometry shown in the inset in Fig. 4. The sample in the shape of a hemi-cylinder can be rota-

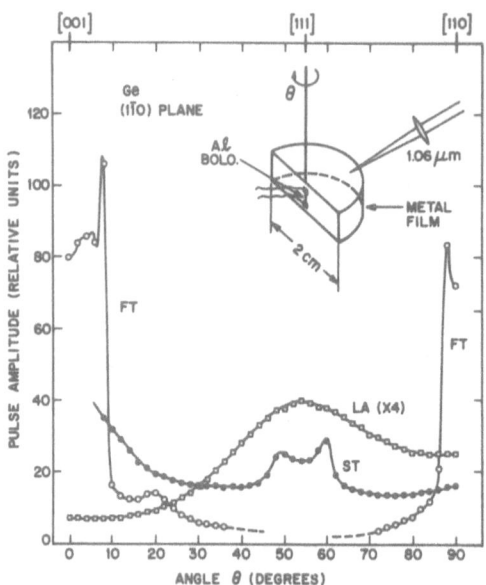

Fig. 4.   Anisotropy of phonon focusing.

ted about its axis by means of a shaft connected to a goniometer at
the top of the cryostat.

An angular distribution in the (1$\bar{1}$0) plane is shown in Fig. 4
for the pulse heights of the three principal acoustic phonon modes:
longitudinal (L), fast transverse (FT) and slow transverse (ST).
The L mode varies smoothly from a minimum along [001] (a defocusing
direction) to a maximum along [111] (a focusing direction). On the
other hand, the behavior of the ST mode is quite extraordinary as one
nears the [001] and [110] axes, where in less than 1° it jumps by an
order of magnitude in intensity giving rise to conical phonon beams
of width ± 7.0° about [001] and ± 2.0° about [110]. The fact that a
considerable fraction of the total TA phonon flux is concentrated
within these cones (particularly [001]) will certainly have ramifica-
tions in the phonon wind pattern. A curious feature in the FT angular distribution is the sharp spikes at the edges of the cones.
These are focusing singularities deriving from critical points
(points of zero curvature) on the constant frequency surface; and
their width is determined by the experimental angular resolution.

The behavior of the ST is unexceptional except for the unusual
structure (at ± 5.5°) about the [111] axis. This represents a
phenomenon called conical refraction[9] which arises from topological
anomalies in the constant frequency surface due to the degeneracy of
the transverse modes in this direction.

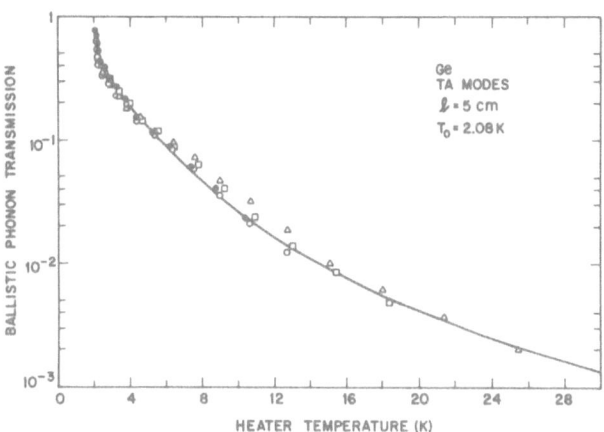

Fig. 5.   Phonon transmission in [111] direction.

MASS-DEFECT SCATTERING OF PHONONS

The energy distribution in a packet of ballistic phonons propagating in Ge will change with distance due to frequency dependent mass-defect (isotope) scattering, i.e. the crystal acts as a low-pass filter, as it were. Phonon scattering can be directly measured by our transmission technique.

The set-up is again as shown in Fig. 1 (without drops), the sample length being increased to 5 cm. TA phonon transmission versus heater temperature T_h is shown in Fig. 5. There is a pronounced fall-off in transmission (approx. 2 1/2 orders of magnitude) in going from 2 K (ambient) to a maximum T_h of $\sim$ 25 K.

The interpretation of these data is straightforward. The phonon transmission can be written

$$T = \int A(\omega) U(\omega, T_h) d\omega \bigg/ \int U(\omega, T_h) d\omega, \tag{1}$$

where $U(\omega, T_h)$ is the phonon distribution function (essentially a Planck distribution) and $A(\omega)$ is an attenuation factor expressed by

$$A(\omega) = \exp\left[-\ell/v_s \tau(\omega)\right] . \tag{2}$$

Here ℓ is the path length, v_s is the sound velocity (for the particular mode in question), and $\tau(\omega)$ is the scattering time. For isotope scattering (Rayleigh scattering) $\tau_{iso}^{-1} = A\omega^4$ where A is a universal constant (independent of phonon polarization and direction) equal to 2.5×10^{-44} sec^3 for Ge. The solid curve in Fig. 5 represents the computer evaluation of Eq. (1) with no adjustable parameters and is seen to fit the data nicely.

Two points need to be made: First, there is a sharp cutoff of high frequency phonons by Rayleigh scattering, as exemplified by the dropoff in transmission in Fig. 5, and this implies a marked diminution of the phonon wind force with distance. This cutoff implies an effective limit to the range of <u>relevant</u> LA phonons ($\lambda_{LA} \sim 1$ mm) but not TA phonons ($\lambda_{TA} \sim 100$ cm). Second, on a purely phonon matter, τ_{iso}^{-1} goes as ω^4 - not q^4 as sometimes mistakenly assumed - this being based on both theoretical[11] and experimental grounds (by comparing, for instance, LA and TA transmission data).

In conclusion, then, we have characterized the main factors underlying the phonon wind, i.e. phonon absorption and phonon propagation. It is our belief that a detailed picture can only be achieved by studies directly on the phonons themselves. On the basis of what we have learned we can understand certain structure in the shape of the e-h droplet cloud in Ge - [111] bulges produced by strong absorption of LA phonons augmented by focusing and [001] "flares" produced by sharply focused TA phonons - as seen in cloud photographs of Greenstein and Wolfe.[12]

We thank F. C. Unterwald and J. P. Garno for technical assistance and T. M. Rice and C. Herring for discussions of theory.

REFERENCES

1. For a review see J. C. Hensel, T. G. Phillips and G. A. Thomas, Solid State Physics <u>32</u>, 87 (1977).
2. V. S. Bagaev, L. V. Keldysh, N. N. Sibeldin and V. A. Tsvetkov, Zh. Eksp. Teor. Fiz. <u>70</u>, 702 (1976) [Sov. Phys. JETP <u>43</u>, 362 (1976).
3. J. M. Worlock and J. Doehler, <u>Proc. Int. Conf. on Lattice Dynamics</u>, 1977 (Flammarion, Paris, 1977), p. 234.
4. J. C. Hensel and R. C. Dynes, Phys. Rev. Lett. <u>39</u>, 969 (1977).
5. L. V. Keldysh, Pis'ma Zh. Eksp. Teor. Fiz. <u>23</u>, 100 (1976) [JETP Lett. <u>23</u>, 86 (1976)].
6. J. C. Hensel and R. C. Dynes, <u>Physics of Semiconductors, 1978,</u> B. L. H. Wilson, ed. (Inst. of Physics, London, 1979), p. 371.
7. G. L. Bir and G. E. Pikus, <u>Symmetry and Strain-Induced Effects in Semiconductors</u>, (Wiley, New York, 1974).
8. L. V. Keldysh and S. G. Tikhodeev, Pis'ma Zh. Eksp. Teor. Fiz. <u>21</u>, 582 (1975) [JETP Lett. <u>21</u>, 273 (1975); D. S. Pan, D. L. Smith and T. C. McGill, Phys. Rev. B<u>17</u>, 3284 (1978); A. Manoliu and C. Kittel, Solid State Commun. <u>21</u>, 641 (1977); R. S. Markiewicz, Phys. Stat. Sol. (b) <u>90</u>, 585 (1978); L. Detweiler and T. M. Rice (unpublished).
9. G. F. Miller and M. J. P. Musgrave, Proc. Roy. Soc. (London) A<u>236</u>, 352 (1956).
10. B. Taylor, H. J. Maris and C. Elbaum, Phys. Rev. B<u>3</u>, 1462 (1971).
11. C. Herring (unpublished work).
12. M. Greenstein and J. P. Wolfe, Phys. Rev. Lett. <u>41</u>, 715 (1978).

EFFECTS OF MAGNETIC FIELD AND UNIAXIAL STRESS ON THERMAL PHONON SCATTERING IN n-Ge

A. Kobayashi and K. Suzuki

Department of Electrical Engineering, Waseda University

Shinjuku, Tokyo 160, Japan

The ground state of Group-V donors in Ge consists of a singlet and a triplet. The former lies lower in energy. Magnetic fields and uniaxial stresses destroy the equivalence of the four conduction band valleys except when they are applied along a [100] axis, and consequently lift completely or partially the degeneracy of the triplet. Therefore these fields are expected to have a considerable effect on the thermal phonon scattering by donors in Ge. Halbo calculated the effect of magnetic fields along a [100] in Sb- and As-doped Ge (1), and that of the [111] stresses was reported in Ref. 2. In this paper we calculate effects of magnetic fields along a [111] axis and of stresses along the [111] and [110] axes in Sb- and As-doped Ge.

First we study effects of magnetic fields. Calculations have been performed under the following assumptions and approximations. (a) All donors are isolated in the concentration under study. (b) The change of the donor ground state by magnetic fields are treated within the model of Lee et al. (3) (c) All envelope functions are identical and have a form $(\pi a^{*3})^{-1/2}\exp(-r/a^{*})$. Magnetic fields give rise to (i) the shrinkage of the wavefunctions or a decrease of a^{*} which increases the scattering strength for high frequency phonons, (ii) the change of the energy difference between the ground levels which alters the population of each level and the resonant frequencies, and (iii) the change in the relative contribution of the indidual valleys to the donor ground state, which reduces strongly some of matrix elements of the donor-phonon interaction. The relative importance of these three depends on the species of donors, the magnitude of fields, and the phonon frequency or the temperature. In Fig. 1 is shown the calculated thermal conductivity K normalized by K_0 (a value in the absence of field) for Sb and As donors together with the experimental data by Halbo and Sladek (4). In calcula-

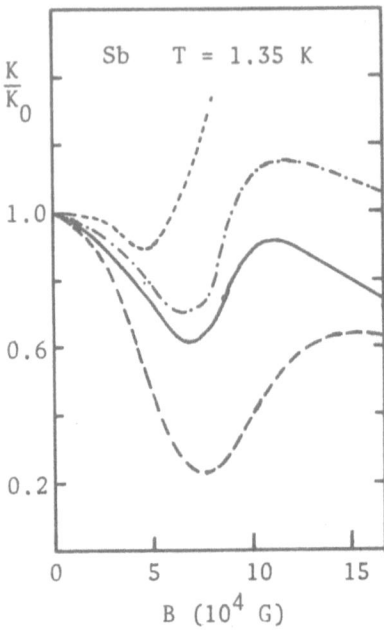

Fig. 1. The dotted line denotes
the experimental data. Other
three lines denote calculated
curves. The dot-dashed line
corresponds to the case where
a* is fixed to a value at zero
field, and the dashed line the
case where the relative contribution of the individual valleys
to a donor state is fixed to
that at zero field.

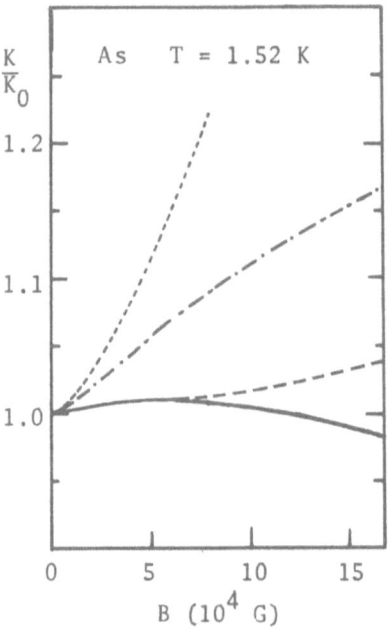

tions we have used 4Δ = 0.32 meV (Sb), 4.23 meV (As), Ξ_u = 16 eV,
L = 0.3 cm, and N = 2 x 10^{16} cm^{-3} (Sb), 4 x 10^{16} cm^{-3} (As) where the
notation and other parameters are the same as in Ref. 5. Effects of
magnetic fields along a [110] axis are expected to be similar to
those of the [111] magnetic field, though the level structures of
the ground state in these two cases are different.

The level structures of the donor ground state under the [111]
and [110] tensile stresses resemble   those under magnetic fields
along the [111] and [110] axes, respectively, though they are different in detail. We have neglected the stress dependence of a*.
Small stresses decrease K. Large [111] tensile and [110] stresses
quench partially the donor-phonon interaction and therefore increase
K, finally giving a **limiting value which is larger than K at zero**
stress,in contrast with the case of magnetic fields in which at high
fields the shrinkage of the wavefunctions or a diminution of a* gives
rise to a decrease of K. Note that compression and tension along a
[110] axis have the same effects. On the other hand, a large [111]
compression quenches completely the phonon scattering by individual
donors and consequently gives K of a pure Ge. For the [111] and
[110] tensile stresses of the same magnitude, the latter has a larger
effect than the former.

The calculated magneto- and piezo-thermal conductivity cannot
explain well the experiments (4, 6, 7). A part of discrepancies will
arise from the fact that the donor concentrations used in the experiments are not so low. We have neglected a modification of the donor
ground state due to internal local fields. This will lead to an un-
derestimation of the phonon scattering at zero stress and at zero
field. It is desirable to carry out an experiment in samples with
the lower donor concentration.

Details of this work will be published elsewhere.

ACKNOWLEDGEMENT
We wish to thank K. Tateishi for his assistance in numerical
calculations.

REFERENCES
1. L. Halbo, Proc. 2nd Int. Conf. on Phonon Scattering in Solids,
   Nottingham, 1975, P. 346.
2. K. Suzuki, phys. stat. sol. (b) 78, K77 (1976).
3. N. Lee et al., J. Phys. Chem. Solids, 34, 1817 (1973).
4. L. Halbo and R. J. Sladek, Proc. 10th Int. Conf. on Physics of
   Semiconductors, Cambridge, Mass., 1970, P. 826.

5. K. Suzuki and N. Mikoshiba, J. Phys. Soc. Jpn., 31, 186 (1971).
6. R. W. Keyes and R. J. Sladek, Phys. Rev. 125, 478 (1962).
7. R. J. Sladek, Proc. Int. Conf. on Physics of Semiconductors,
   Exeter, 1962, P. 35.

PHONON SCATTERING IN N-TYPE SILICON

B. B. Touami and D. V. Osborne

School of Mathematics and Physics
University of East Anglia
Norwich, England

The scattering of phonons by imperfections in solids has been extensively studied through measurements of thermal conductivity. At low temperatures, the intrinsic phonon scattering length often becomes very long, and may easily exceed the size of the specimen. In this case boundary scattering contributes substantially to the observed thermal resistance, and it may become difficult to analyse the results in such a way as to deduce reliably the comparatively small scattering cross-section of the imperfections.

The ballistic transmission of heat pulses through crystalline solids at low temperatures occurs because of the very fact that phonon scattering processes are weak. There is, however, some scattering, and it is possible to detect the scattered phonons separately from the directly propagated pulse. We here report preliminary measurements of this type and believe that they can in principle provide rather direct information about the single scattering cross-section.

In order to explore the technique, we have examined a series of three phosphorus-doped silicon crystals having phosphorus concentrations of $1 \cdot 8 \times 10^{16}$ cm^{-3} ($0 \cdot 34$ Ωcm), $2 \cdot 5 \times 10^{15}$ cm^{-3} ($2 \cdot 0$ Ωcm) and 8×10^{12} cm^{-3} (620 Ωcm). At our ambient temperature of 2K, the donor atoms are not ionised. The outer electron of each phosphorus atom has a wave function extending over a radius of about 15 Å. Phonon scattering in such a crystal will be due to the outermost electron or each donor atom and to the differing masses of the atomic species present (92% ^{28}Si, 5% ^{29}Si, 3% ^{30}Si, < 10^{-4}% ^{31}P).

In order to measure the scattering we have used the geometry shown in figure 1. The heater H is of constantan and the detector

D is a superconducting tin-silver bolometer working at its transition
temperature. If a heat pulse be sent out from H, the signal at the
detector contains contributions from scattering and from reflections
as shown in figure 1. The form of the signal is shown in figure 2,
where S is the scattered signal and R are the reflected signals for
various modes. The scattered signal starts at a time appropriate
for transverse waves. We have measured the height S_m of the maximum
of S as a function of transient heater temperature T_H. S_m depends
on (a) geometrical factors and anisotropy, (b) the scattering cross-
section of the imperfection at which scattering occurred, (c) the
probability that phonons will be further scattered out of the path
HSD, and (d) the heater power W partly because a higher power
produces more phonons and partly because a higher power gives rise
to a higher heater temperature T_H, and hence to phonons of higher
energy which are more strongly scattered.

The measured S_m for the three Si(P) specimens described above
are shown in figure 3 as functions of T_H. We see at once that the
addition of minute quantities of phosphorus makes a very substantial
difference to the scattered signal, out of all proportion to any
possible effect of the mass anomaly of the added atoms. We therefore
believe that for the two impure specimens, most of the observed
scattering is due to the bound electron attached to each P atom.

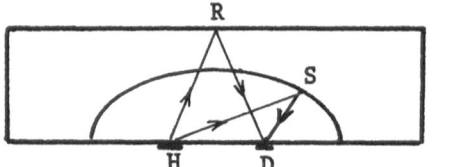

H = heater
D = detector
R = reflection
S = scattering

Figure 1 Propagation in silicon crystal

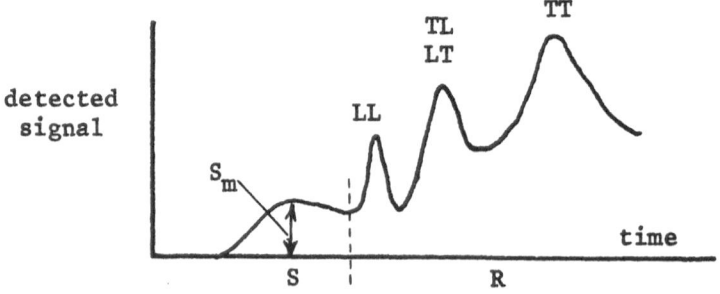

Figure 2 Typical signal

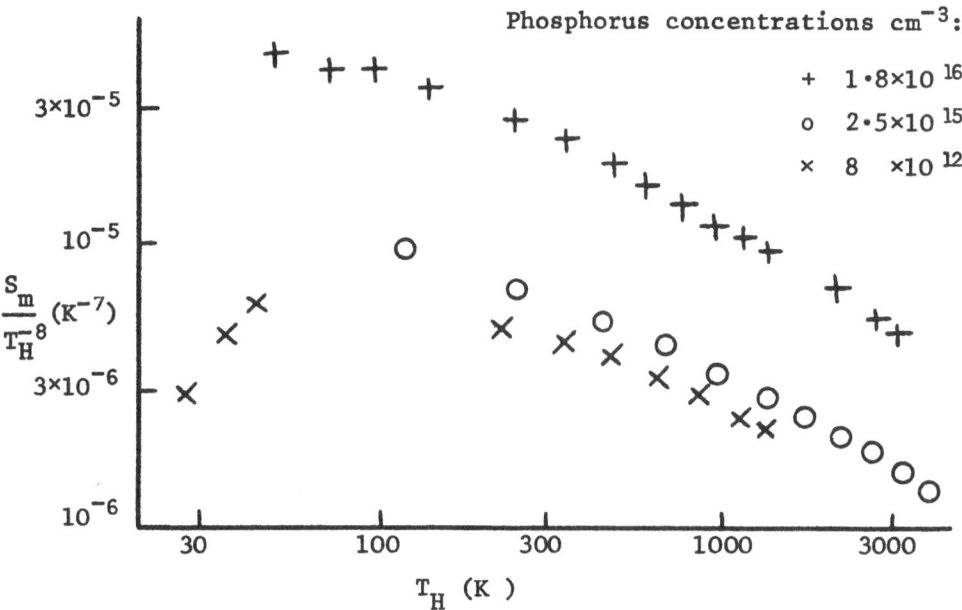

Figure 3   Scattered signal as a function of T_H

The comparison of experiment with theory is not yet quantitative, partly because insufficient measurements have been taken, and partly because detailed calculations allowing for the anisotropy of the silicon have not yet been completed. We have however used a simplified model assuming a single scattering process as in figure 1. We have allowed for loss of phonons due to further scattering out of the path HSD, but have not so far included additional phonons traversing from H to D by multiplying scattered paths which do not reach the ellipsoid on which S lies.

The application of this simple model leads to a scattering length of 42 cm for 7K transverse phonons in undoped silicon, in rough agreement with an estimate of 53 cm (Holland 1963) derived from thermal conductivity measurements. The scattering lengths for similar phonons in the doped specimens appear to be about 30 cm for a phosphorus concentration of $2 \cdot 5 \times 10^{15}$ cm^{-3} and 10 cm for a concentration of $1 \cdot 8 \times 10^{16}$ cm^{-3}. These lengths are not yet wholly dependable, since the model does not give a very good fit to the upper curves of figure 3. Experimental and theoretical work is continuing.

REFERENCE

Holland, M. G., Phys. Rev. 132 2461 (1963).
****

DISCUSSION

J. D. N. Cheeke: Were these heater temperatures calculated?

D. V. Osborne: These heater temperatures were calculated in the same manner as Weis wrote up in some detail some years ago, using the power and some mismatch and the emissivity.

J. D. N. Cheeke: There is the possible danger of getting back-scattering into the heater, so the heater temperature if you have a lot of defects would be hotter than theory would predict.

D. V. Osborne: That is certainly possible.

J. D. N. Cheeke: Another danger is that if the heater film is very thin, then the theoretically calculated temperature might not be appropriate.

D. V. Osborne: All I can rely on is the fact that Weis got reason-able agreement of temperature with heat power in the same experi-ments, and ours are not so different in general character of heater and substrate as his, so I think the temperatures are really not unreliable.

ULTRASONIC ATTENUATION IN p-InSb AT LOW TEMPERATURES

J.D.N. Cheeke and G. Madore

Département de physique,
Université de Sherbrooke,
Sherbrooke, Qué, J1K 2R1

Previous work on ultrasonic attenuation in p-type semiconductors at low temperatures has shown good agreement with theory for localised carriers while few results are available for the non-localised limit. Effects similar to those observed for two level systems have been seen and qualitatively explained in the former case. In the present work, we present preliminary results on Ge doped p-InSb where it is possible to study both regimes at relatively small impurity concentrations by use of the magnetic field dependence of the effective Bohr radius a*.

InSb is of the zinc blende structure and at low temperatures we are here concerned with states near the top of the valence band; the four fold degenerate Γ_8 level is split by Δ_0 ($\sim$ 0.8 eV) from the doubly degenerate Γ_7 ($m_3 \sim 0,11\ m_o$) level at $k = 0$. For $k \neq 0$, Γ_8 splits into two doubly degenerate bands (heavy holes, $m_1 \sim 0,4\ m_o$, and light holes $m_2 \sim 0,016\ m_o$). The traditional view of p-InSb of Hilsum and Rose-Innes (HRI)[1] is that for $p < 10^{14}$ cm^{-3} electrical conduction is by hopping (R_H undefined, $\Delta\rho/\rho \sim \rho_0^3 H^2$) [2] for $10^{14} < p < 10^{17}$ there is a progressively broadening impurity band (IB) (Max in R_H(T), low T saturation of ρ) and for $p > 10^{17}$ the IB merges into the valence band. For each concentration, one can in principle identify ε_1, ε_2 and ε_3 activation energy regimes for electrical conductivity as a function of temperature, corresponding to hopping, IB and ionised impurities. More recent work confirms this general picture as rather large values of a* $\sim$ 100–300 Å are found [3], as well as three similar regimes in the piezoresistance [4].

The present experiments were carried out in a superconducting solenoid using Matec 6600 and 2470 equipment and a PAR 164 gated

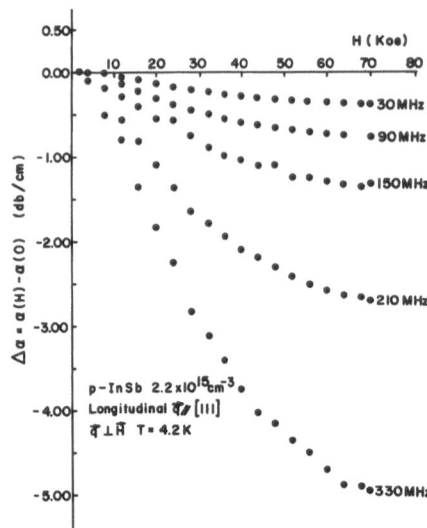

Fig. 1:Attenuation vs perpendicular magnetic field for the 2.10^{15}
cm^{-3} specimen. Similar results were
obtained for parallel fields.
Again, the general trend was similar but with proportionally smaller variations for the 10^{14} specimen, while no field dependence was
observed to within $\pm$ 0,2 db for
the 8.10^{16} specimen.

boxcar integrator. Quartz or $LiNbO_3$ transducers were bonded to the
crystals by NONAQ: the results were quite reproducible and there
was no evidence of transducer-induced dislocation effects at these
frequencies. All of the measured attenuation values were found to
be amplitude independent. Three specimens with concentrations 10^{14},
2.10^{15} and 8.10^{16} cm^{-3} were measured; these should correspond roughly to partially localised, IB, and almost degenerate carriers according to the HRI scheme. Magnetic field results are shown for the
2.10^{15} specimen in Fig. (1). The electrical conductivity of the specimen of Fig. 1 was found to follow a very similar variation in
field, and it has a rather small thermal activation energy of
$\varepsilon_2 \sim 0.22$ meV between 2 and 4 K. The temperature dependence of the
same specimen is shown in Fig. 2; again, similar results were obtained for the 10^{14} specimen, and we find $T_{max} \sim \omega$.

The peak near 10 K is almost certainly a relaxation peak due
to the localised acceptor ions, as has been observed by many other
workers. The origin of the magnetic field dependence does not seem
to be unambiguously determined, as it could in principle be due to
carrier freezout or to the Zeeman effect of the localised states.
The HRI picture would tend to suggest a freezout interpretation,
which is consistent with both α and σ depending linearly on a decreasing carrier concentration. Clustering of impurities or very weak
impurity banding could then account for the small residual effect
seen at $10^{14} cm^{-3}$. However, in view of the large mean impurity separation (1000 Å for $N \sim 10^{14} cm^{-3}$) the existence of an IB would seem
doubtful for an uncompensated specimen; ours is certainly compensa-
ted but we do not know the mean impurity-impurity separation.

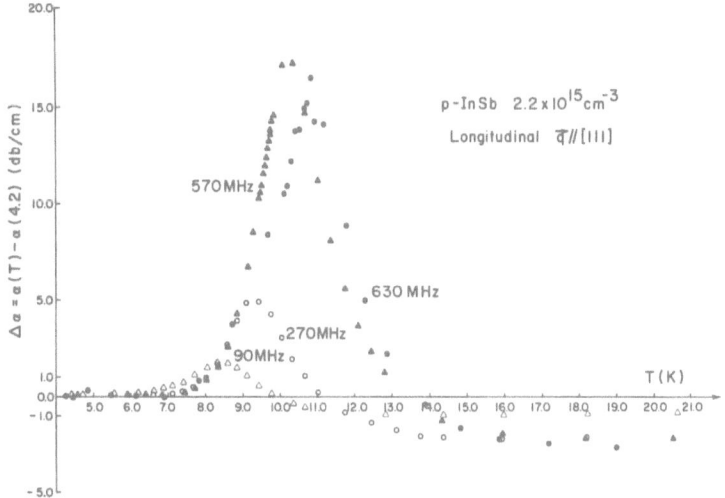

FIGURE 2

Further, Hall effect and resistivity results are needed to settle
these questions as well as concurrent resistivity and attenuation
measurements in both parallel and perpendicular fields.

The results would also be compatible with a resonance interaction with localised states, diminished by the Zeeman splitting in
magnetic fields. While the concentration dependence is compatible
with this interpretation, further work is needed on the frequency,
temperature and power dependence at higher frequencies in order to
verify this hypothesis.

This work was supported by the National Research Council of
Canada. We wish to thank F.R. Ladan for the crystals and J.-P.
Maneval for many fruitful discussions.

REFERENCES

1. C. Hilsum and A.C. Rose-Innes, "Semiconducting III-V Compounds",
        Pergamon, Oxford (1961).
2. N. Mikoshiba, Rev. Mod. Phys., 40, 833 (1968).
3. E.M. Gershenzon, I.N. Kurilenko and L.B. Litvak-Gorskaya, Sov.
        Phys. Semicond., 8, 768 (1974).
4. D.I. Aladashvili, V.V. Galavanov and S.A. Obukhov, Sov. Phys.
        Semicond., 7, 1349 (1974).

                          ****

DISCUSSION

Voice: On the low temperature side of the relaxation peak the
attenuation is higher than on the high temperature side. Do you
have any explanation for this?

P. J. King: There are several systems, in fact, where this is a
normal event. If you take a typical expression for a relaxation
peak containing $1/T$ times the usual sort of expression $\omega^2\tau/(1 +
\omega^2\tau^2)$, if τ^{-1} contains something like a direct process AT then you
get a finite attenuation in the low temperature limit, so there are
many systems where the attenuation is higher on the low temperature
side.

ULTRASONIC HOLE-BURNING OF ACCEPTORS IN GERMANIUM

H. Tokumoto, T. Ishiguro, and K. Kajimura

Electrotechnical Laboratory
Tsukuba Research Center, Sakuramura
Ibaragi 300-31, Japan

The ground state of lightly doped acceptors in Ge can be expressed by randomly distributed two-level system. The ultrasonic waves are absorbed resonantly associated with hole transitions within the ground state.[1] The relaxation of the excited holes to the thermal equilibrium is characterized by the longitudinal relaxation time T_1 and the transverse relaxation time T_2. The relaxation times were measured by the ultrasonic hole-burning technique in Ge sample doped with Ga of $3.5 \times 10^{15} cm^{-3}$ at liquid helium temperatures.

Two ultrasonic pulses with duration of ~ 0.5 μsec were used: A preceeding pulse ("saturating pulse") with frequency f_1 has acoustic intensity enough to saturate the two-level system and a following pulse ("probing pulse") with frequency f_2 has weak intensity which does not alter the population of the levels appreciably. For the measurement of T_1, the attenuation of the probing pulse was measured as a function of delay time τ from the saturating pulse for $f_1 = f_2$. To measure T_2, the attenuation of the probing pulse was measured as a function of f_2 at τ≈0.

Figure 1 shows the temperature dependence of the derived T_1 without a magnetic field. The linear dependence of T_1 to T^{-1} indicates that the direct process is dominant for the longitudinal relaxation.

Figure 2 (a) shows the attenuation of the probing pulse ($f_2 = 1.008$ GHz) with the fast-transverse mode as a function of f_1 at 1.4 K without the magnetic field. A fairly broad minimum is seen around $f_1 = f_2$ in the absorption line, which is called a "hole" burned by the saturating pulse. The full-width at half-minimum

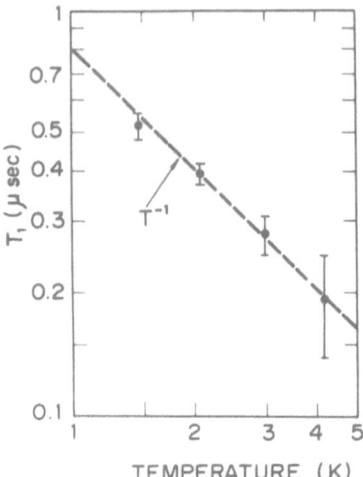

Fig. 1.   Longitudinal relaxation time T_1 as a function of
           temperature T.   Longitudiual waves of 1.022 GHz were
           propagated along the [001] direction.

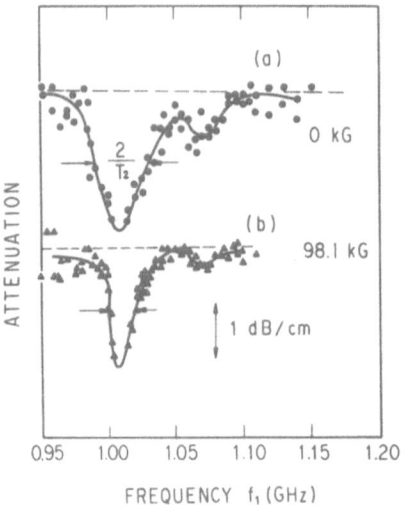

Fig. 2.   Attenuation of the probing pulse of 1.008 GHz as a
           function of the saturating pulse frequency at 1.4 K
           under H=0 kG (a) and H=98.1 kG (b).   Fast-transverse
           waves were propagated along the [110] direction.

$(2/T_2)$ of the hole is ~ 40 MHz ($T_2 \approx 8$ nsec), which was little affected by the change of the saturating pulse intensity. The width is appreciably greater than the spectrum width of the saturating pulse (≈ 4 MHz) and is much smaller than T_1 (≈ 0.2 µsec). By using the magnetoacoustic resonance absorption peaks[1], we obtained the burned hole under strong magnetic fields H as shown in Fig. 2(b). The width $(2/T_2)$ at 98 kG is ~ 19 MHz ($T_2 \approx 17$ nsec), which is very close to the calculated value (≈ 20 nsec) neglecting the interaction among acceptors.[1] These results suggest that the interaction among acceptors is important in the relaxation process in the absence of the magnetic field.

There are several contributions to the width of the hole burned by the saturating pulse; spectral diffusion, the relaxation process $(2T_1)^{-1}$, the spectrum width of the saturating pulse, and the strain field of the saturating pulse. The possibilities except the spectral diffusion are ruled out by the experimental facts. The spectral diffusion takes place through (1) the exchange interaction, (2) the magnetic dipolar interaction,[2] and (3) the elastic dipolar interaction.[3] The width of the spectral diffusion is proportional to $|J_{ij}|$; $|J_{ij}|$ is the exchange energy for (1) and it is proportional to the concentration of acceptors for (2) and (3). For the present sample, the contributions of (2) and (3) are too small to explain the results because of low concentration. On the other hand, the exchange energy of a nearest neighbor acceptor pair leads to the diffusion width of ~ 10 MHz, which is in good agreement with the experimental results.

Under strong magnetic fields, the exchange interaction is suppressed remarkably due to the magnetic field-induced shrinkage of wave functions and therefore we can ignore the contribution of the spectral diffusion to the transverse relaxation.

We wish to acknowledge helpful conversations with Professor K. Suzuki and Professor L. J. Challis.

REFERENCES:

1. H. Tokumoto and T. Ishiguro, Magnetoacoustic resonance study in Ga doped Ge, Phys. Rev. B 15:2099 (1977)
2. J. R. Klauder and P. W. Anderson, Spectral diffusion theory in spin resonance experiments, Phys. Rev. 125:912 (1962)
3. J. L. Black and B. I. Halperin, Spectral diffusion, phonon echoes, and saturation recovery in glasses at low temperatures, Phys. Rev. B 16:2879 (1977)

****

DISCUSSION

M. Wagner: There should also be an interaction between the two-level systems which should come from the exchange of virtual phonons. Did you consider this also?

T. Ishiguro: Yes, but the effect if very small because our concentration is very low.

L. J. Challis: Is the exchange of virtual phonons the same as the elastic dipole-dipole interaction?

M. Wagner: Yes, I think it's the same.

L. J. Challis: You have now got agreement between your estimate for T_2 for acceptor-acceptor interaction with a figure of around about 20 MHz. Is that consistent with Dr. Lassmann's suggestion that the acceptor-acceptor interaction is not responsible for his broad distribution? He is observing widths of 1 GHz, which is very much bigger than your result of 20 MHz.

T. Ishiguro: I think this is very different.

PHONON SCATTERING AT ACCEPTORS WITH

Γ_8 GROUND STATES IN SEMICONDUCTORS

Ernst Sigmund and Kurt Lassmann

Institut für Theoretische Physik and Physikalisches
Institut, Universität Stuttgart, Pfaffenwaldring 57,
D - 7000 Stuttgart 80, W-Germany

The ground state of acceptors in cubic semiconductors is four-fold degenerate and the interaction with the lattice vibrations leads to the possibility of a Jahn-Teller effect[1], causing a dynamical splitting of the electronic levels. However, due to the extended nature of the wavefunction of the defect, random internal fields may have dominant influence to destroy the J.T. effect. Such an influence should be smaller for deeper acceptors, where the wavefunctions are more concentrated.

In fact different types of experiments indicate a resonance energy in the meV range for the deeper acceptors GaAs(Mn), GaP(In), Si(In) and Si(B), which have been studied by the following methods:
α) Ultrasonic experiments and analysis of the relaxation attenuation[2,3]
β) Thermal conductivity measurements[4]
γ) Quasimonochromatic phonon scattering experiments[5]
δ) Luminescence measurements[6]
ε) Raman spectroscopy[7].

All these detected resonance energies are much larger than the splittings due to random internal fields, which may be of the order of 10 to 100 μeV.

The acceptor-hole-lattice interaction Hamiltonian can be written as

(1) $\quad H_I = \sum\limits_{q\lambda} \Lambda^{q\lambda} (b_{q\lambda} + b^+_{q\lambda})$

where the electronic part $\Lambda^{q\lambda}$ contains Dirac's 4x4 matrices $\hat{\rho}_i$ and $\hat{\sigma}_j$ (i,j=1,2,3):

$$(2) \quad \Lambda^{q\lambda} = D^{\epsilon}(\hat{\rho}_1 r_1^{q\lambda} + \hat{\rho}_2 r_2^{q\lambda}) + D^{\tau}\hat{\rho}_3 (\hat{\sigma}_1 s_1^{q\lambda} + \hat{\sigma}_2 s_2^{q\lambda} + \hat{\sigma}_3 s_3^{q\lambda})$$

This form is equivalent to the usual one[8] expressed in angular momentum operators. It is useful for our treatment. The following abbreviations are used

$$(3a) \quad r_1^{q\lambda} = (\frac{\omega_{q\lambda}}{2Mc_\lambda^2})^{1/2} f(q) \frac{1}{3}\left[2\hat{q}_z e_{\lambda z} - \hat{q}_x e_{\lambda x} - \hat{q}_y e_{\lambda y}\right]$$

$$(3b) \quad r_2^{q\lambda} = (\frac{\omega_{q\lambda}}{2Mc_\lambda^2})^{1/2} f(q) \frac{1}{\sqrt{3}}\left[\hat{q}_x e_{\lambda x} - \hat{q}_y e_{\lambda y}\right]$$

$$(3c) \quad s_1^{q\lambda} = (\frac{\omega_{q\lambda}}{2Mc_\lambda^2})^{1/2} f(q) \frac{1}{\sqrt{3}}\left[\hat{q}_z e_{\lambda y} + \hat{q}_y e_{\lambda z}\right]$$

$$(3d) \quad s_2^{q\lambda} = (\frac{\omega_{q\lambda}}{2Mc_\lambda^2})^{1/2} f(q) \frac{1}{\sqrt{3}}\left[\hat{q}_z e_{\lambda x} + \hat{q}_x e_{\lambda z}\right]$$

$$(3e) \quad s_3^{q\lambda} = (\frac{\omega_{q\lambda}}{2Mc_\lambda^2})^{1/2} f(q) \frac{1}{\sqrt{3}}\left[\hat{q}_x e_{\lambda y} + \hat{q}_y e_{\lambda x}\right]$$

Here the long wavelength approximation is used, i.e. only acoustic phonons are coupled. $b_{q\lambda}$ and $b_{q\lambda}^+$ are the annihilation and creation operators for the phonon with wavevector $\vec{q}$ in the branch λ. $\omega_{q\lambda}$ is the angular frequency and c_λ the velocity of sound. $\hat{q}$ is the unit vector along $\vec{q}$ and $\vec{e}_\lambda$ is the polarization vector of the phonon. M is the mass of the crystal. $D^{\epsilon}(=D_{\hat{H}}^a)$ and $D^{\tau}(=D_{u'}^a)$ are the deformation potential constants for a $[1,0,0]$ and $[1,1,1]$ strain respectively. The extended nature of the defectstate is reflected in the cut-off function

$$(4) \quad f(q) \cong \left[1 + \frac{1}{4} a^{*2}q^2\right]^{-2}$$

where a^* is the effective Bohr radius.

The scattering rate (or inverse life time) of a phonon q,λ is given by

$$(5) \quad \tau_{q\lambda}^{-1} = -\omega_{q\lambda}^{-1} \lim_{\epsilon \to 0} Im \ T(\omega_{q\lambda} + i\epsilon)_{q\lambda, q\lambda}.$$

$T(\omega_{q\lambda}+i\epsilon)$ is the T-matrix defined by the phonon Green's functions $(G=\tilde{G}_0-G_0TG_0)$ of the unperturbed (G_0) and perturbed (G) crystal. Following refs. 9 and 10 the phonon Green's function can be replaced by Green's functions between the electronic operators $\hat{\rho}_i$ and $\hat{\sigma}_j$:

$$(6) \quad \tau_{q\lambda}^{-1} = 4\pi \lim_{\epsilon \to 0} \text{Im} <<\Lambda^{q\lambda};\Lambda^{q\lambda}>>.$$

We consider a longitudinally polarized phonon travelling in a $[1,0,0]$ direction. This simplifies our further treatments considerably, because in this case we are only concerned with the "spin-spin" Green's function $<<\hat{\rho}_1;\hat{\rho}_1>>$.

Using the equation of motion method the Green's function hierarchy can be expanded up to the fourth order. In second order we factorize the quadratic forms of the phonon operators. In fourth order we close the hierarchy taking only the inhomogeneous part of the Green's functions into account.
The thermal expectation values are calculated by use of the exponential transformation

$$(7) \quad U = \exp \{\sum_{q\lambda} \omega_{q\lambda}^{-1} \Lambda^{q\lambda} (b_{q\lambda}-b_{q\lambda}^+)\}$$

which guarantees that they are exact at least up to the second order in the coupling parameters.
Introducing an isotropic model for the crystal the mean scattering rate can be calculated analytically for zero temperature. As a final result we get the Lorentzian-like form

$$(8) \quad <\tau^{-1}>(\omega) = \frac{P(\omega)}{(\omega-\Delta(\omega))^2+\Gamma(\omega)^2}$$

P,Δ and Γ are rather lengthy expressions, therefore we have omitted

TABLE I: Experimental and theoretical values of resonance energies. The small greek letters indicate the method of measurement (see text)

	GaAs (Mn)	GaP (In)	Si (In)	Si (B)
experimental values (meV)	3.1 ± 0.3 α, β	2.7 ± 0.3 ϵ	4.2 ± 0.2 $\alpha,\beta,\gamma,\delta,$	~2. α,β,γ
theoretical values (meV)	2.6	2.3	4.0	1.9

them. $\Gamma(\omega)$ represents the "width" of the resonance. The formal resonance frequency is given by the relation $\omega - \Delta(\omega) \equiv 0$. The results for several systems are given in table I. They are in good agreement with the experimental values.

REFERENCES

1. T.N. Morgan, Phys. Rev. Lett. 24:887 (1970)
2. K. Lassmann, Hp. Schad, Solid State Commun. 18:449 (1976)
3. Hp. Schad, K. Lassmann, Phys. Rev. Lett. 56A:409 (1976)
4. A. de Combarieu, K. Lassmann, "Phonon Scattering in Solids" Plenum Publishing Corp. p.340 (1976)
5. H. Schenk, W. Forkel, W. Eisenmenger, Frühjahrstagung DPG Freudenstadt 1978
6. R. Sauer, W. Schmid, J. Weber, Sold. State Commun. 27:705 (1978)
7. L.L. Chase, W. Hayes, J. Ryan, J. Phys. C 10:2957 (1977)
8. see e.g. G.L. Bir, E.J. Butikov, G.E. Pikus, J. Phys. Chem. Solids 24:1467 (1963)
9. M.V. Klein, Phys. Rev. B 186:839 (1969)
10. M. Rueff, E. Sigmund, M. Wagner, Phys. Stat. Sol. (b)81:511 (1977)

EFFECT OF INTERNAL STRAINS ON LOW TEMPERATURE THERMAL CONDUCTIVITY

OF Li DOPED Ge

A. Adolf, D. Fortier, J.H. Albany

Laboratoire de Physique des Matériaux, Service de Chimie
Physique, Centre d'Etudes Nucléaires de Saclay,
91190 Gif sur Yvette, France

K. Suzuki

Dept. of Electrical Engineering, Waseda University,
Shinjuku, Tokyo, Japan

M. Locatelli
Service des Basses Temperatures, Centre d'Etudes
Nucléaires de Grenoble, France

In a previous paper (1), we determined the valley-orbit
splitting 4Δ of Li donors in Ge from the measurement of the thermal
conductivity K down to 0.4 K and its analysis. Here the measurement of K has been extended down to 0.06 K. The result shows that
the ground state of Li donors has not a "normal" structure, but
an "inverted" one, that is, the triplet lying below the singlet.
The experimental data at T <0.3 K (see Fig. 1) cannot be explained only by scattering mechanisms considered in Ref. 1. In
order to explain the lower temperature behaviour of K, we must
take account of effects of random internal strains on the ground
state and therefore on the donor-phonon interaction.

An analysis of K is proceeded as follows: Stresses along the
<111> and <110> axes split the triplet into two levels and into
three levels, respectively, while a <100> stress does not. For
simplicity, internal strains are assumed to be those related
to the former stresses. Also a distribution of the magnitude of
random stress X, P(X), is assumed to have a Gaussian or a
Lorentzian form. We adopt a "two sites" model proposed by Challis
et al. (2), in which a part of the impurities occupy distorted
sites and the others undistorted sites. As the splitting of the
energy levels within the ground state distributes over a certain

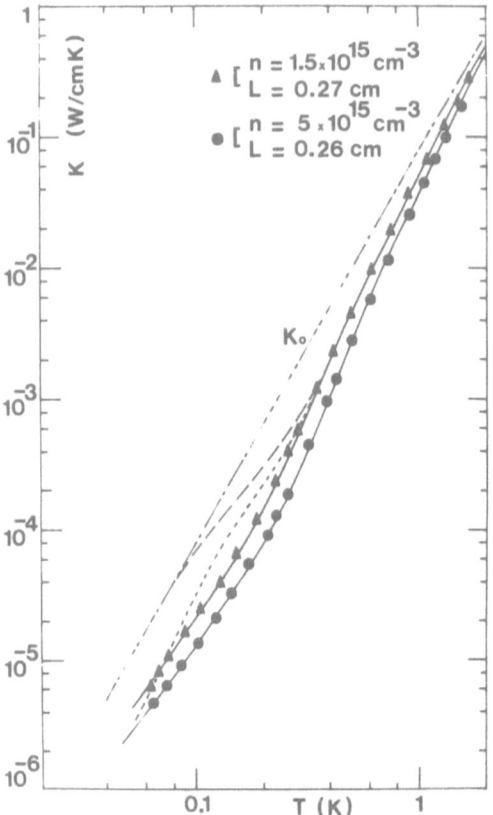

Fig. 1. Thermal conductivity of Li doped Ge. Experimental points
         and   . Calculated curves:
         ____.____.__ __   undoped sample
         _____ Li doped Ge taking into account resonant ab-
sorption scattering with internal strains along the <111>
axis and an "inverted" electronic level. The adjustable
parameters are N_R/N: 0.09 and $\sigma = 5 \times 10^6$ dynes/cm^2. The
other parameters are the same as in Ref. 1 except $4\Delta =$
0.10 meV instead of 0.12 meV.
--------------- Effect of internal strains along the <110>
axis.
         __ __ __ __ __ Calculated thermal conductivity neglecting
the effect of internal strains (Ref. 1). It is worth
noticing that this curve gives also the calculated thermal
conductivity taking into account a "normal" electronic
structure including the effect of internal strains.

range reflecting P (X), the corresponding resonant absorption of low frequency phonons contributes considerably to the thermal resistance at lower temperatures. However, the resonant absorption due to the electron transition between the singlet and the split-off triplet states is neglected. The expressions of the relaxation rates of phonons due to elastic and inelastic scattering by donors in undistorted sites are also used, for simplicity, for those by donors in distorted sites. The donor density in distorted sites, N_R, the half width of P(X), σ, and 4Δ are treated as adjustable parameters.

An excellent fit with the experiment has been obtained for the inverted ground state, a Gaussian distribution, and stresses along the <111> axis with $4\Delta = 0.1$ meV, $N_R/N = 0.09$, and $\sigma = 5 \times 10^6$ dynes/cm^2 where N is the total donor concentration. This value of 4Δ is somewhat smaller than 0.12 meV, the value obtained in Ref. 1.

REFERENCES

(1) A. Adolf et al., Phys. Rev. Lett. __41__, 1477 (1978)
(2) L.J. Challis et al., Phys. Rev. Lett. __39__, 558 (1977)

MAGNETIC FIELD DEPENDENCE OF BALLISTIC HEAT-PULSE ATTENUATION

IN Sb DOPED Ge

Tatsuro Miyasato, Masao Tokumura and Fumio Akao

The Institute of Scientific and Industrial Research,
Osaka University, Yamadakami, Suita Osaka 565 JAPAN

ABSTRACT
     The magnetic field and the frequency dependence of the attenu-
ation of the ballistic heat-pulse propagating in Sb $(0.4-0.9 \times 10^{15}/$
$cm^3)$ doped Ge were measured at liquid He temperatures up to 60 kG.
We used the CdS thin film bolometer which is useful in the magnetic
field. The heat-pulse was propagated along the [111] or [100]
axis, and the magnetic field $\vec{H}$ was applied along the direction of
the wave vector $\vec{q}$. In the former case ($[111]//\vec{H}//\vec{q}$), the Zeeman
splitting of the ground states of the neutral donor electrons reflects directly on the attenuation. In the latter case ($[100]//\vec{H}//$
$\vec{q}$), the magnetic field dependence of the attenuation mainly arises
from the shrinkage effect of the donor wave function.

INTRODUCTION
     The neutral shallow donor states in n-type Ge is expressed by
the hydrogenic state, and the interaction between the donor electrons are neglected if the concentration of donors is low and the
condition $a* \ll R$ (where, $a*$ is the effective Bohr radius of donor
wave function, $R=(3/4\pi n)^{1/3}$ and n is the concentration) is satisfied.
The localized donor electrons are strongly coupled with the lattice
vibrations (acoustic phonons, thermal phonons). This effect is
observed as a large increase of the acoustic attenuation and thermal
resistivity. These phenomena result from the fact that the donor
ground state of the V group donor in Ge is composed of the lower
singlet and the higher triplet (the energy separation between these
levels is expressed by 4Δ and called the valley orbit splitting) and
the fact that the many-valley semiconductors such as Ge have considerably large deformation constants.
     When the magnetic field is applied on these ground states, it

is expected that these states are influenced by the Zeeman effect
and the shrinkage effect of the donor wave function.   These effects
are observed as the magnetic field dependence of the acoustic at-
tenuation(1) or that of the thermal conductivity(2).

The purpose of the present experiment is to observe the effect
by the ballistic heat-pulse attenuation in lightly Sb doped Ge.
When the magnetic field is applied along the [100] axis, the magnetic field does not break the equivalence of the four conduction
band valley, and as a consequence the ground states do not split.
But the shrinkage of the effective Bohr radius a* should be taken
into account.   When the magnetic field is applied along the [111]
axis, the triplet states split by the Zeeman effect, and the level
width of these levels and the electron population are changed.

EXPERIMENTAL TECHNIQUE AND PROCEDURE

Sample: Sb-doped single crystal Ge is polished and cut in a
shape of rectangular parallelepiped of $6 \times 8 \times 8-10$ mm^3.   The concentration of Sb is $0.4-0.9 \times 10^{15}$/cm^3.   CdS bolometer(3): CdS thin film
(1 micron) is evaporated on one face of the sample(Ge) after the SiO
thin film (0.5 micron) is evaporated.   The interdigital electrode
is evaporated on the CdS film through the mask.   Au-heater: Au-
thin film (60Å) is evaporated as a generator onto the other face
after the SiO thin film (0.5 micron) is evaporated.   The dimension
of the generator is 1×1.2 mm^2 and the resistance is 50 ohm at liq.
He temperatures.   The magnetoresistance of the Au-heater is less
than 0.2% up to 60 kG, so the generator is not affected by the field.

About 200 ns duration of dc pulse is fed to the Au-heater and
the detected echo signal by the CdS bolometer is gated and amplified
by the Box-car integrator,and then recorded by the X-Y recorder as a
function of the magnetic field.

The mode of the signal echo is determined by measuring the
time of flight of the signal echo.   In the present experiments, L-
mode is not observed in both ([100] and [111]) cases.

The sample temperature is 1.8 K, and the supplied power to the
heater is from 100 to 1000 mW/mm^2.   The heat pulse temperature is
estimated by the formula(4), and the estimated heater temperature T_h
is 4-7.2 K.

EXPERIMENTAL RESULTS

The magnetic field dependence of the signal echo (T mode)
height is shown in Figs.1 and 2, as a function of the magnetic field
at various heater temperatures.   The signal echo height is normalized to that at zero field.

In the case of H∥q∥[100] configuration, the signal echo height
increases with magnetic field for all heater temperatures, and as the
heater temperature T_h increases, the rate of the increase of the
echo height increases.   In the case of H∥q∥[111] configuration, the
signal echo height increases with magnetic field at fields less than
20-30 kG, and shows peak at 25-35 kG, then decreases with the field

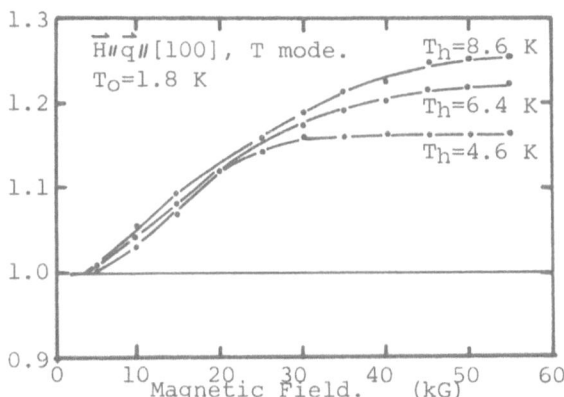

Fig. 1.   Magnetic field dependence of T mode ($\vec{e}$//[010] or [001] at various heater temperatures. Magnetic field $\vec{H}$ and wave vector $\vec{q}$ is parallel to [100] axis. Temperature of sample (Ge) T_o is 1.8 K. Concentration of Sb is 0.4 x 10^{15}/cm^3.

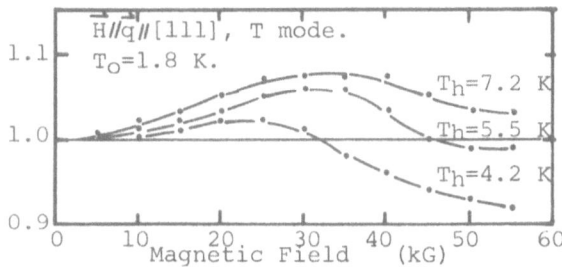

Fig. 2.   Magnetic field dependence of T mode ($\vec{e}$//[11$\bar{2}$] or [1$\bar{1}$0]) at various heater temperatures. Magnetic field $\vec{H}$ and wave vector $\vec{q}$ is parallel to [111] axis. Temperature of sample (Ge) is 1.8 K. Concentration of Sb is 0.9 x 10^{15}/cm^3.

at higher field.   Also in this case, the rate of the increase of echo height is large as the heater is heated up.

CONCLUSIONS
     The experimental results for the magnetic field dependence of the heat-pulse transmission seem to show the opposite behavior from the present theoretical predictions and other experimental results such as the magnetothermal conductivity.   But, e.g., the experiments of the magnetothermal conductivity by Halbo and Sladek(2) is for high Sb doped (10^{16}–10^{17}/cm^3), while, in the present experi-

ment by the heat-pulse transmission for $\vec{H}/\!\!/\vec{q}/\!\!/[100]$ case, the condition a*≪R is satisfied fully and the magnetic field dependence of
the valley orbit splitting is observed directly.   Accordingly, the
discrepancy between the present results and those of Halbo is natural.   To clarify these problems, further experiments are continued.

ACKNOWLEDGEMENT

     We express our thanks to Dr. Katsuo Suzuki for his helpful
discussions and encouragements.

References

(1)T.Miyasato and F.Akao: J.Phys.Soc.Japan 41(1976)502.
(2)L.Halbo and R.J.Sladek: Proceeding of the Tenth International
    Conference on the Physics of Semiconductors, 1970, P.826.
(3)T.Ishiguro et al: appl.Phys.Lett.25(1974)533.
(4)T.Ishiguro et al: J.Phys.Soc.Japan 39(1975)1547.

SYMMETRY OF A DONOR RELATED (DX) CENTER IN $Al_xGa_{1-x}As$

V. Narayanamurti, M. A. Chin and R. A. Logan

Bell Laboratories
600 Mountain Avenue
Murray Hill, NJ 07974

The phenomena of low temperature persistent impurity photo-conductivity in compound semiconductors has attracted considerable recent attention.[1,2] For the case of $Al_xGa_{1-x}As$ it involves a non-effective mass like donor-complex labeled a "DX" center. The unoc-cupied DX center is believed to be resonant with the conduction band, yet sufficiently localized to produce a large lattice relaxation[1] (see top of Fig. 1 which shows the configuration coordinate diagram proposed by Lang et al). In this figure the defect activation energy, E_O (~ 0.1 eV), is small compared to the optical activation energy, E_{op} (~ 1.5 eV). When $kT < E_O$, persistent photoconductivity is possi-ble. Also shown in Fig. 1 is a crystal model of GaAs showing possible distortions of As site donors (such as Te) and Ga site donors (such as Sn).

In this paper we report on the attenuation of ballistic phonons generated in bulk GaAs and propagating through thick ($\sim 10\mu$) epitaxial layers of $Al_xGa_{1-x}As$ doped with 10^{18} Sn or Te donors. These mea-surements yield the symmetry of the DX centers and imply the exis-tence of donor related resonant states in the conduction band.

The geometry of the experiment is shown in the inset of Fig. 1 (bottom right-hand corner). The epilayers were grown by liquid phase epitaxy (LPE). Granular Al bolometers and/or tunnel junctions were used as phonon detectors. The experiments were done either in the dark or in the presence of light generated by a tungsten lamp. Suitable absorbing filters could be inserted between the lamp and crystal for effective illumination of the $Al_xGa_{1-x}As$ epilayers. The detectors were sufficiently thin to allow transmission of the light. For clarity we present data for x=0.5 only and compare these results to those obtained with several different Al compositions in the x=0

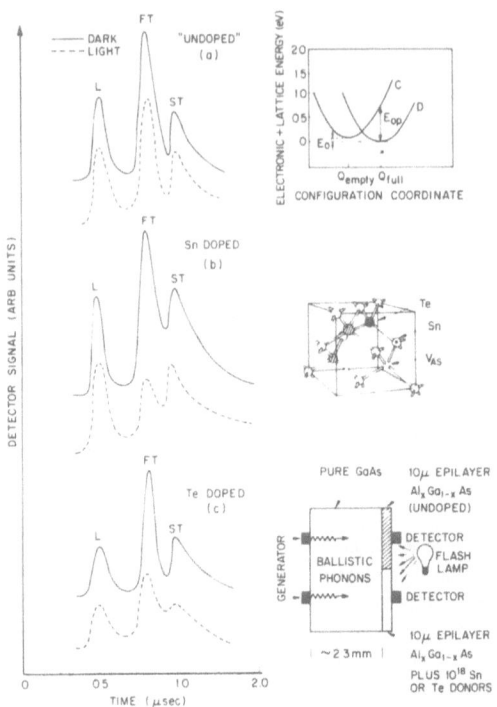

Fig. 1 Ballistic phonon intensities for three different samples of $A\ell_{.5}Ga_{.5}As$ (a) nominally undoped, (b) $\sim 10^{18}$ Sn donors and (c) $\sim 10^{18}$ Te donors. Solid lines are data taken in the dark whilst the dashed lines are taken after photoexcitation. Propagation direction [110]. Also shown in the figure are a typical configuration coordinate for donor related (DX) centers in AℓGaAs, a crystal model and a schematic experimental arrangement.

to x=0.5 range.

In Fig. 1(a) to 1(c) we show typical [110] ballistic phonon signals taken with the sample in the dark (solid curves) and after illumination (dashed lines). In all cases the heater power density was very low and typically ~ 0.03 W/mm^2. All three data (undoped, Sn doped and Te doped) are very similar in the dark and the intensities of the ballistic phonons can semiquantitatively be described by phonon focussing in GaAs.

The ballistic phonon intensities, however, change markedly with photoexcitation. In the case of the nominally undoped sample [case (a)] we observe a small decrease of the FT mode signal relative to the other modes. In the case of the Sn doped sample the FT mode intensity drops dramatically between a factor of 2 and 3 while the

ST mode remains virtually unaffected. In the case of the Te doped
sample we see strong attenuation of both FT and ST modes with the
latter attenuation being somewhat stronger. The attenuation persists even after the light is switched off as long as the low temperature is maintained. The effects described here have been observed with several different samples and does not depend on Aℓ composition in the range x=0.3 to 0.5. In this composition range the
band structure of $Aℓ_xGa_{1-x}As$ changes from direct to indirect. Thus
the effects we describe cannot be explained by free carrier or band
structure effects due to ordinary piezoelectric or deformation potential scattering. For x<.3 we observe strong attenuation in the
dark with little sensitivity to light. In this range, the activation
energy E_0 for DX centers is negative and the DX centers are always
ionized.

   Since the unusual data we have observed parallels the behavior
of the DX center in $Aℓ_xGa_{1-x}As$ studied by photocapacitance[1] we interpret our data as follows: The strong sensitivity to photoexcitation
and persistence of the phonon attenuation after the light is switched
off (in the x=.3 to .5 range) and the variations we have observed
with Sn and Te doping is unequivocal evidence that the attenuation
is to be associated with ionized DX centers. Furthermore, from the
known selection rules[3] for anisotropic defects interacting with the
phonon stress field we identify the symmetry of the Sn DX center to
be trigonal and the Te DX center to be orthorhombic and suggest the
existence of donor related energy states in the conduction band of
GaAs. The small attenuation we observe in the undoped sample is
also to be identified with trigonal centers. Since the "undoped"
LPE material has a background donor concentration in the mid 10^{16}
to 10^{17} range and is grown in quartz tubes and in carbon boats, the
most likely donors are Si or C which may be expected to take a Ga
site similar to the case of Sn as is observed.

   We would like to thank H. G. White for the epilayer growth.

REFERENCES

1. D. V. Lang, R. A. Logan and M. Jaros, Phys. Rev. B 19, 1015
    (1979); D. V. Lang and R. A. Logan, Phys. Rev. Lett. 39, 635
    (1977).
2. R. J. Nelson, Appl. Phys. Lett. 31, 351 (1977).
3. A. S. Nowick and W. R. Heller, Advan. Phys. 14, 101 (1965).

****

DISCUSSION

R. Bray: How does the charge state of the impurity influence how
it will respond to phonons in the crystal compared to the uncharged
state?

V. Narayanamurti: One doesn't know. We seem to see the attenuation
of the ionized center. The neutral central presumably has undergone
such a large lattice distortion that basically it is fixed in position. The acoustic wave will induce either resonance or relaxational absorption. Those low lying energy states probably don't
exist for the occupied DX center. Of course one doesn't know a
priori because if the thing could move around (the configurational
part of the diagram should have orientational structure - it's like
a Jahn-Teller effect) you could get attenuation from the neutral
DX center as well.

BRILLOUIN SCATTERING BY INJECTED ACOUSTIC DOMAINS IN ZnS

Y. Itoh, M. Fujii, S. Adachi and C. Hamaguchi

Department of Electronics, Osaka University
Suita City, Osaka 565, Japan

In this paper we report resonant Brillouin scattering in ZnS studied by using the acoustic domain injection method[1-4], where phonon domains amplified through acoustoelectric effect in CdS are injected into end-bonded ZnS. We used two kinds of ZnS, cubic and polytype, crystals. According to the birefringent measurement,the latter shows 10 % stacking faults parallel to the (111) plane. Samples were cut in the form of parallelepipeds and optical flat surfaces were obtained by polish and chemical etching. Indium layer was deposited by vacuum evaporation onto the end surfaces of CdS and ZnS, and they were bonded together by heating the indium layer. The details of the experimental setup are described elsewhere[2-4].

Figure 1 shows experimental data of the transmission efficiency of the acoustic domains from CdS into ZnS samples in the frequency range of 0.3 - 1.0 GHz (90 % efficiency is achieved at 0.3 GHz in one of the samples). Theoretical value[1] of the efficiency is 98 % without frequency dependence. The reduction of the efficiency in the higher frequency region is mainly due to the acoustic mismatch of the waves in the three layer structure CdS-In-ZnS. We found that the wide frequency range and high transmission efficiency provide many advantages in investigating phonon-phonon[1] and electron-phonon-photon[2-6] interaction in semiconductors. Especially resonant behavior of the Brillouin scattering cross section is investigated in detail in the region of photon energy below the fundamental absorption edge.

In Fig. 2, we present the experimental data of the resonant Brillouin scattering cross section σ_B, which shows resonant cancellation (357 nm in cubic crystal and 360 nm in polytype at room temperature) and resonant enhancement at shorter wavelengths. The results show a good agreement with the theoretical relation

$\sigma_B \propto |R_{1s}+R_0|^2$, where R_{1s} is the resonant term arising from the excitation of the electrons in the conduction band and holes in the three valence bands (exciton formation and damping effect are taken into account) and R_0 is the non-dispersive term arising from

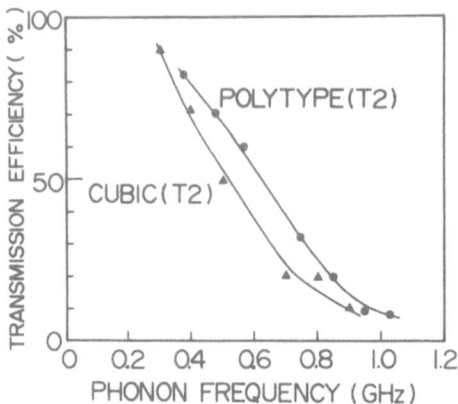

Fig. 1.   Frequency dependence of the transmission efficiency of the acoustic phonon domains from CdS into ZnS samples.

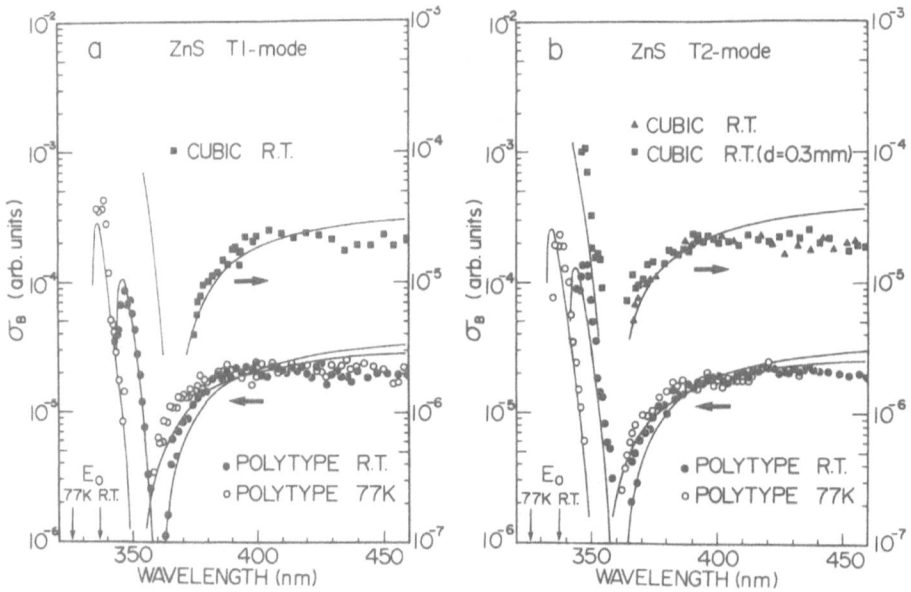

Fig. 2.   Dispersion curves of the Brillouin scattering cross sections for 0.3 GHz a) T1-, and b) T2-mode phonons measured at room temperature and 77 K. Solid curves are calculated by the Brillouin scattering theory. Unless otherwise noted, the thickness of the sample is 1.2 mm.

the remote band contribution. Resonant cancellation is well
explained by a sign reversal of $R_{1s}+R_0$. A new maximum and
reduction of the scattering cross section near the band edge are
presumably due to the damping of the excited electronic states by
impurity, acoustic and optical phonon scatterings[4].
      Figure 3 shows the dispersion curves of the photoelastic
coefficients P_{44} and $P_{11}-P_{12}$ determined by the relation $(P_{11}-P_{12})^2$
$\propto \sigma_B(T1)$ and $(P_{44})^2 \propto \sigma_B(T2)$ [2-5]. Since the present method does
not give sign and absolute values of the photoelastic coefficients
but the relative values, they are adjusted to the value of Yu and
Cardona[7] at 400 nm. The solid curves were calculated by the

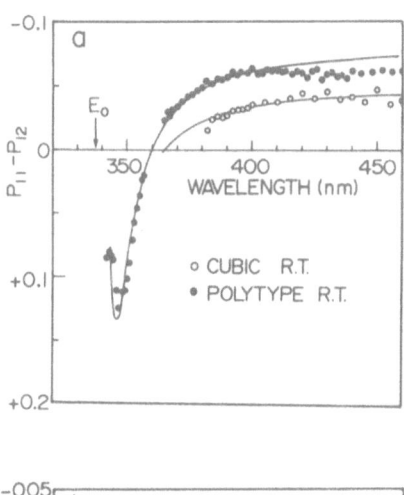

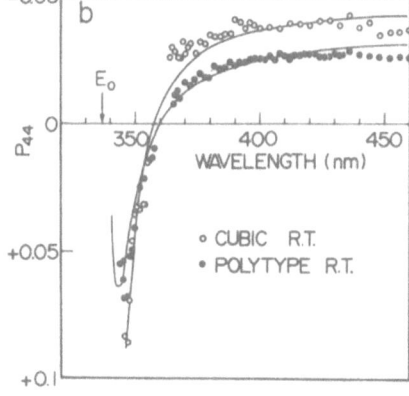

Fig. 3.   Dispersion curves of the photoelastic coefficients
           a) $P_{11}-P_{12}$ and b) P_{44} obtained from the data shown in
           Fig. 1. Solid curves are calculated by the PB theory.

piezobirefringence (PB) theory taking account of the damping
effect in cubic[3,4,8] and hexagonal[5,6,9] materials for cubic
and polytype ZnS, respectively. We find in Fig. 3 that the
theoretical curves obtained from PB analysis show a good agreement
with the experimental data in the region observed.

REFERENCES

[1]   K. Yamabe, K. Ando and C. Hamaguchi, Japan J. Appl. Phys.
           16, (1977) 747.
[2]   K. Ando and C. Hamaguchi, Phys. Rev. B11, (1975) 3876.
[3]   K. Ando, K. Yamabe, S. Hamada and C. Hamaguchi, J. Phys.
           Soc. Japan. 41, (1976) 1593.
[4]   S. Adachi and C. Hamaguchi, Phys. Rev. B11, (1979) 938.
[5]   R. Berkowicz and T. Skettrup, Phys.Rev. B11, (1975) 2316.
[6]   Y. Itoh and C. Hamaguchi, in preparation.
[7]   P. Y. Yu and M. Cardona, J. Phys. Chem. Solids. 34, (1973)
           29.
[8]   Y. Itoh, S. Adachi and C. Hamaguchi, Phys. Stat. Sol. 93,
           (May 1, 1979).
[9]   Y. Itoh, K. Yamabe, S. Adachi and C. Hamaguchi. J. Phys.
           Soc. Japan. 46, (1979) 542.

ELASTIC PROPERTIES AND VALENCE-FORCE-FIELD THEORY

IN ZERO-GAP SEMICONDUCTORS

Kenji Kumazaki

Hokkaido Institute of Technology, Sapporo 061-24, Japan

INTRODUCTION

The valence-force-field theory has the advantage of the most phenomenological and economical description of the short-range valence-forces in the tetrahedrally coordinated crystals. This theory has been modified[1] by taking account of the combined effects of Coulomb forces and internal strains. The regularities on the elastic constants of zero-gap semiconductors[2] (β-HgS, HgSe and HgTe) are well explained by the Phillips ionicity as well as those of other zincblende (Zb) crystals.[1] However, the application of this modified valence-force-field (MVFF) theory to the zero-gap semicon-ductors[2] (ZGS) was not as satisfactory as in other Zb crystals.

The purpose of this work is to make clear the cause of the deviation from MVFF thory in ZGS. The additional aim is to discuss the phase transition in CdSe-HgSe system.

EXPERIMENTAL PROCEDURES

The single crystals of HgSe and CdSe-HgSe are grown by the Bridgman method. The elastic constant of these crystals are meas-ured by the pulse-echo overlap method.

RESULTS AND DISCUSSION

The reduced elastic constants obtained from our experiment are used for comparison between elastic properties. The reduced shear modulus of ZGS and other Zb crystals[1] are plotted against the Phillips ionicity as shown in Fig.1. The reduced elastic con-

stants decrease linearly as the ionicity increases.   This decrease
is related to that of the bond-bending forces as the ionicity increases.   The same linear relation is shown in other reduced elastic constants and the ionicity.

On the other hand, the applicability of MVFF theory to ZGS must
be checked by the Martin's relation.[1]   The values of the Martin's
relation are 1.220 for β-HgS, 1.260 for HgSe and 1.238 for HgTe.

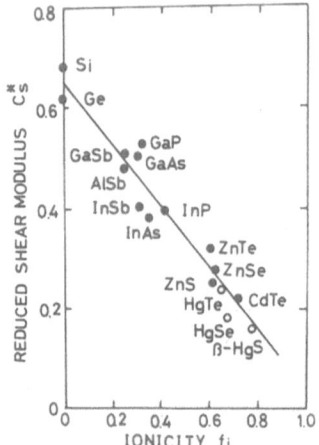

Fig. 1. Reduced shear modulus vs Phillips ionicity.

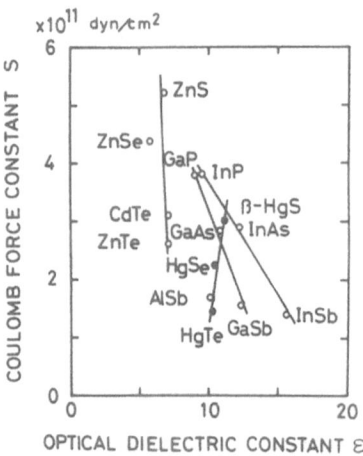

Fig. 2. Coulomb force parameter vs optical dielectric constant.

The values for other Zb crystals[1] are within 10%, even for ionic
crystals.   The large deviation from MVFF theory in ZGS is caused
mainly by the small Coulomb forces, which are also coupled with the
internal strain term.   The Coulomb forces are strongly dependent on
the optical dielectric constants.   The relation between the Coulomb
force parameter[1] and the optical dielectric constant is shown in
Fig.2.   The linear trend in both constants is seen in Zn-, Ga- and
In-compound semiconductors.   But the different characteristics is
recognized in ZGS.

The dynamic dielectric function[3,4,5] in ZGS consists of the
sum of Γ_8-Γ_8 interband, intraband and phonon contributions.   The
effect of the interband contribution[4] on the plasmon-LO phonon coupling and on the frequency of LO phonon is significant among them.
The optical dielectric constants in the plasma region are reduced
by the plasma term.[3]   The magnitude of the plasma frequency is in
the order of β-HgS, HgSe and HgTe.   Therefore the Coulomb forces
are decreasing in the order of HgTe, HgSe and β-HgS as seen in
Fig.2.   If the Coulomb forces are completely described by the MVFF
theory, the optical dielectric constants of HgSe will be the high
value of 23 at low temperature.   This value is larger by a factor
of two than that[4] determined by the optical reflectivity at 2 K.
There is the discrepancy between the experimental and the calculated spectra in HgTe[3] and HgSe.[4]   It is pointed out[4,5] that one
of possible origins for this discrepancy is the adiabatic approximation breaking down of the dielectric function in ZGS.   A more
suitable valence-force-field theory in addition to the Coulomb
forces is needed for ZGS.

The elastic constants of CdSe-HgSe crystals are measured only
in the [110] direction because of difficulty of obtaining the large
single crystals.   The relation between the reduced elastic constant

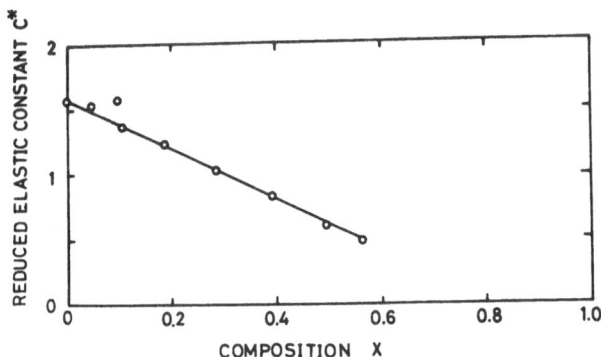

Fig. 3. Reduced elastic constant vs CdSe composition.

C^* and the composition of CdSe x is shown in Fig.3, where $C^* = (C_{11} + C_{12} + C_{44})/C_0$ and C_0 the normalization constant.   The value of C^* decreases with x and becomes zero at $x = 0.81$ if the line in Fig.3 is extrapolated out to higher composition.   A solid with a zero elastic constant cannot be stable and the crystal structure must be changed.   Namely the structure of CdSe-HgSe crystal changes from the zincblende to the wurtzite structure.   In fact, the phase diagram[6] shows that the cubic phase (zincblende) exist for $0 < x < 0.77$ and hexagonal phase (wurtzite) for $0.81 < x < 1.0$.   The mixed phase of both structures is observed for $0.77 < x < 0.81$.

REFERENCES

1.   R. M. Martin, Elastic Properties of ZnS Structure Semiconductors, Phys. Rev. B 1: 4005 (1970).

2.   K. Kumazaki, Elastic Properties and Ionicity of Zero-Gap Semi-conductors, phys. stat. sol. (a) 33: 615 (1976).

3.   W. Szuszkiewicz, A. M. Witowski, and M. Grynberg, The Dynamic Dielectric Function in HgSe and HgTe, phys. stat. sol. (b) 87: 637 (1978).

4.   A. Manabe, H. Noguchi, and A. Mitsuishi, Dielectric Function in HgSe at 2, 95 and 300 K, phys. stat. sol. (b) 90: 157 (1978).

5.   M. Grynberg, R. Le Toullec; and M. Balkanski, Dielectric Func-tion in HgTe between 8 and 300 K, Phys. Rev. B 9:517 (1974).

6.   A. Kalb and V. Leute, The Miscibility Gap of the System CdSe-HgSe, phys. stat. sol. (a) 5: K 199 (1971).

BRILLOUIN AND RAMAN SCATTERING IN LAYER SEMICONDUCTORS GaS AND GaSe

C. Hamaguchi, K. Wasa and M. Yamawaki

Department of Electronics,
Osaka University
Suita, Osaka 565, Japan

Brillouin and Raman scattering experiments have been made to obtain the five elastic constants and the interlayer force constants in layer semiconductors GaS and GaSe. Resonant enhancement of the Brillouin scattering in GaSe has been observed for the first time by using thermal tuning of the fundamental absorption edge at 6328A.

Single crystals of β-GaS and ε-GaSe were grown by Bridgman method. Brillouin scattering measurements were made by using He-Ne laser and details of the experimental setup are given elsewhere[1]. Raman scattering was observed with conventional technique using Ar ion and He-Ne laser. Identification of the Brillouin scattering signals with the phonon modes was done with the help of formula derived by Hamaguchi[2]. The elastic constants are determined from the best fitting[1,3], which are listed in Table 1. Also given in Table 1 are the values obtained from the neutron scattering[4,5].

In order to determine the force constants Raman scattering data

Table 1. Elastic constants of GaS and GaSe $[10^{10}\mathrm{Nm}^{-2}]$

elastic constants	GaS		GaSe	
	our work	neutron scattering[4]	our work[1]	neutron scattering[5]
C_{11}	15.7	15.5	10.5	9.4
C_{12}	3.32	—	2.75	—
C_{13}	1.50	—	1.22	—
C_{33}	3.58	3.64	3.57	3.18
C_{44}	0.81	1.33	1.05	1.25

are combined with the Brillouin scattering data. From the Raman
scattering we determined the energy of rigid layer mode E_{2g} to be
22.7 cm^{-1} for GaS and E' mode to be 19.5 cm^{-1} for GaSe. The present
results are summarized in Table 2. The energy of the B_{2g} mode is
estimated to be 42.5 cm^{-1}[4] for GaS and A" mode to be 36.0 cm^{-1}[6]
for GaSe from the neutron scattering. Using these values we calculated the dispersion curves which are shown in Fig. 1 along with the
neutron scattering data[4,6]. We find in Fig. 1 that the upper
branch is in good agreement with the neutron scattering data, while
the lower branch isn't. The discrepancy is due to the difference in
C_{44} between Brillouin and neutron scattering data(Table 1). The
interlayer force constants determined from the present work are
listed in Table 3.

Resonant behavior of the Brillouin scattering cross section has
been investigated near the band edge in GaSe by virtue of thermal
tuning of the band gap through the incident radiation at 6328Å.

Table 2. Raman vibration in β–GaS and ε–GaSe at room temp.[cm^{-1}]

Representation	E_{2g}^{2}	E_{1g}^{1}	A_{1g}^{1}	E_{1g}^{2}	E_{2g}^{1}	A_{1g}^{2}
GaS	22.7	74.7	188.4	291.8	295.5	360.6
Representation	E'	2E"	$A_{1}^{'}$	E"	E'	$2A_{1}^{'}$
GaSe	19.5	59.2	134.8	252.3 249.4	212.6	307.8

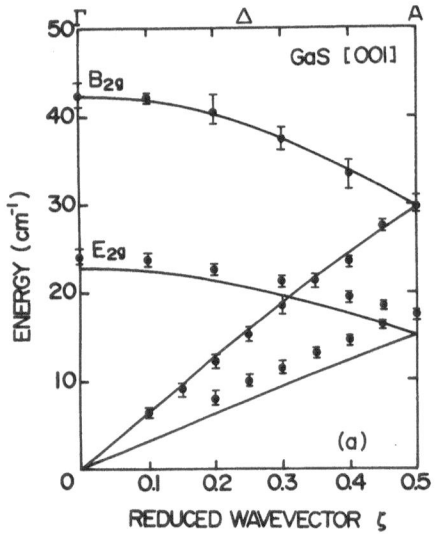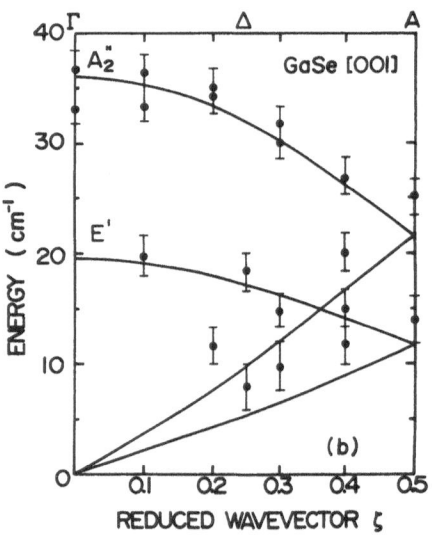

Fig. 1   Dispersion of the phonon propagating along the
c-axis in (a) β–GaS (b) ε–GaSe

Table 3. Interlayer force constants [Nm^{-1}]

	C_{1L}	C_{2L}	C_{1T}	C_{2T}
GaS	5.42	-0.06	1.56	-0.10
GaSe	5.72	-0.82	1.68	-0.24

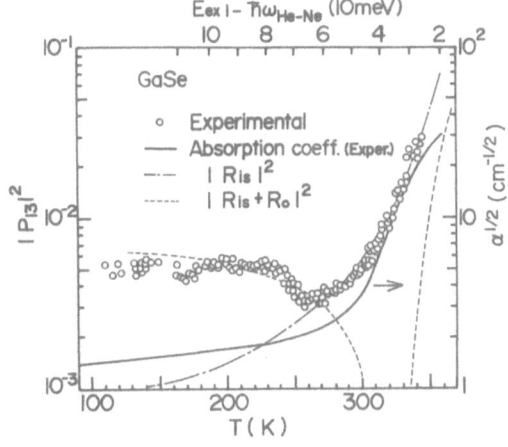

Fig.2   Resonant Brillouin scattering in ε-GaSe.

The result is shown in Fig.2 by the open circles, where the solid curve shows the absorption coefficient and the dot-dashed curve is calculated from Loudon's light scattering theory[7] by taking into account discrete and continum excitons with the binding energy 20 meV and constant broadening factor Γ=12 meV. Dotted curve represents calculated result assuming an existence of cancellation as observed in many semiconductors[8]. Agreement between the experimental data and theoretical curves in the whole region is not obtained. Contribution from the indirect edge should be considered.

References

[1] M.Tanaka, M.Yamada & C.Hamaguchi,J.Phys.Soc.Japan,38:1708(1975).
    Revised data are given in; M.Yamada, K.Wasa & C.Hamaguchi,J.
    Phys.Soc.Japan,40:1778(1976).
[2] C.Hamaguchi,J.Phys.Soc.Japan,35:832(1973).

[3] C.Hamaguchi, K.Wasa & M.Yamawaki,Sci.Repts.Res.Inst.Tohoku Univ.,Ser.A,Suppl.27:43(1979).

[4] B.M.Powell, S.Jandl, J.L.Brebner & F.Levy,J.Phys.C,Solid St. Phys.,10:3039(1977).

[5] J.L.Brebner, S.Jandl & B.M.Powell,Nuovo Cim.,38:263(1977).

[6] J.L.Brebner, S.Jandl & B.M.Powell,Solid St.Commun.,13:1555(1973)

[7] R.Loudon,Proc.Roy.Soc.,London,A275,218(1963).

[8] S.Adachi & C.Hamaguchi,Phys.Rev.,B19:938(1979) and see references therein.

QUANTUM EFFECT ON TRANSVERSE MAGNETORESISTANCE OF DEGENERATE

PIEZOELECTRIC SEMICONDUCTORS IN STRONG MAGNETIC FIELDS *

Chhi-Chong Wu and Anna Chen

Department of Applied Mathematics
National Chiao Tung University
Hsinchu, Taiwan, 300, China

The transverse magnetoresistance for nondegenerate semiconductors with the isotropic parabolic energy bands has been investigated for the case where acoustic phonons are the dominant scattering mechanism.[1] It was shown that the transverse magnetoresistance increases with the magnetic field in the quantum limit. Some experimental results[2] for the inelastic scattering mechanism showed that the transverse magnetoresistance depends strongly on the magnetic field. In this paper we calculate the transverse magnetoresistance of degenerate piezoelectric semiconductors like n-type InSb with isotropic parabolic energy bands throughout the strong-field region in which the splitting of Landau levels is much greater than the average carrier energy. In a crystal with low symmetry, the piezoelectric interaction may become important. This is the interaction between acoustic phonons and electrons through the polarization induced by the piezoelectricity of the crystal. Therefore we investigate the quantum effect of the transverse magnetoresistance for the inelastic scattering of acoustic phonons from the deformation-potential and piezoelectric couplings. The scattering is treated in the Born approximation for strong magnetic fields. We assume that the inelasticity is the dominant mechanism in resolving the divergence, and the cutoff energy due to the inelastic scattering does not change appreciably with the temperature.

For the scattering due to acoustic phonons, the dissipative current lying in the direction of the total electric field is given by[1-3]

$$J_d = (\pi |e| L^2 v_H / k_B T) \sum_{\vec{kn}, \vec{k}'n', q} |C(q)|^2 q_y^2 N_q (N_q + 1) (m^*/q_z \hbar^2) \exp(-\tfrac{1}{2} L^2 q_\perp^2)$$

$$\times \ \{[f(E_{\vec{k}n} - \hbar\omega_q) - f(E_{\vec{k}n})] \ \Big(\frac{n!}{n'!}\Big) \ (\tfrac{1}{2}L^2q_{\perp}^2)^{n'-n} [L_n^{n'-n} (\tfrac{1}{2}L^2q_{\perp}^2)]^2 \delta[k_z$$

$$- \ \tfrac{1}{2}q_z - \frac{m^*\omega_c}{q_z\hbar}(n'-n) - \frac{m^*\omega_q}{q_z\hbar}]\delta(k_z - q_z - k_z') - [f(E_{\vec{k}n} + \hbar\omega_q)$$

$$- \ f(E_{\vec{k}n})] \ \Big(\frac{n'!}{n!}\Big) \ (\tfrac{1}{2}L^2q_{\perp}^2)^{n-n'} [L_{n'}^{n-n'} (\tfrac{1}{2}L^2q_{\perp}^2)]^2 \delta[k_z + \tfrac{1}{2}q_z - \frac{m^*\omega_c}{q_z\hbar}(n$$

$$- \ n') - \frac{m^*\omega_q}{q_z\hbar}]\delta(k_z + q_z - k_z')\} \ , \hspace{2cm} (1)$$

where $\omega_c = |e|B/m^*c$ is the cyclotron frequency of electrons, $\vec{L} = (\hbar/m^*\omega_c)^{\tfrac{1}{2}}$ is the classical radius of the lowest Landau level, $\vec{v}_H = c(\vec{E} \times \vec{B})/B^2$ is the Hall velocity with the applied electric field $\vec{E}$, $f(x)$ is the Fermi-Dirac distribution function, $q = (i, \vec{q})$ denotes collectively the branch and wave vector for the phonon mode with the energy $\hbar\omega_q$, $N_q = [\exp(\hbar\omega_q/k_BT) - 1]^{-1}$ is the Planck distribution function for the phonons in thermal equilibrium, $C(q)$ is the elec-

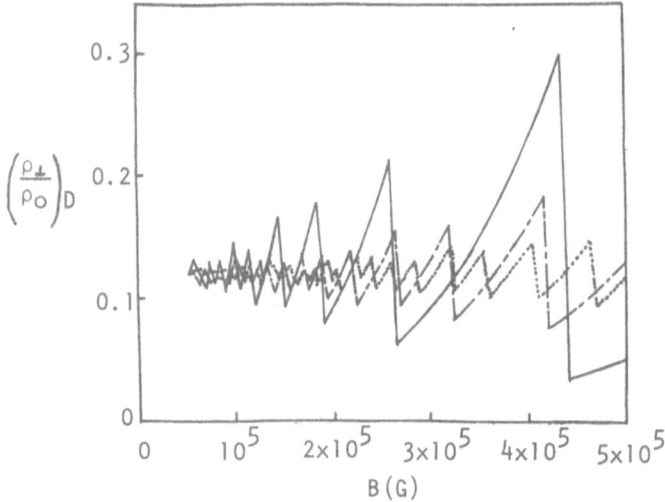

Fig. 1. Transverse magnetoresistance $(\rho_{\perp}/\rho_o)_D$ as a function of mag-
         netic field B in n-type InSb due to the deformation-poten-
         tial coupling. The electron density: full curve: 3×10^{18}
         cm^{-3}; chain curve:$10^{19}cm^{-3}$; broken curve:3×10^{19} cm^{-3}.

tron-phonon coupling constant, $\delta(x)$ is the Dirac δ function, $L_m^p(x)$ is the associated Laguerre polynomial[4] and q_z, $q_\perp$, and q_y are the components of the phonon wavevector directed parallel to the magnetic field, normal to the magnetic field, and in the $\vec{B} \times \vec{E}$ direction, respectively.

Using the same method as our previous works,[3-5] we can have the expression for the transverse magnetoresistance due to the deformation-potential coupling in the degenerate case as $(\rho_\perp/\rho_o)_D$ and the expression for the transverse magnetoresistance due to the piezo-electric coupling in the degenerate case as $(\rho_\perp/\rho_o)_P$. From our numerical results shown in Figs. 1 and 2, it can be seen that the transverse magnetoresistance oscillates with the magnetic field for both deformation-potential and piezoelectric couplings. In Fig.1, the transverse magnetoresistance for the deformation-potential coupling oscillates about the constant value with the magnetic field. These oscillations will be enhanced while increasing the magnetic field. However, in Fig. 2, the transverse magnetoresistance for the piezoelectric coupling oscillates and increases its amplitudes with the magnetic field from the small value of $(\rho_\perp/\rho_o)_P$. Moreover, it is shown that these two coupling mechanisms play the same important

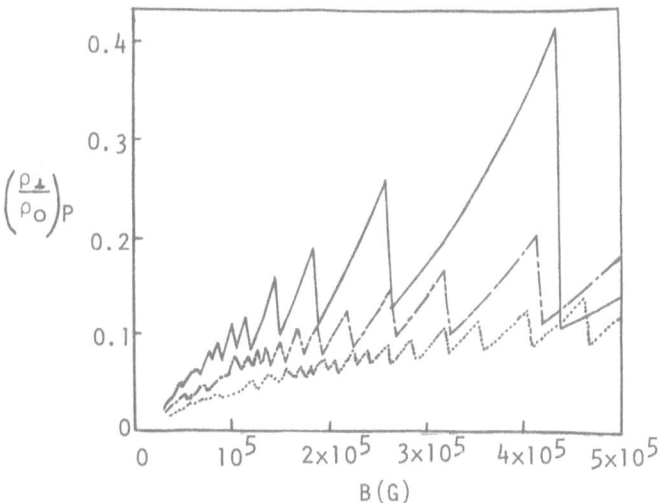

Fig. 2. Transverse magnetoresistance $(\rho_\perp/\rho_o)_P$ as a function of magnetic field B in n-type InSb due to the piezoelectric coupling. The electron density: full curve: 3×10^{18} cm^{-3}; chain curve: 10^{19} cm^{-3}; broken curve: 3×10^{19} cm^{-3}.

role for the transverse magnetoresistance in strong magnetic fields. These results are quite different from those of our previous works[5] for nondegenerate semiconductors in which the acoustic-phonon scattering due to the piezoelectric coupling contributes very insignificantly to the transverse magnetoresistance. As the magnetic field increases, the number of oscillations for both couplings decreases.

* Partially supported by National Science Council of China in Taiwan.

REFERENCES

1. Cassiday D.R. and Spector H.N., Phys. Rev. B9, 2618 (1974).
2. Aliev M.I., Askerov B.M., Agaeva R.G., Daibov A.Z., and Ismailov I.A., Fiz. Tekh. Poluprovodn. 9, 570 (1975)[Sov. Phys.-Semiconductors 9, 377 (1975)].
3. Roth L.M. and Argyres P.N., in "Semiconductors and Semimetals", (Academic, New York, 1966), Vol. 1, p. 159.
4. Erde'lyi A., Magnus W., Oberhettinger F., and Tricomi F.G., "Higher Transcendental Functions" (McGraw-Hill, 1953),Vol.2, 188.
5. Wu C. and Chen A., Phys. Rev. B18, 1916 (1978).

NONEQUILIBRIUM PHONONS IN THREE- AND TWO-DIMENSIONAL

CHARGE TRANSPORT IN SEMICONDUCTORS

Peter Kocevar

Inst. f. Theor. Physik d. Univ. Graz

Universitätsplatz 5, A-8010 Graz, Austria

Under favourable circumstances (low temperatures, strong electron-phonon (el-ph) coupling, weak phonon dissipation mechanisms) phonon amplification by drifting hot carriers in semiconductors can become very large and therefore of practical importance.

Although the "phonon-maser", based on resonant excitation of a narrow band of ph-modes, has not yet been realized as cheap and easily tunable microwave generator of adequate efficiency, there are many practical aspects of nonequilibrium phonons (NEP's).

In particular NEP-effects in the energy-dissipation of hot carriers can modify speed, carrier-trapping, -generation and -recombination, laser-action and optical properties of electronic devices[1].

Putting aside the many phenomena connected with inhomogenous charge and field distributions (e.g. acousto-electric effects, domains etc.) we restrict our following discussion to bulk-effects and the corresponding Boltzmann regime with well defined collisions between electrons and phonons as ballistic particles.

We assume that "single-mode" 3-ph-relaxation towards the thermal equilibrium Planck-distribution N_0 dominates the nonelectronic NEP-losses, the corresponding relaxation time τ_{nonel} defined as in the analogous case of ultrasonic attenuation of externally induced hyperthermal phonons.

Any progress in the solution of the ph-Boltzmann-
equation seems at present only be achievable through the
use of the electron temperature model [2] or its two
modifications, the displaced and heated Maxwellian (HDM)[3]
and the "two-temperatures" (TT) model[4], which have also
been found adequate to understand ordinary 3- and 2-
dimensional high field transport . In all these models
one finds closed expressions for the ph-el collision
integrals, allowing us to write the ph-Boltzmann equation
for a given mode of wave vector $\vec{q}$ as

$$\frac{\partial N(\vec{q},t)}{\partial t} = - \frac{N(\vec{q},t)-N_{el}}{\tau_{ph-el}} - \frac{N(\vec{q},t)-N_0}{\tau_{nonel}} \quad , \tag{1}$$

where τ_{ph-el} and N_{el} are given (not necessarily positive)
functions of $\vec{q}$, the el-parameters and the lattice temperature T_L, and where $\tau^{-1} = \tau_{ph-el}^{-1} + \tau_{nonel}^{-1}$ obviously
plays the role of an amplification- or a damping-coeffi-
cient, depending on whether

$$\tau_{ph-el}^{-1} < 0 \qquad \text{and} \qquad |\tau_{ph-el}^{-1}| > \tau_{nonel}^{-1} \tag{2}$$

or not. If immediately after the application of an electric field pulse, when the ph-system is still in equilibrium, conditions (2) are fulfilled for certain modes,
the corresponding N's increase until either higher order
nonelectronic losses begin to stabilize them or the
viscous action of the amplified modes on the carriers
reduces the electron parameters from their initial resonance
regime to stationary values for which $|\tau_{ph-el}^{-1}| = \tau_{nonel}^{-1}$
(see refs. (2) and (3) for the case of n-Ge and n-InSb
at T_L=4.2K and 2oK).

Equ.(1) shows that any quantitative analysis of ph-
instabilities or NEP-effects requires a detailed know-
ledge of $\tau_{nonel}(\vec{q},T_L)$ for the relevant ph- modes with $|\vec{q}|$
of the order of the mean thermal el-wave-vector. The
corresponding acoustic frequencies range from several
hundred Gcy to approximately 1 Tcy, lying well within
the regime $\hbar\omega \approx kT_L$, for which no quantitative theories
or systematic measurements exist!

In addition to measurements of $\tau_{nonel}(\vec{q},T_L)$ for these
acoustic and the optic ph-modes in the technologically

most important crystalline and amorphous semiconductors,
similar attenuation measurements for acoustic surface-
waves and bulk modes at interfaces or in the layered
structures of the standard MOSFETs and hetero-junction
devices would be very revealing.

Finally, as a possible starting point for a quantitative analysis of the recently intensified experimental
work on optically excited hot carriers[1], we show the ph-
el collision term for the TT-model which should give a
good description for the frequently encountered case of
dominant el-el scattering and strong energy dissipation
through optical ph emission[4]. The TT-distribution function is given by

$$f(E) = A\, e^{-W/kT'}\, e^{(W-E)/kT} \qquad\qquad E < W$$

$$= A\, e^{-E/kT'} \qquad\qquad E > W .$$

For Fröhlich-coupling to LO modes in polar materials we
have $W = \hbar\omega_{LO}$. In the usual effective mass approximation
we get

$$\left.\frac{\partial N}{\partial t}\right)_{ph\text{-}el} = \frac{2\, m_{eff}^2\, e^2\, kT\, \omega_{LO}}{\hbar^4\, q^3}\left(\frac{1}{\epsilon_\infty} - \frac{1}{\epsilon_0}\right) A\, r .$$

$$\cdot\left[e^{-r\left(\frac{x^2 + z^2}{2z}\right)^2}(N+1) - \left(\Theta(x_- - z) + \Theta(z - x_+)\right) e^{-r\left(\frac{x^2 - z^2}{2z}\right)^2} N +\right.$$

$$\left. + \Theta(z - x_-)\Theta(x_+ - z)\, e^{-rx^2}\left(r - 1 + r\, e^{-\left(\frac{x^2 - z^2}{2z}\right)^2 + x^2}\right) N\right], \quad (3)$$

$$r = T/T', \qquad z = \hbar q\,(2m_{eff}\,kT)^{-1/2}, \qquad \Theta = \text{step func.}$$

$$x = (W/kT)^{1/2}, \quad x_\pm = (2^{1/2} \pm 1)x, \qquad \begin{array}{l} \epsilon_\infty = \text{optic diel. fct.} \\ \epsilon_0 = \text{static diel. fct.} \end{array}$$

For $(kT/W)^3 \ll 1$:

$$A \approx \frac{n_{el}\hbar^3}{2} (2m_{eff}kT')^{3/2} \, \bar{r}^{1/2} \left[r + \frac{2}{\pi^{1/2}} e^{-rx^2}(1-r)(x+\frac{1+r}{2xr}) \right]^{-1},$$

where n_{el} is the free carrier concentration.

Inserting (3) into (1) and assuming immediate adaption of the carriers to the instantaneous ph-distribution, one could stepwise integrate (1) and obtain T(t) and T'(t) through the two moment-equations of ref.(4) at each time-instant, analogous to the calculations of refs.(2) and (3).

REFERENCES

1. Lectures of J.C.Hearn, K.Hess, A.Smirl and R.Ulbrich in Proc.NATO ASI on Physics of Nonlinear Transport in Semiconductors, Urbino 1979, Plenum Press (t.b.p.).
2. N.Perrin and H.Budd, Phys.Rev.B 9,3454(1974).
3. P.Kocevar, J.Phys.C 5,3349(1972).
4. G.Persky and D.J.Bartelink, Phys.Rev.B 1,1614(197o).
****

DISCUSSION

K. F. Renk: Which technique do you propose for measuring phonon lifetimes?

P. Kocevar: I think for acoustic phonons it should be the heat pulse technique which should at least get one further up in frequency. It is, of course, not necessary to have very exact numbers so every directional dependence should be roughly given, but numbers which one can trust in estimating non-equilibrium phonon effects for device performances.

V. Narayanamurti: We have been doing experiments in gallium arsenide. At high amplitudes you can actually see very strong amplification depending upon the mobility in the sample. This is a quantum regime with very high frequency phonons, there are very strong anisotropies, it depends on the mode direction, and it depends on the quality of the sample.

THE EFFECTS OF HOT PHONON IRRADIATION ON THE SUPERCONDUCTING

TRANSITION OF THIN ALUMINIUM FILMS

†S.L. Chan, S.J. Rogers, C.J. Shaw and H.D. Wiederick*

Physics Laboratory,
University of Kent at Canterbury,
Kent CT2 7NR,
England

Substantial deviations from the equilibrium superconducting state can be easily generated in thin film samples. The pair breaking effect of incident photons has been observed in such films [1], and in the present work we have looked for its analogue for the case of phonon irradiation. We have so far been concerned only with Aℓ films on sapphire substrates. The sapphire crystals were cylindrical, and the Aℓ films were typically 100-200 nm thick. Phonon pulses were generated electrically in a constantan heater ∿ 20 nm thick.

The sapphire serves as a "vacuum" which transmits the phonon pulse from the heater to the thin film sample. The spectral composition of this pulse is determined by the temperature rise in the heater, but, regardless of its spectral composition, the pulse will cause a temperature rise in the irradiated film which can easily take the temperature of the Aℓ from that of the substrate at say 0.1K to the superconducting transition temperature ∿ 1.5K. We have tried to establish whether the response of the superconductor, when it is in this way driven to its normal transition, can be adequately represented in terms of simple radiative heating; if there is preferential pair breaking by energetic phonons one might expect to see effects attributable to differences in spectral composition for a given radiation intensity. In seeking for such effects we have made a study of the "threshold radiation energy" required to make a small voltage appear across the superconducting film when it is biassed with a small current. In the discussion we shall be concerned only with the effects of the degenerate transverse pulse propagated along the c axis. We consider in turn the various factors that influence the threshold energy for this pulse.

(1) <u>Pulse Width.</u> For relatively narrow pulses in the range 0.05 to 1 μs, to a very good approximation, the threshold voltage (V_{th}) varies inversely with the square root of the pulse width; this implies that what is important is the total pulse energy rather than its power. This is true for pulse widths considerably greater than the intrinsic width of the pulse as observed when the Aℓ film is used as a bolometer at its transition temperature. However, the amplitudes of such detected pulses were found to vary relatively little with width for pulses longer than ∿ 0.15 μs.

(2) <u>Crystal Length.</u> For simple heating, to compensate for the effects of the inverse square law (for matched heaters) we should expect V_{th} to scale with crystal length. The variation is, in fact, more rapid than that; observations on crystals of three different lengths are well represented by $V_{th} \propto (\text{length})^{4/3}$. On the basis of the acoustic mismatch model [2], if the substrate temperature is 0.1K the observed V_{th} values for 0.2 μs pulses are such as to give rise to heater temperatures (T_h) of 5.2K and 2.5K respectively in our 9.5 mm and 3.12 mm samples. The corresponding calculated temperature rises in the Aℓ films, 1.4K and 1.1K, are reasonably in accord with the temperature at which the transition to normal resistance starts to occur (∿ 1.35K for these films). As regards the T_h values, the emitted phonon spectrum will have its maximum at 3.5 kT_c, the B.C.S. value for 2Δ, when $T_h = 1.25T_c$.

It is difficult in practice to achieve a physical situation in which the maximum in the phonon spectrum does not occur at an energy somewhat greater than 2Δ, but for the two cases above it can be said that whereas for the longer sample nearly all of the energy is carried by hot phonons with energies > 2Δ, for the shorter sample a significant fraction of the energy is carried by less energetic phonons.

(3) <u>Heater Area and Substrate Temperature.</u> In some of the experiments the effect of the power density in the heater was directly examined by switching a fixed input pulse between matched heaters differing in area by a factor of 10. It was generally found that smaller values of V_{th} were observed for the smaller heater. This is seen, for example, in Fig. 1a which shows the variation in V_{th} with substrate temperature (T) for such a matched pair. For simple radiative heating we should expect:

$$V_{th}(T) = V_{th}(0)\left[1 - (T/T_c)^4\right]^{1/2}.$$

The solid curves in Fig. 1a are of this form and obviously fit the data well.

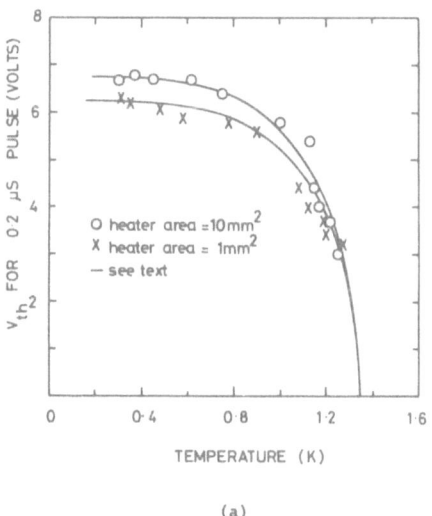

(a)

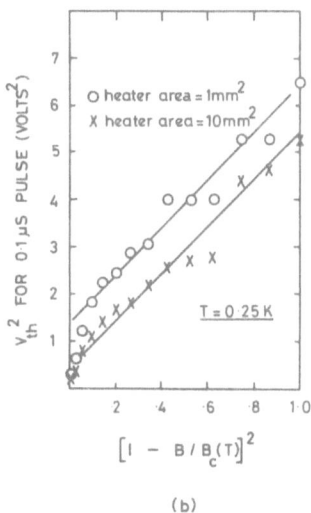

(b)

Figure 1   (a)   The effect of temperature for a 9.5 mm crystal
          (b)   The effect of magnetic field for a 3.12 mm crystal

(4) <u>Applied Magnetic Field</u>. We have examined the effect of changing the S.C. transition temperature by the application of an external magnetic field. Fig. 1b shows data for a pair of matched heaters as described in (3) above. If V_{th} is a measure simply of the heat flux needed to raise the temperature of the film to the new transition temperature, we should expect to find a linear relationship when $V_{th}^2 (B)$ is plotted against $(1 - B/B_c)^2$ in this way. This appears to be the case, but the difference in intercept in the data for the two heater areas is not easily explained in geometric terms.

To sum up, although the observations are broadly in accord with the simple radiative model, some of the results suggest that changes in the spectral composition of the phonon pulses may have an effect on the "normal" response of the S.C. Aℓ films. There are a number of unresolved aspects which may be clarified by further experimental work particularly on thinner films.

*   Permanent address: R.M.C., Kingston, Ontario, Canada.

†   Permanent address: Universiti Malaysia, Malaya.

REFERENCES

[1] L.R. Testardi, Phys. Rev. B4, 2189 (1971)

[2] O. Weis, J. Phys. (Paris) 33, Colloq. C4 - 49 (1972)

PHONON EMISSION FROM A PHOTOEXCITED

SUPERCONDUCTING DOMAIN

B. Pannetier and J.P. Maneval

Groupe de Physique des Solides de l'E.N.S.
24 rue Lhomond, 75231 Paris Cedex 05, France

Photon absorption in a superconductor leads, after fast decaying of the primary excitation, to a longer-lived system of phonons and quasiparticles with energies comparable to the superconducting gap 2Δ. Various theoretical approaches to the steady-state problem have been proposed [1,2]. We report here an experimental investigation using the bulk of the superconductor as a time-of-flight analyzer of the excess excitations.

A tin crystal at low reduced temperature ($T/T_c \ll 1$) is irradiated by a light pulse (20 nsec duration) from an Ar^+ laser while, on the other side, an evaporated tin tunnel junction serves as a magnetically tunable detector [3]. For the sake of brevity, we restrict ourselves in this communication to a description of the phonon system, a typical distribution of which is represented in Fig. 1 :

Fig. 1 : Phonon distribution (schematic). $2\Delta_b$: bulk gap. $h\nu_d$: detector threshold energy. B : magnetic field. Along [001], $2\Delta_b$ is smaller than $h\nu_d$(0).

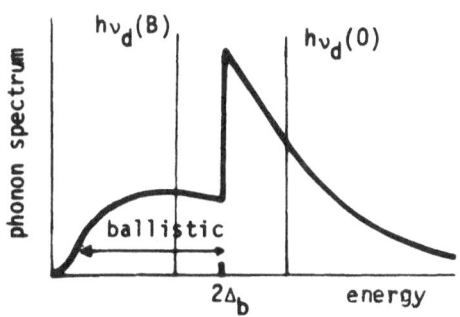

In the spectrum, we have to distinguish two regions : (i) the subgap phonons ($h\nu < 2\Delta_b$), weakly coupled to the electrons and able to escape ballistically (at the sound velocity) into the crystal, and (ii) the phonons of energy larger than $2\Delta_b$, evolving in strong interaction with the excited quasiparticle gas. The characteristic velocity of the coupled quasiparticle phonon system (the mixed mode [4]) is about ten times the sound velocity.

In complement to the time-of-flight study, a fine frequency analysis is performed in the near-gap region (240-270 GHz) by magnetic tuning of the detector gap across $2\Delta_b$. The results are shown in Fig. 2.

At 0.92 kelvin, the detected signals show the decoupling of the excitations. Their interpretation can be made as follows : the M.M. peak is due to <u>recombination</u> phonons ($h\nu \gtrsim 2\Delta_b$) emitted by the quasiparticle gas. It is observed to vanish at very low temperatures (not shown) as the recombination rate decreases. On the other hand, it is almost insensitive to tuning of $h\nu_d$, which is in agreement with prediction of Fig. 2.

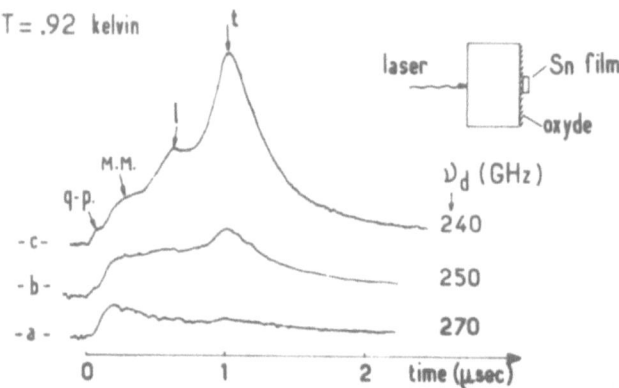

Fig. 2 : Detected signal versus time in bulk tin : length 1.5 mm, direction : [001]. Magnetic fields are a) 0, b) 173 gauss, c) 210 gauss. The symbols l and t stand respectively for longitudinal and transverse ballistic phonons. M.M. is for mixed mode.

The ballistic l and t phonon pulses, corresponding to <u>relaxa-tion</u>,or bremstrahlung-phonons,are essentially absent in zerofield, because $h\nu_d > 2\Delta_b$. They appear at opening of the frequency window and their amplitude increases almost linearly with $(2\Delta_b - h\nu_d)$, indi-cating that the spectrum is approximately flat below $2\Delta_b$.

At higher fields(Fig.2c), an initial peak(named q.p.for quasi-particles) is detected by the tunnel diode, at a velocity which cannot be distinguished from the Fermi velocity. Contrary to the above mentioned phonon signals, the q.p. peak does not show if the detecting tunnel diode is electrically insulated from the crystal by a SiO layer. We delete this question till another publication.

Analysis of the low frequency modes can be performed thanks to bolometric detection (this corresponds to the case of extreme tuning $h\nu_d =0$). The departure from near-equilibrium conditions is made evident by the contrast between photoexcitation and (low-power) heat pulse irradiation (Fig. 3). A heat pulse only provides low-energy ballistic phonons (lower trace) ; in fact, due to limited mean free paths at finite temperatures[3] , the frequencies contained in the l and t peaks are estimated to be $\leqslant$ 50 GHz. Laser irradiation (upper trace) results in a composite signal, virtually void of phonon peaks, but rather containing a large amount of recombination phonons (at

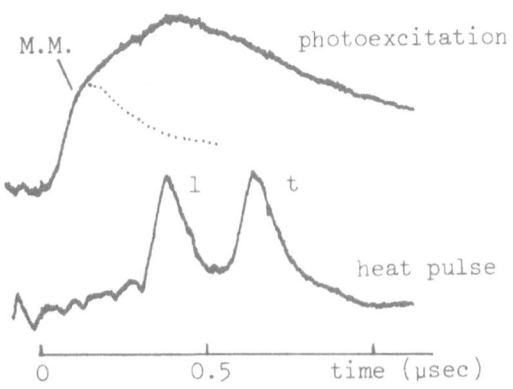

Fig. 3 : Bolometric signals in tin at 1.3 K $\cdot$ [110] direction.   Length 11 mm. Dotted curve corresponds to the normal state.

the front edge), and a tail of low-energy phonons emitted inside
the bulk of the sample. Propagation of the latter is not allowed in
the normal state (dotted curve).

Therefore, it is clear that the phonon distribution resulting
from photoexcitation is displaced towards the high energies.
Explanation is by considering the spontaneous emission from a hot
quasiparticle gas, which favours population of energetic modes
compared to thermal equilibrium population.

REFERENCES

1. J.J. Chang and Scalapino, Phys. Rev. B 15, 2651 (1977).
2. N. Perrin, to appear in J. de Physique (Paris).
3. B. Pannetier et al, Phys. Rev. Letters 39, 646 (1977).
4. J.P. Maneval et al, J.de Phys.(Paris), Col. 39, C6-1559(1978).

ELECTRON-PHONON SCATTERING RATES IN Cu, Ag,

AND Au FROM MAGNETOACOUSTIC EXPERIMENTS[*]

P. B. Johnson and J. A. Rayne

Physics Department
Carnegie-Mellon University
Pittsburgh, PA   15213

The temperature dependence of the electron-phonon scattering rate on the (100) central belly orbit and the (110) "dog's bone" orbit in copper, silver, and gold has been determined from magnetoacoustic measurements.

The samples used in this study were spark cut from high purity oxygen annealed single crystal ingots. Two samples of each material were prepared with normals along [001] and [011] directions. All of these samples were approximately 1 cm thick except for the [001] gold specimen which was only about 0.6 cm thick. A separate set of copper samples 0.3 cm thick was used at ultrasonic frequencies above 210 MHz. These samples were necessary because of the high zero-field electronic attenuation in copper.

Magnetoacoustic measurements up to 270 MHz were made in transmission with longitudinal waves generated by 30 MHz X-cut transducers. For the thinner copper samples a Z-cut delay line was used to provide adequate time separation between the transmitter pulse and the received echoes. Data were obtained with a 12-inch Varian electromagnet, using a PAR Model 162 boxcar signal averager to obtain the amplitude of the transmitted acoustic signal. The temperature of the sample was controlled electronically and measured with a calibrated germanium thermometer.

The oscillatory component of the acoustic attenuation for a given extremal orbit is given by the relation

$$\alpha = C_q^{-1/2} \cos(qD - 3\pi/4)/\sinh(\pi\nu/\omega_c) \qquad (1)$$

[*]Work supported by the National Science Foundation.

In this equation C is a constant depending on the characteristics
of the Fermi surface, q is the phonon wave vector, D is the orbit
diameter in real space, $\omega_c=eH/m^*c$ is the cyclotron frequency for
the orbit and v is the orbital average of the carrier scattering
rate. The value of D is given by $D=(\hbar c/eH)\Delta k$ where Δk is the extremal diameter of the Fermi surface. It follows that the oscillations due to an extremal orbit are periodic in 1/H and that

$$\Delta k = (e/\hbar c)[\lambda/\Delta(1/H)] \qquad (2)$$

where λ is the sound wavelength and $\Delta(1/H)$ the period in 1/H.

The average scattering rate, v, can be represented by
$v=v_I+v_p(T)$ where v_I is the temperature independent impurity scattering rate and $v_p(T)$ the scattering rate due to thermal phonons. In
the limit where $\pi v/\omega_c \gg 1$ the magnitude of the attenuation in
Eq.(1) becomes

$$\alpha = \alpha_o \exp[-(\pi/\omega_c)v_p(T)] \quad ,$$

where α_o is the attenuation at absolute zero for a fixed field.
Hence from suitable plots of α vs T, values of $v_p(T)$ can be obtained.

Figure 1 shows typical magnetoacoustic data. The oscillations
are periodic in 1/H, and using Eq.(2) values of the extremal Fermi
surface radius $\Delta k/2$ were calculated for each measured orbit. In
all cases these calculated values agreed well with accepted values
for the noble metals.

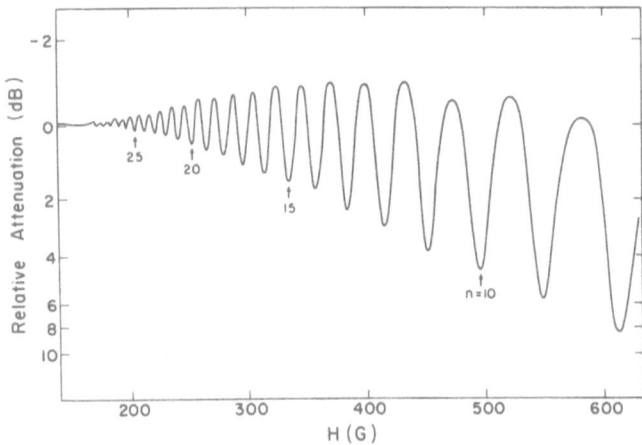

Fig. 1.   Magnetoacoustic oscillation in copper for 152 MHz
           longitudinal waves propagating along [011]. The
           temperature is 4.65K and H is parallel [100].

For the measured orbits plots of $\ell n \alpha$ vs T^3 yielded linear results for all measured harmonics. From the slope of these plots, values of ν_p/T^3 were calculated. So long as the values of ω_c corresponding to various harmonics were in the limit where $\pi\nu/\omega_c \gg 1$, the calculated values of ν_p/T^3 were in good agreement. These values are summarized in Table 1. Also included in the table are results from other experiments and theory. In those cases where comparison is possible, the results are all in good agreement.

Table 1. Electron-Phonon Scattering Rates in the Noble
          Metals (10^6 sec^{-1} K^{-3})

	$\hat{n} \parallel [011]$, $H \parallel [100]$ (Central Belly Orbit)	$\hat{n} \parallel [001]$, $H \parallel [110]$ (Dog's Bone Orbit)
Copper		
Magnetoacoustic (this work)	2.7 ± 0.1	6.2 ± 0.2
RFSE[1]	2.72^a	6.37^a
Theory[2]	2.3 ± 0.5	6.7 ± 1.7
Cyclotron Resonance[3]	2.9 ± 1.2	
Magnetoacoustic[4]	2.9 ± 0.2^b	6.0 ± 0.3^b
Silver		
Magnetoacoustic (this work)	4.2 ± 0.2	13.7 ± 0.4
RFSE[5]	4.1 ± 0.2	
RFSE[6]	4.2^a	13.6^a
Gold		
Magnetoacoustic (this work)	14.2 ± 0.5	26.5 ± 2.0

^{a}These data have been obtained from an extrapolation technique outlined in Ref. 7. Values of $\nu_I = 2.97 \times 10^9$ sec^{-1} in copper and $\nu_I = 1.3 \times 10^9$ sec^{-1} in silver were used in the extrapolation.

^{b}In their work the investigators have mistakenly assigned these scattering rates to different orbits on the Fermi surface.

REFERENCES
1. V. F. Gantmakher and V.A.Gasparov, Sov.Phys.-JETP 37, 864 (1973).
2. David Nowak, Phys. Rev. B 6, 3691 (1972).
3. P. Haussler and S.J.Welles, Phys. Rev. 152, 675 (1966).
4. D.S.Khatri and J.R.Peverley, Phys. Cond. Matter 19, 67 (1975).
5. P.B.Johnson and R.G.Goodrich Phys. Rev. B 14, 3286 (1976).
6. V. A. Gasparov, Sov. Phys.-JETP 41, 1129 (1976).
7. P. B.Johnson, J.C.Kimball, and R.G.Goodrich, Phys. Rev. B 14, 3282 (1976).

*****

DISCUSSION

K. Fossheim: Koch and his co-workers developed the surface cyclotron resonance method whereby they could map out point by point the scattering rate. I believe they have done this also in copper. Could you just mention how the results compare?

J. A. Rayne: I didn't mean to imply that this was a unique tool. It's true that the radio frequency size effect has also given in the case of copper a fairly complete analysis of how the scattering rate varies over the entire Fermi surface. However, I should point out that to the best of my knowledge these are the only data which have given results for gold. It's very difficult to get data using the other techniques simply because of the problem of producing electro-polished samples that are sufficiently thin to give well-defined surface size-effect measurements or the Koch-type of experiment. The nice part of the magnetoacoustic effect measurements is that you use a bulk sample so there's no question that you are looking at a real bulk electron-phonon scattering. So it's rather satisfying to see that in the case of copper everying fits together finally.

J. P. Maneval: In the case of radio frequency size effect measurements there are directions which exhibit divergence of the collision rates. Is your method more appropriate to get closer to this?

J. A. Rayne: No, I would say not. There's no question that the radio frequency size effect can give you more detailed information, because the essence of this technique is that you have to be able to make an unambiguous identification of a single orbit. In the case of the radio frequency size effect they are automatically discriminated by the value of the field at which you observe the anomaly in the surface impedance. However, to the best of my knowledge nobody has succeeded in doing these things in the case of gold.

QUANTUM OSCILLATIONS EFFECT ON THE LATTICE THERMAL CONDUCTIVITY OF

GRAPHITE UNDER VERY HIGH MAGNETIC FIELDS

C. AYACHE and M. LOCATELLI

Centre d'Etudes Nucléaires de Grenoble, S.B.T./L.C.P.

85 X - 38041 GRENOBLE-CEDEX (France)

We have extended previous thermal conductivity measurements under magnetic fields[1][2] on graphite to very high fields ranges. These correspond to the quantum transport regime by electrons and holes (i.e. $\hbar\omega > kT$) and the higher experimental field is about the extreme quantum limit for electrons (i.e. $\hbar\omega \sim \eta$)[3]. In this way, we got evidence for variations of the lattice thermal conductivity with the magnetic field at low temperatures (1.5 to 20 K).

EXPERIMENTAL CONDITIONS

The experiments were performed on highly oriented pyrolytic graphite (HOPG) samples obtained from Union Carbide Corporation[4]. The thermal conductivity (K) is measured parallel to the basal planes while the magnetic field (B) is directed along the c-axis. A differential technique is used which avoids the problems of thermometry under a magnetic field and provides a very good resolution (typically $\Delta K(B)/K(0) \sim 10^{-3}$). A superconductive coil supplies the magnetic fields up to 8 Teslas.

EXPERIMENTAL RESULTS

Figure 1 shows the B-dependence of the ratio $\dot{K}(B)/K(0)$ at 4 K. When B is just raised, K strongly decreases and then nearly remains constant. The total change (as measured at 1 T) is about 34 % of K(0). If B is raised more some oscillations become apparent. The first one has a small dip about 1.9 T with a very small amplitude (~ 1 % of K(0)). The second one has again a dip around 3.2 T but it is more extended and larger (~ 6 % of K(0)). Finally, between 4 and 8 T another depression in K(B) can be distinguished with a smaller amplitude (~ 4 % of K(0)).

When the temperature is raised all the above B-dependences become less and less pronounced and disappear between 10 and 20 K.

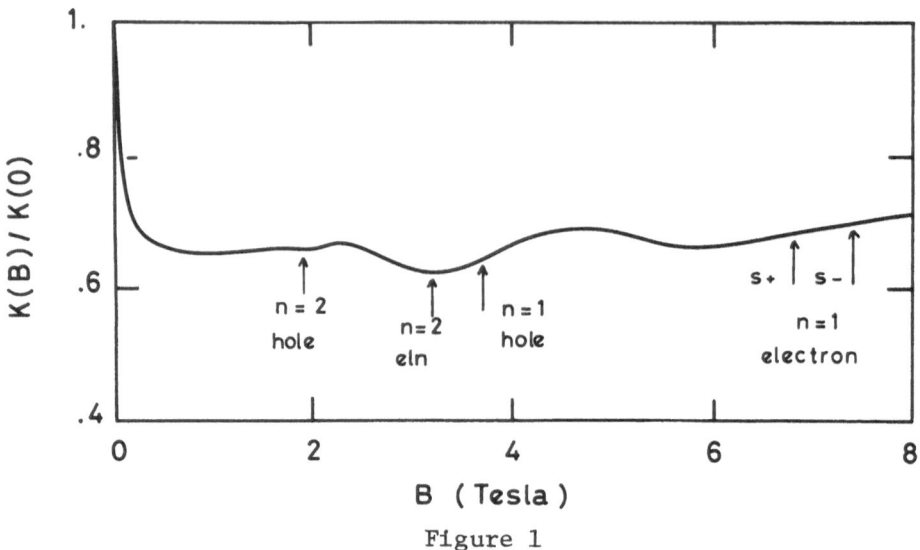

Figure 1

DISCUSSION

        In graphite K results from the electronic (K_e) and lattice
(K_1) contributions[6]. The lower field region decrease is then
readily understood as the lowering of K_e with B[1][2]. So, at about
1 T, we can consider that we are left with K_1 only. As a consequence the observed variations at higher fields are variations of the
lattice contribution. This result is in opposition with the scattering of phonons by grain-boundaries which is, on another side,
well consistent with the K temperature dependence[6]. Another scattering mechanism which becomes effective under high magnetic fields
must thus be introduced. The most natural one is the electron-phonon interaction.

        Our study is essentially concerned with fields corresponding
to the quantum regime for electrons and holes[3]. Charge carriers
are then distributed on Landau levels characterized by the quantum
number n. The field values corresponding to the different levels
crossing the Fermi level have been determined through different
quantum oscillations experiments[3]. Arrows on Figure 1 represent
some of them. One can see that a correspondence is readily established between the 3T dip and the n=1-hole and n=2-electron level
on the other hand. However, the correspondence is not so clear
between the higher field effect and the n=1-electron level

        These correspondences may be confirmed on a theoretical basis
as the electron-phonon interaction goes through a logarithmic divergence whenever a phonon connects two electronic states with
the same n[7]. The effect can thus be compared to the magneto-pho-
non oscillations effect but instead of the electronic one it is
the phonon mean free path which is affected here.

Up to now we only considered changes in the relaxation mechanism under high magnetic fields. However, the phonon spectrum also can be affected under such conditions connected with the strong enhancement of the electron-phonon interaction[9]. So the experimental curve can also express such processes.

Finally, we can mention that the observed effect seems to be related to the very rapid disappearance of quantum oscillations in graphite and also with an anomalous maximum which has been observed in the resistivity versus temperature curve under a high fixed magnetic field [10].

REFERENCES

[1]   M.G. HOLLAND, C.A. KLEIN, and W.D. STRAUB, J.Phys.Chem.Sol.27 (1966) 903

[2]   C.K. CHAU and S.Y. LU, J.of Low Temp.Phys.15 (1974) 447

[3]   I.L. SPAIN in Chemistry and Physics of Carbon (P.L. WALKER and P.A. THROWER, eds) 8 (1973) 1

[4]   A.W. MOORE in Chem.Phys.Carbon, op.cit., 11 (1973) 69

[5]   M. LOCATELLI, the XVth Intern.Conf. on Low Temp.Phys.Grenoble (1978)

[6]   B.J. KELLY in Chem.Phys.Carbon, op.cit., 5 (1969) 119

[7]   P.G. HARPER, J.W. HODBY, R.A. STRADLING, Rep. Prog. Phys. 36 (1973) 1

[8]   A.A. ABRIKOSOV, Sol.Stat.Phys.suppl.N° 12, Academic Press (1972)

[9]   J.D. AXE in Electron-Phonon Interactions and Phase Transitions (T. RISTE, ed.) Plenum Press (1977) 50

[10] C. AYACHE, Thèse de Doctorat d'Etat (1978), Grenoble and references therein.

LIST OF DELEGATES

D. A. Ackerman   Department of Physics, University of Illinois, Urbana, Illinois 61801

A. C. Anderson   Department of Physics, University of Illinois, Urbana, Illinois 61801

C. H. Anderson   RCA Laboratories, Princeton, New Jersey 08540

H. Araki   Metals Research Laboratory, Brown University, Providence, Rhode Island 02912

B. H. Armstrong   IBM Scientific Center, 1530 Page Mill Road, Palo Alto, California 94304

W. Arnold   Max-Planck-Institut for Festkorperforschung, Heisenbergstrasse 1, 7 Stuttgart 80, WEST GERMANY

W. Bausch   Fachhochschule Darmstadt, Fachbereich Mathematik u. Naturwissenschafte, D 6100 Darmstadt-Schöfferstr. 3, WEST GERMANY

D. J. Bishop   Bell Laboratories, 600 Mountain Avenue, Murray Hill, New Jersey 07974

J. L. Black   Brookhaven National Laboratory, Upton, New York 11973

R. I. Boughton   Department of Physics, Northeastern University, Boston, Massachusetts 02115

R. Bray   Department of Physics, Purdue University, West Lafayette, Indiana 47907

W. E. Bron   Department of Physics, Indiana University, Bloomington, Indiana 47401

M. A. Brown   Department of Physics, University of Technology, Loughborough, Leicestershire, ENGLAND

469

R. Calemczuk   Centre d'Etudes Nucleaires de Grenoble,
                      Avenue des Martyrs, 38041 Grenoble, FRANCE

L. J. Challis   Department of Physics, University of
                      Nottingham, University Park, Nottingham
                      NG7 2RD, ENGLAND

W-C. Chang   Department of Physics, Purdue University,
                      West Lafayette, Indiana 47907

L. D. Chapman   Department of Physics, Purdue University,
                      West Lafayette, Indiana 47907

J. D. N. Cheeke   Department de Physique, Université de
                      Sherbrooke, Sherbrooke P Q J1K 2R1,
                      CANADA

B. B. Chick   Metals Research Laboratory, Brown Univer-
                      sity, Providence, Rhode Island 02912

M. A. Chin   Bell Laboratories, Murray Hill, New
                      Jersey 07974

R. W. Cline   Department of Physics, Massachusetts
                      Institute of Technology, Cambridge,
                      Massachusetts 02139

R. Colella   Department of Physics, Purdue University,
                      West Lafayette, Indiana 47907

A. M. de Goër   Centre D'Études Nucléaires De Grenoble,
                      Avenue Des Martyrs, 38041 Grenoble, FRANCE

F. de la Cruz   Centro Atómico Bariloche, S. C. de
                      Bariloche, Rio Negro, ARGENTINA

W. Dietsche   Physik Department E10, TU München, D 8046
                      Garching, WEST GERMANY

E. R. Dobbs   Department of Physics, Bedford College,
                      Regents Park, London NW1 4NS, ENGLAND

K. Dransfeld   Max-Planck-Institut für Festkorper-
                      forschung, 7000 Stuttgart 80, Heisen-
                      bergstrasse 1, WEST GERMANY

R. C. Dynes   Bell Laboratories, Murray Hill, New
                      Jersey 07974

W. C. Egbert   Department of Physics, University of
                      Georgia, Athens, Georgia 30602

W. Eisenmenger   Physikalisches Institüt der Universität,
                      Stuttgart 80, Pfaffenwaldring 57, WEST
                      GERMANY

C. Elbaum   Department of Physics, Brown University,
                      Providence, Rhode Island 02912

D. Fortier	Service de Chimie Physique, Centre D'Etudes de Saclay, 91190 Gif-Sur-Yvette, FRANCE
K. Fossheim	Department of Physics, Norwegian Inst. of Tech., 7034-NTH Trondheim, NORWAY
J. W. Gardner	Department of Physics, University of Illinois, Urbana, Illinois 61801
Affan Ghazi	Department of Physics, University of Nottingham, Nottingham, NG7 2RD, ENGLAND
B. Golding	Bell Laboratories, 600 Mountain Avenue, Murray Hill, New Jersey 07974
D. Goodstein	Division of Physics & Astronomy, California Institute of Technology, Pasadena, California 91125
C. Gough	Department of Physics, University of Birmingham, Birmingham B15 2TT, ENGLAND
J. E. Graebner	Bell Laboratories, 600 Mountain Avenue, Murray Hill, New Jersey 07974
M. Greenstein	Department of Physics, University of Illinois, Urbana, Illinois 61801
W. Grill	Institut f. Festkorperforschung, Technische Hochschule Darmstadt, 61 Darmstadt, Hochschulestr. 2, WEST GERMANY
J. M. Grimshaw	Department of Physics, University of Nottingham, University Park, Nottingham NG7 2RD, ENGLAND
F. Guillon	Department of Physics, Queen's University, Kingston, Ontario, CANADA K7L 3N6
G. Guillot	Department de Physique, Université de Sherbrooke, Sherbrooke P Q J1K 2R1, CANADA
C. Hamaguchi	Department of Electronics, Osaka University, Suita City, Osaka 565, JAPAN
J. P. Harrison	Department of Physics, Queen's University, Kingston, Ontario, CANADA K7L 3N6
B. Hebral	CRTBT, CNRS, Avenue des Martyrs, 38042 Grenoble, FRANCE
J. Heiserman	Department of Applied Physics, Stanford University, Stanford, California 94305

J. C. Hensel	Bell Laboratories, 600 Mountain Avenue, Murray Hill, New Jersey 07974
A. Hikata	Metals Research Laboratory, Brown University, Providence, Rhode Island 02912
Y. Hiki	Tokyo Institute of Technology, Department of Applied Physics, Oh-okayama, Meguro-ku, Tokyo 152, JAPAN
F. R. Hope	Department of Physics, University of Exeter, Exeter, ENGLAND
D. Houde	Department de Physique, Université de Sherbrooke, Sherbrooke, P Q J1K 2R1, CANADA
T. Huber	Department of Physics, Brown University, Providence, Rhode Island 02912
S. Hunklinger	Max-Planck-Institut für Festkorperforschung, 7000 Stuttgart 80, Heisenbergstrasse 1, WEST GERMANY
A. Ikushima	The Institute for Solid State Physics, The University of Tokyo, Roppongi, Minato-ku, Tokyo 106, JAPAN
T. Ishiguro	Electrotechnical Laboratory, Tanashi Branch, 5-4-1 Mukodai-Machi, Tanashi-Shi, Tokyo, JAPAN
I. Iwasa	Department of Physics, University of Tokyo, Bunkyo-ku, Tokyo, JAPAN
D. P. Jones	Cavendish Laboratory, Madingley Road, Cambridge CB3 OHE, ENGLAND
J. M. Katerberg	Department of Physics, University of Illinois, Urbana, Illinois 61801
R. S. Katiyar	I.F.-Unicamp, C.P. 1170, 13.100-Campinas, S.P., BRASIL-S.A.
V. Keith	Department of Physics, Queen's University, Kingston, Ontario, CANADA K7L 3N6
H. Kinder	Physik-Department der Technischen Universität München, 8046 Garching, Institut für Festkorperphysik, WEST GERMANY
P. J. King	Department of Physics, University of Nottingham, University Park, Nottingham, ENGLAND

P. G. Klemens	Department of Physics, University of Connecticut, Storrs, Connecticut 06268
P. Kocevar	Institute of Theoretical Physics, Universitaetplatz 5, A 8010 Graz, AUSTRIA
Y. Kogure	Department of Physics, University of Illinois, Urbana, Illinois 61801
K. Kumazaki	Department of Engineering Science, Hokkaido Institute of Technology, Teine, Sapporo 061-24, JAPAN
H. Kwun	Department of Physics, Brown University, Providence, Rhode Island 02912
C. Laermans	Department of Physics, Kath Universiteit Leuven, Celestijnenlaan 200D, 3030 Heverlee, BELGIUM
J. C. Lasjaunias	Centre De Recherches Sur Les Très Basses Températures, Avenue des Martyrs, B.P. 166, Centre de Tri, 38042 Grenoble, Cedex, FRANCE
K. Lassmann	Physikalisches Institut der Universitat 7, Stuttgart 80, Pfaffenwaldring 57, WEST GERMANY
M. Lax	Bell Laboratories, 600 Mountain Avenue, Murray Hill, New Jersey 07974
D. G. Legrand	General Electric Company, Schenectady, New York 12301
R. G. Leisure	Laboratorie d'Ultrasons, Tour 13, 4 Place Jussieu, 75230 Paris, Cedex 05, FRANCE
J. E. Lewis	Department of Physics, State University of New York, Plattsburgh, New York 12901
P. Lindenfeld	Department of Physics, Rutgers University, New Brunswick, New Jersey 08903
F. Lipschultz	Department of Physics, University of Connecticut, Storrs, Connecticut 06268
M. Locatelli	Centre D'Études Nucléaires De Grenoble, Avenue Des Martyrs, 38-Grenoble, FRANCE
A. R. Long	Department of Natural Philosophy, University of Glasgow, Glasgow G12 8QQ SCOTLAND
K. Luszczynski	Department of Physics, Washington University, St. Louis, Missouri 63130

R. A. MacDonald — Thermophysics Division, National Bureau of Standards, Washington, D. C. 20234

G. Madore — Department de Physique, Université de Sherbrooke, Sherbrooke, P Q J1K 2R1, CANADA

J. P. Maneval — ENS Physique des Solids, 24 rue Lhomond, 75231, Paris 5, FRANCE

M. B. Manning — Department of Physics, Brown University, Providence, Rhode Island 02912

H. J. Maris — Department of Physics, Brown University, Providence, Rhode Island 02912

D. Matsumoto — Department of Physics, University of Illinois, Urbana, Illinois 61801

A. K. McCurdy — Worcester Polytechnic Institute, Worcester, Massachusetts 01609

R. S. Meltzer — Department of Physics & Astronomy, University of Georgia, Athens, Georgia 30602

F. Michard — D.R.P. Université P. et M. Curie, 4 Place Jussieu, Paris, FRANCE

T. Miyasato — Institute of Scientific and Industrial Research, Osaka University, Yamadakami, Suita, Osaka 565, JAPAN

D. Monk — Department of Physics, University of Nottingham, Nottingham NG7 2RD, ENGLAND

F. Moss — Department of Physics, University of Missouri, 8001 Natural Bridge Road, St. Louis, Missouri 63121

Y. Narahara — Institute of Physics, University of Tsukuba, Sakuramura, Ibaragikeu 300-31, JAPAN

V. Narayanamurti — Bell Laboratories, 600 Mountain Avenue, Murray Hill, New Jersey 07974

G. A. Northrup — Department of Physics, University of Illinois, Urbana, Illinois 61801

D. V. Osborne — School of Mathematics and Physics, University of East Anglia, Norwich, NR7 7TJ, ENGLAND

W. C. Overton, Jr. — Los Alamos Scientific Laboratory, Los Alamos, New Mexico 87545

J. H. Page	Department of Physics, Queen's University, Kingston, Ontario, CANADA
B. Pannetier	Groupe de Physique des Solides de l'Ecole Normale Supérieure, 24 rue Lhomond, 75231 Paris Cedex 05, FRANCE
G. Park	Department of Physics, Brown University, Providence, Rhode Island 02912
B. Perrin	D.R.P. Université P. et M. Curie, 4 Place Jussieu, Paris, FRANCE
C. Peters	Department of Physics, Queen's University, Kingston, Ontario, CANADA K7L 3N6
R. O. Pohl	Department of Physics, Cornell University, Ithaca, New York 14853
V. W. Rampton	Department of Physics, University of Nottingham, University Park, Nottingham NG7 2RD, ENGLAND
C. A. Ratsifaritana	Department of Physics, University of Connecticut, Storrs, Connecticut 06268
A. K. Raychaudhuri	Department of Physics, Cornell University, Ithaca, New York 14853
J. A. Rayne	Department of Physics, Carnegie-Mellon University, Pittsburgh, Pennsylvania 15213
K. Renk	Fachbereich Physik der Universität, 84 Regensburg, WEST GERMANY
A. Ridner	Department of Physics, Brown University, Providence, Rhode Island 02912
S. J. Rogers	The Physics Laboratory, The University, Canterbury, Kent, England·
J. E. Rives	Department of Physics & Astronomy, University of Georgia, Athens, Georgia 30602
B. Salce	Centre D'Études Nucleaires De Grenoble, Avenue Des Martyrs, 38-Grenoble, FRANCE
D. Salin	Laboratorie d'Ultrasons, Tour 13, 4 Place Jussieu, 75230 Paris, Cedex 05, FRANCE
W. M. Saslow	Department of Physics, Texas A & M University, College Station, Texas 77843
G. Seidel	Department of Physics, Brown University, Providence, Rhode Island 02912

F. W. Sheard	Department of Physics, University of Nottingham, Nottingham NG7 2RD, ENGLAND
N. S. Shiren	IBM Watson Research Center, P.O. Box 218, Yorktown Heights, New York 10598
E. Sigmund	Institut für Theoretische Physik, D-7000 Stuttgart, Pfaffenwaldring 57, WEST GERMANY
G. P. Srivastava	New University of Ulster, Coleraine, County Londonderry, NORTHERN IRELAND BT52 ISA
H. L. Stormer	Bell Laboratories, 600 Mountain Avenue, Murray Hill, New Jersey 07974
K. Suzuki	Department of Electrical Engineering, Waseda University, Shinjuku, Tokyo 160, JAPAN
P. Taborek	Department of Physics, California Institute of Technology, Pasadena, California 91125
C. Tannous	Department de Physique, Université de Sherbrooke, Sherbrooke, P Q J1K 2R1, CANADA
A. Thellung	Institut für Theoretische Physik der Universität Zürich, 8001 Zürich Schonberggasse 9, Zurich, SWITZERLAND
R. L. Thomas	Department of Physics, Wayne State University, Detroit, Michigan 48202
J. U. Trefny	Department of Physics, Colorado School of Mines, Golden, Colorado 80401
J. W. Tucker	Department of Physics, Sheffield University, Sheffield 537RH, ENGLAND
R. Vacher	Max-Planck Institut für Festkörper-forschung, Heisenberger Str. 1, D-7000 Stuttgart 80, WEST GERMANY
J. W. Vandersande	Department of Physics, University of the Witwatersrand, Johannesburg, SOUTH AFRICA
J. C. A. Van Der Sluijs	Department of Physics, University College of North Wales, Bangor LL57 2UW WALES
W. M. Visscher	Los Alamos Scientific Laboratory, Los Alamos, New Mexico 87545

M. Wagner   III Institute for Theoretical Physics,
                              University of Stuttgart, Pfaffenwaldring
                              57, 7000 Stuttgart 80, WEST GERMANY

D. G. Walmsley   New University of Ulster, Coleraine,
                              NORTHERN IRELAND BT52 1SA

H. Weinstock   Department of Physics, Illinois Institute
                              of Technology, Chicago, Illinois 60616

O. Weis   Exp.-Ph. IV, Universitat Ulm, D-7900
                              Ulm-Donau, WEST GERMANY

K. Weiss   Philips Research Wy-3, Eindhoven,
                              NETHERLANDS

R. M. Westervelt   Division of Applied Sciences, Harvard
                              University, Cambridge, Massachusetts
                              02138

J. K. Wigmore   Department of Physics, University of
                              Lancaster, Lancaster LA1 4YB, ENGLAND

J. L. Wilson   Department of Physics, Brown University,
                              Providence, Rhode Island 02912

J. P. Wolfe   Department of Physics, University of
                              Illinois, Urbana, Illinois 61820

C.-C. Wu   Department of Applied Mathematics,
                              National Chiao Tung University, Hsinchu,
                              45 Po Al Street, Taiwan, REPUBLIC OF CHINA

A. F. G. Wyatt   Department of Physics, University of
                              Exeter, Exeter, ENGLAND

M. Wybourne   Department of Physics, University of
                              Nottingham, Nottingham, ENGLAND

M. Yaqub   Department of Physics, Ohio State Univer-
                              sity, Columbus, Ohio 43210

INDEX

479